ADAMS OF FLEET STREET,
INSTRUMENT MAKERS TO KING GEORGE III

Frontispiece This elaborate and unusually large (16 × 10½ inches) trade card was used around 1800 by the youngest member of the Adams family, Dudley Adams, apparently in an attempt to draw attention to the antiquity of the business as well as its wide scope and royal appointments. The instruments depicted were current when the business was commenced by his father, George Adams senior, in 1734. Most of them were obsolete by the end of the eighteenth century, when Dudley Adams was trading at 60 Fleet Street. (*British Museum, London, Department of Prints and Drawings, Heal Collection 105.1*)

Adams of Fleet Street, Instrument Makers to King George III

John R. Millburn

with the kind support of the

Scientific Instrument Society
London

Ashgate

Aldershot • Burlington USA • Singapore • Sydney

© 2000 Museum of the History of Science, Oxford

All rights reserved. No part of this publication may be reproduced, stored in a retrieval system, or transmitted in any form or by any means, electronic, mechanical, photocopied, recorded or otherwise, without the prior permission of the publisher.

The author has asserted his right under the Copyright, Designs and Patents Act, 1988, to be identified as the author of this work.

Published by

Ashgate Publishing Limited
Gower House, Croft Road,
Aldershot, Hampshire GU11 3HR
England

Ashgate Publishing Company
131 Main Street
Burlington VT 05401
USA

Ashgate website: http://www.ashgate.com

ISBN 0 7546 0080 7

British Library Cataloguing-in-Publication Data
Millburn, John R.
 Adams of Fleet Street, Instrument Makers to King George III
 1. Adams (Family)—History. 2. Scientific apparatus and
 instruments industry—England—London—History—18th
 century. I. Title.
 338.4'76817'09421'09033

US Library of Congress Cataloging-in-Publication Data
Millburn, John R.
 Adams of Fleet Street : Instrument Makers to King George III / John R.
 Millburn.
    p. cm. Includes bibliographical references and index.
 1. Adams family. 2. Scientific apparatus and instruments—England—London
 —History—18th century. 3. Instrument manufacture—England—London—
 History—18th century. 4. Instrument manufacture—England—Biography.
 I. Title.
 Q185.M48 2000   99-056586
 681'.75'09224212–dc21
 [B]

This volume is printed on acid free paper.

Typeset by Manton Typesetters, Louth, Lincolnshire, UK and printed in Great Britain by MPG Books Limited, Bodmin, Cornwall.

Contents

List of Figures vii
Foreword xi
Preface and Acknowledgements xiii
Notes on Units and Prices xvii

Part I George Adams Senior

1 Family Background 3
2 Foundation and Development of the Fleet Street Business 14
3 Royal Connections 75
4 Adams's Globes 107
5 George Adams Senior's Last Few Years, 1767–72 133

Part II George Adams Junior

6 Continuation of the Business 159
7 Essays and Lectures 198
8 Instruments for van Marum 250
9 Hannah Adams and the Succession 263

Part III Dudley Adams

10 Globe Maker and Instrument Maker 273
11 Bankrupt: the End of the Adams Instrument Business 307
12 Electrician and Political Reformer 318

Notes 333
Appendix I George Adams Senior's Catalogue 1766 362

Appendix II	George Adams Junior's Last Catalogue 1795	367
Appendix III	Aids to Dating Adams Instruments and Publications	383
Appendix IV	Short-Title List of Publications by the Adams Family	386
Bibliography		399
Index		407

List of Figures

Dudley Adams's trade card, *c.* 1800		*Frontispiece*
Adams Family Tree		xx–xxi
1.1	The Fleet Street–Shoe Lane junction in the mid-eighteenth century	6
1.2	The Fleet Street–Shoe Lane junction in 1986	7
2.1	A Wright-type grand orrery	16
2.2	Grand orrery by G. Adams, pre-1752	17
2.3	Frontispiece, *Micrographia Illustrata* 1746	22
2.4	Compound microscope, '1746' pattern	23
2.5	Compound microscope, shown in a) compound and b) simple configurations	28–9
2.6	All-brass solar microscope	35
2.7	Detail of signature on a sector by G. Adams	39
2.8	St Bride's church, Fleet Street, in the mid-eighteenth century	43
2.9	Frontispiece of B. Cole's *New Sea Quadrant* 1748	46
2.10	Cole sea quadrant, 1748 pattern	47
2.11	Frontispiece of Adams's *New Sea Quadrant* 1748	50
2.12	Title-page of Adams's *New Sea Quadrant* 1748	51
2.13	Standing universal ring dial	55
2.14	Set of geometrical solids	58–9
2.15	Grocers' Hall	64
2.16	Universal equinoctial sundial	67
2.17	Set of Napier's Bones	71
2.18	Draughtsman's tool	73
3.1	Frontispiece of Adams's tract on Hadley's quadrant, *c.* 1757	77
3.2	Plane table, upper surface	80

3.3	Plane table, underside	81
3.4	'Prince of Wales' microscope	84
3.5	A trunnion-mounted microscope	86
3.6	The constellation 'Microscopium' on an Adams celestial globe	87
3.7	Cuff-type microscope by Adams	88
3.8	Architectonic sector	90
3.9	Perambulator	93
3.10	Map of London, 1761	96–7
3.11	Large double-barrelled air-pump	99
3.12	Elaborate silver microscope with eyepiece micrometer	103
4.1	Title-page of Adams's book on the globes, 1766	110
4.2	Terrestrial globe plate in Adams's book on the globes, 1766	111
4.3	18-inch terrestrial globe in the King George III Collection	116
4.4	18-inch celestial pair to the terrestrial globe in figure 4.3	117
4.5	12-inch celestial globe, in a plainer stand than shown in figures 4.3 and 4.4	119
4.6	Template for use with Adams's globes	122
4.7	Detail of the inscription on a 12-inch celestial globe	125
4.8	3-inch terrestrial globe in a pocket case	128
5.1	George Adams's trade card, *c.* 1765	134
5.2	Title-page, *Micrographia Illustrata*, fourth edition, 1771	136
5.3	Adams's new 'pocket' microscope, 1771	140
5.4	'Variable' microscope from Hill's *Construction of Timber*, 1770	141
5.5	Hill's cutting engine (microtome)	142
5.6	Section of timber, from Hill	143
5.7	The silver 'variable' microscope in its case	145
5.8	Gunner's calipers, from a magazine case of instruments	148
5.9	Gunner's calipers, with a pocket case	149
5.10	Gunner's perpendicular or gunner's level	150
5.11	Stick barometer, post-1766	154
5.12	Detail of the head of the stick barometer in figure 5.11	155
5.13	Magnetic compass in a silver box	156
6.1	Stick barometer with glazed door	160
6.2	Pocket case of drawing instruments	165
6.3	Hydrostatic balance	170
6.4	Combined perpendicular and quadrant	175

6.5	2½-inch telescope with brass draw-tubes	178
6.6	Trade card, Christopher Stedman senior	182
6.7	Stick barometer, *c.* 1790	185
6.8	Theodolite with vertical arc and rackwork	187
6.9	Compound microscope by George Adams junior	190
6.10	Magazine case of drawing instruments and paints	194–5
6.11	Adams's bill for electrical equipment	196
7.1	Electrical machines, plate I of *Electricity*	202
7.2	Electrical apparatus, plate V of *Electricity*	203
7.3	Frontispiece, fourth edition of *Electricity* 1792	205
7.4	Newspaper advertisement for *Essays on the Microscope*	208
7.5	Examples of microscopic organisms	210
7.6	Lucernal microscope, plate from *Hall's Encyclopaedia*	212
7.7	Lucernal microscope as made by Jones	214
7.8	Compound microscopes as made by Jones	215
7.9	Plate XVII from *Astronomical and Geographical Essays*, a compound orrery assembled as a planetarium	219
7.10	The compound orrery in figure 7.9 with the tellurian attachment fitted	220
7.11	Compound orrery assembled as a tellurian	221
7.12	A similar instrument to that in figure 7.11 with an extra dial on the arm	222
7.13	Compound orrery assembled as a planetarium	223
7.14	Hybrid orrery (combined planetarium and orrery)	224
7.15	Armillary spheres	225
7.16	Improved equatorial mounting with telescope	226
7.17	Equatorially-mounted telescope, plate from *Hall's Encyclopaedia*	227
7.18	Extant equatorial mounting matching that shown in figure 7.17	228
7.19	Small wooden octant with engine-divided scale	232
7.20	Detail of scale from figure 7.19	233
7.21	Complex drawing instruments, plate XI of *Geometrical and Graphical Essays*	234
7.22	Military use of a plane (plain) table	237
7.23	Plate XV: circumferentor, theodolite with pivot	238
7.24	Plate XIV (Jones): common theodolite, 4-inch theodolite	239
7.25	Plate XVI: 'best' theodolites, Adams's and Ramsden's patterns	240
7.26	4-inch theodolite with limited elevation movement	241

x List of Figures

7.27	4-inch theodolite packed in its case	242
7.28	Detail of scale of 4-inch theodolite	243
7.29	Perambulator dial	244
8.1	Seconds pendulum escapement for Atwood's machine	253
8.2	Terrestrial globe at Teyler's Museum	254
8.3	Detail of inscription on terrestrial globe	255
8.4	Detail of inscription on globe mounting	256
8.5	Celestial globe at Teyler's Museum	257
8.6	Detail of inscription on celestial globe	258
10.1	Inscription on thermometer by Dudley Adams, Charing Cross	274
10.2	Trade card, Dudley Adams (detail)	275
10.3	Reflecting telescope by Dudley Adams	278
10.4	Patent spectacles in use	281
10.5	Patent portable telescope, and detail of inscription	282
10.6	Larger patent telescope with ten draw-tubes	283
10.7	Detail of inscription on celestial globe, 1797	286
10.8	Detail of inscription on terrestrial globe, Dudley Adams, n.d.	287
10.9	Adams perpendicular	288
10.10	Adams perpendicular, and detail of signature	289
10.11	Adams proportional compasses, and detail of signature	290
10.12	Adams 360-degree protractor with vernier	291
10.13	Adams surveyor's level	294
10.14	Miscellaneous surveying instruments, including level with telescope	295
10.15	Detail of head of Adams barometer	296
10.16	Adams universal inclining sundial	297
10.17	Adams large compound microscope	298
10.18	Dudley Adams's billhead with royal arms, 1808	299
10.19	Title-page of Dudley's 'thirtieth' edition of *Globes* 1810	300
10.20	12-inch celestial globe, '1813'	301
10.21	Detail of inscription on the globe shown in figure 10.20	302
11.1	60 Fleet Street in modern times	311
11.2	Advertisement for Howes, Hart & Hall at 60 Fleet Street, 1821	314
11.3	Title-page of a tract by Francis West, 1829	315
12.1	Title-page, *Electricity is the Fountain*	321
12.2	Francis Lowndes's medical electrical apparatus	323
12.3	Title-page, *A Trifle … Materia Medica*	324

Foreword

Jim Bennett

The rise in the eighteenth century of what has been called 'public science' has recently become an object of sustained interest for historians, no longer content to restrict their attention to theoretical developments set in the context of the careers of leading figures in universities and learned societies. To appreciate the scientific culture of the day in broader terms, in the way it was experienced by the great majority of participants, it has been necessary to bring into the narrative coffee houses, public lectures and demonstrations, museums and rational entertainments, and popular books. Prominent in this wider world of experimental philosophy and practical mathematics was the vigorous commercial trade in scientific instruments, whose visibility made such an impression on foreign visitors to London.

Through two generations, the instrument makers Adams played a prominent part in this world and in the development of this public scientific culture. Their shops were famous, their instruments were prized in England and abroad, and their popular books had a widespread influence in creating and maintaining an audience for natural philosophy. Adams was a commercial success in a fiercely competitive world, while at the same time maintaining a reputation for sound books and reliable instruments.

That John Millburn has turned his attention to this story is a source of satisfaction and anticipation for everyone involved in the field of instrument history. Through his monographs on Benjamin Martin and James Ferguson, Millburn has established a formidable reputation for meticulous scholarship, unflagging application to painstaking research, and the secure grounding of his conclusions in literary and material evidence. This was just as well, for the vagaries of the modern business of publishing conspired to obstruct the appearance of the book for some years: at times it seemed that it might never be published at all. The Museum of the History of Science became involved through the author's generous transfer of the text to the Museum's keeping and control, and despite the lapse of years, it was clear that its scholarly grounding was such that it could still be printed in essentially its 1994 form. No substantial additions or revisions to the text were necessary, but readers should realize that the bibliography has not been updated to the year 2000. An exception has been made for inventory numbers of instruments in the Museum of the History of Science, where a new numbering system has been introduced; these references have been updated to avoid confusion.

Adams makes a natural sequel to Martin and Ferguson in Millburn's *oeuvre*, but a particularly challenging one, on account of the longevity of the business and the variety of its output in instruments and publications. Success sustained over such a period came only from the kind of adaptable opportunism that results in a complex story whose traces are much more difficult to bring together into a connected narrative than those of the career of an institutional scientist. The elements are, for the most part, found not in the familiar archives and libraries, but in popular publications and ephemeral tracts and advertisements, as well as in surviving objects. Of all instrument historians active today, probably only John Millburn could have completed such a work, built as it is from a wealth of instance and a richness of scattered evidence. We are grateful to everyone concerned, first to the author, but also to Ashgate for their willingness to undertake publication and to the Scientific Instrument Society for their encouragement and support. Through their combined efforts a landmark work in the history of instruments, of natural philosophical commerce, and of 'public science' has finally appeared.

Preface and Acknowledgements

In some respects this account of the Adams family forms a companion volume to my earlier books on Benjamin Martin (1976) and James Ferguson (1988). Martin, Ferguson, and the elder George Adams were all about the same age, and in the middle of the eighteenth century all three lived in Fleet Street or its environs. Their paths must have crossed on many occasions, though their backgrounds were rather different, and it is not known whether they met socially as well as in the course of their business transactions with each other. None of them left a large personal manuscript archive for potential biographers to sift through. Some letters and other relevant manuscripts do exist, but they are few and far between, and serve rather to fill in peripheral details than to form the backbone upon which a biographer of a more famous historical figure might expect to build. For my two earlier books, information on Martin's and Ferguson's activities was derived largely from their publications and public lectures; George Adams senior, on the other hand, produced only two major books, on globes and on microscopes (though his son George junior was rather more prolific in the 1780s). He was first and foremost a tradesman, whose products (scientific instruments of all types) have survived in fairly large numbers and often appear in the marketplace today. Consequently, it is his products, and his relationships with customers and rival instrument makers, that provide the basic material for part I of this book, taking the story up to the death of George Adams senior in 1772.

But the Adams story, unlike those of Martin and Ferguson, does not end with the principal subject's death. George Adams senior married twice and had many children. He was succeeded by two sons, who successively carried on the business into Regency times, until the younger son (Dudley) crashed into bankruptcy in 1817. The final chapter, on the latter's venture into 'medico-electrical therapeutics' in the 1820s, has little to do with instrument making but was necessary to bring the story to a close.

George Adams senior and his two surviving sons were all citizens of London and members of the Grocers' Company, one of the 'top twelve' City livery companies. The archives of this company, deposited in the Guildhall Library, London, provided much useful information on the family's participation in the affairs of their guild. Similarly, the archives of the Loriners' Company, though relatively fragmentary, enabled George's father Morris Adams to be traced and the background to George's early life filled in from the archives of the relevant parish, St Bride's, Fleet Street. I am indebted to the Librarian of the Guildhall Library for permission to quote numerous excerpts from guild and parish archives, and also from the archives of Christ's Hospital deposited there. For permission to quote in full one of Dudley Adams's letters I am indebted to the Trustees of the British Library.

For a substantial part of the eighty-three years during which the Adams business flourished in Fleet Street, a major customer was the Office of Ordnance; so far as military equipment suppliers were concerned, this was roughly comparable with the Ministry of Defence today. The voluminous financial and administrative records of this government department yielded an unexpectedly large amount of detailed information on the drawing, surveying, and gunnery instruments supplied by the Adams business from 1748 onwards. Excerpts from these and other crown copyright records at the Public Record Office are quoted by permission of the Controller of Her Majesty's Stationery Office.

Choosing illustrations for this book presented some problems, as I did not wish to reproduce yet again numerous illustrations of Adams instruments which are available elsewhere, for example in museum catalogues. On the other hand, as this is intended to be a comprehensive reference book on the Adams family and their products, clearly it should include major known instruments which readers might expect to find here. I have therefore compromised by including a few of the latter – such as specific instruments in the King George III Collection – but have also used photographs of many minor products of the type that instrument collectors may still find in the marketplace. The sources of these illustrations, which are all acknowledged in the captions, include auction houses, instrument dealers in Britain and America, and some private collectors, as well as the principal science museums.

Here I am grateful to the Trustees of the British Library and the British Museum; the Librarian and Keeper of Prints and Maps of the Guildhall Library; and the Science and Society Picture Library of the Science Museum for their kind permission to reproduce illustrations for which copyright is held by these institutions. I am particularly indebted to Jeremy P. Collins, of Christie's South Kensington, London, for allowing me to search his photographic files and select numerous photographs of Adams instruments sold there during the past ten years, and for providing copies for use in my book; and also Dr David Coffeen, of Tesseract, Hastings-on-Hudson, New York, for similarly providing copies of illustrations which had appeared in his catalogues.

Many instruments by Adams at the Museum of the History of Science, Oxford, had not previously been photographed: I am indebted to A.V. Simcock, Librarian, for obtaining new negatives where necessary, and for supplying information from the museum's files on the instruments concerned. Professor Gerard L'E. Turner kindly made available some of his own photographs of instruments at Oxford and Haarlem. Dr J.A. Bennett, then Curator of the Whipple Museum, Cambridge, similarly provided photographs of instruments and other items held there. Several private owners also had instruments in their collections specially photographed at my request, including Andrew Alpern (New York), Silvio A. Bedini (Washington, DC), Howard Dawes, and Peter J. McSloy.

Most of the reproductions of plates in publications by the Adams family and other authors are taken from the relevant volumes in my own library, especially the 'essays' on various subjects produced by George Adams junior between 1784 and 1794 (chapter 7), but for certain rare items I am indebted to the British Library and the Whipple Museum Library.

The Adams surname died out (in this particular dynasty) in 1830, but George Adams senior had numerous descendants through his eldest surviving daughter, Mrs Sarah Blunt (1738–1812), and possibly through other married daughters as well. Arthur Graham Blunt, a past Master of the Grocers' Company and a great-great-great-great-grandson of George Adams senior, kindly provided me with genealogical information on the Blunt family of Charing Cross and their cousins the Rogers family, latterly of Bristol, to whom some Adams family portraits and papers (unfortunately not traced) passed on the death of George senior's youngest daughter, Isabella (later Isabel) Adams, in 1830.

In addition to those mentioned above, other individuals to whom I am indebted for information or assistance include: Dr J.H. Appleby,

Norwich; Miss C. Armet, Mount Stuart; Dr J.A. Chaldecott, Eastbourne; V.K. Chew, Science Museum, London; Dr Gloria C. Clifton, Project SIMON; the late Michael A. Crawforth and his wife Diana, Oxford; Peter Delehar, London; Maya Hambly, London; Dr Anita McConnell, London; Dr A.Q. Morton, Science Museum, London; Hugh Orr, British College of Optometrists, London; Dr R.K. Smeltzer, Princeton, NJ; the late Peter J. Wallis and his wife Ruth, Newcastle upon Tyne; Alice Walters, Oaklands, California; Janice Wilson, Wellcome Institute, London; Dr H.J. Zuidervaart, Middelburg, Holland.

John R. Millburn
Aylesbury, November 1994

Notes on Units and Prices

The Adams family and their contemporaries measured their world in feet and inches, and priced their products in guineas, or in pounds, shillings, and (old) pence. Since their own publications or catalogues are quoted in this book, measurements are necessarily given in imperial units. For the benefit of readers more familiar with metric units, 1 inch is 2.54 centimetres.

As it is now over twenty years since the ancient English monetary system was replaced by decimal division of the pound, and guineas have practically disappeared from the language, younger readers may like to be reminded that the pound sterling used to represent twenty shillings of twelve (old) pence each; that is, 240 (old) pence made one pound. The symbols generally used in print during the eighteenth century were ℓ (libra = pounds), s (shillings), and d (from denarius = penny, subdivided into halves and quarters). For typographical convenience the modern symbol £ has been used throughout for pounds. The guinea, which in the seventeenth century fluctuated in value, had been fixed at twenty-one shillings by the time George Adams senior commenced trading. (The shilling, not the pound, was the fundamental unit on which the English coinage had been based since Elizabethan times.) As there was no coin of value £1 until the sovereign was introduced in the early nineteenth century, prices of more than a few shillings were often expressed in guineas, or fractions (half and quarter) of a guinea, for which gold coins were available. For accounting purposes, however, these were always converted into pounds, shillings, and pence. Thus, the price of (say) a microscope, quoted to a customer as 5½ guineas, would be entered in the books as £5 15s 6d. Conversely, a price which seems an odd value in pounds, shillings, and pence will often be found to represent a round figure in guineas.

While English measures of length and so on can be expressed precisely in equivalent metric units, it is very misleading to attempt to do the same with money. Numerically a shilling (twelve old pence) is exactly five new pence, or £0.05, but the purchasing power of the eighteenth-century shilling was, of course, very much larger than that of 5p today. Just how much larger is a question to which there is no simple answer, because not only has inflation (mainly during the last few decades) made our present money relatively worthless, it has affected different goods and services in different ways. Also, technological developments during the past two centuries often make comparisons between prices for eighteenth-century and modern products difficult, even though their names (for example microscope, theodolite) may be the same.

It is nevertheless desirable, when reading about the activities of eighteenth-century tradesmen such as the Adamses, to bear in mind what any prices quoted meant to the persons concerned. The price of bread, which is often used to derive a cost-of-living index for the working classes, was largely irrelevant to a fairly prosperous family not dependent on the land for their income. Of more direct interest to them was the price of labour employed in their business or by their subcontractors. Although no trading records have survived for the Adams firm, some information on wages in specific trades can be obtained from bills mentioning labour costs, and from announcements in contemporary newspapers. For example, in 1764 the rate for journeyman tailors in the City of London was fixed at 2s 6d per day (this was for a fourteen-hour day in the summer, from 6 a.m. to 8 p.m., and an hour less in winter). An experienced instrument maker could expect to earn about twice as much as this. The extensive financial records of the Office of Ordnance provide the following daily rates:

Manual labourer	1s 6d
Mason	2s 0d
Bricklayer	2s 6d
Carpenter	2s 4d
Draughtsman	1s 6d to 5s 0d in five grades
Instrument maker	5s 0d (approximately, on subcontract work)

At the same period (mid-eighteenth century) an Ordnance engineer, who was equal in status to an army captain, earned 10s per day, while the basic salary of the Chief Master of the Royal Military Academy was £200 per year (equivalent to about 11s per day).

Comparison of the instrument maker's 5s per day with current wage rates for skilled industrial workers gives a multiplying factor of around 200. In other words, as a very rough guide one may think of the mid-eighteenth century shilling as being equivalent to £10 in today's money. If a more exact comparison is required, one must take into account the actual costs of the goods or services on which the money is or was spent. Personal travel, for example, was relatively much more expensive in the eighteenth century than today (a post-chaise cost about 1s a mile), but very few instrument makers would have spent much of their income in this area. On the other hand, domestic servants, who today are scarce and expensive, were much more affordable then, and a middle-aged instrument maker with a family would probably have employed several.

The figures quoted above for the mid-eighteenth century remained almost constant for the whole of George Adams senior's working life (1730s to 1770s), and for most of George junior's as well; in other words, annual inflation would have been a meaningless term to them. It was only in the last decade of the eighteenth century that prices of the necessities of life began to increase significantly (due to the Napoleonic wars), leading to increases in the cost of labour and hence of manufactured goods. Of the Adams family, only Dudley and his surviving sisters had to cope with a rapidly increasing cost of living, and its effects on a manufacturing and retail business.

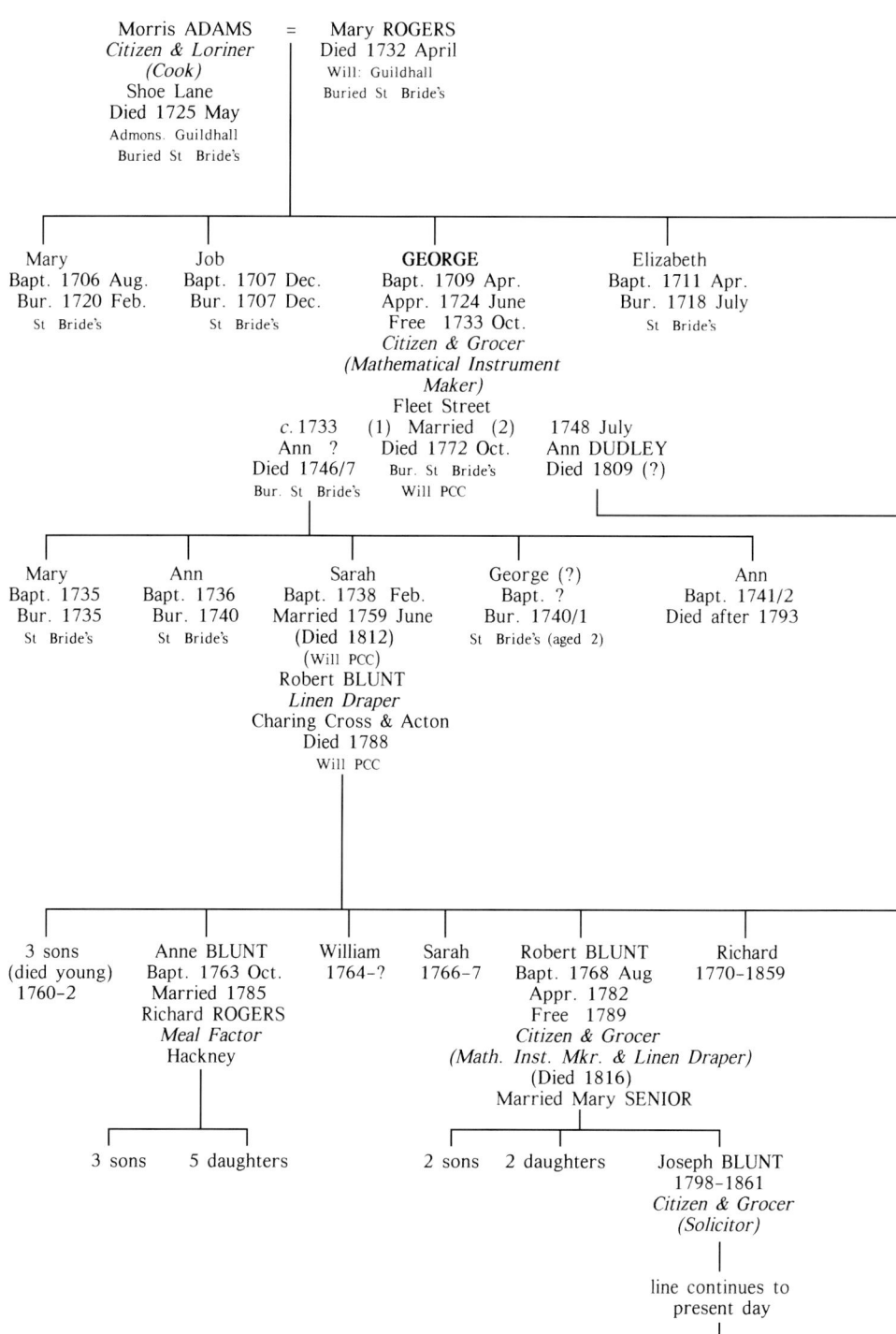

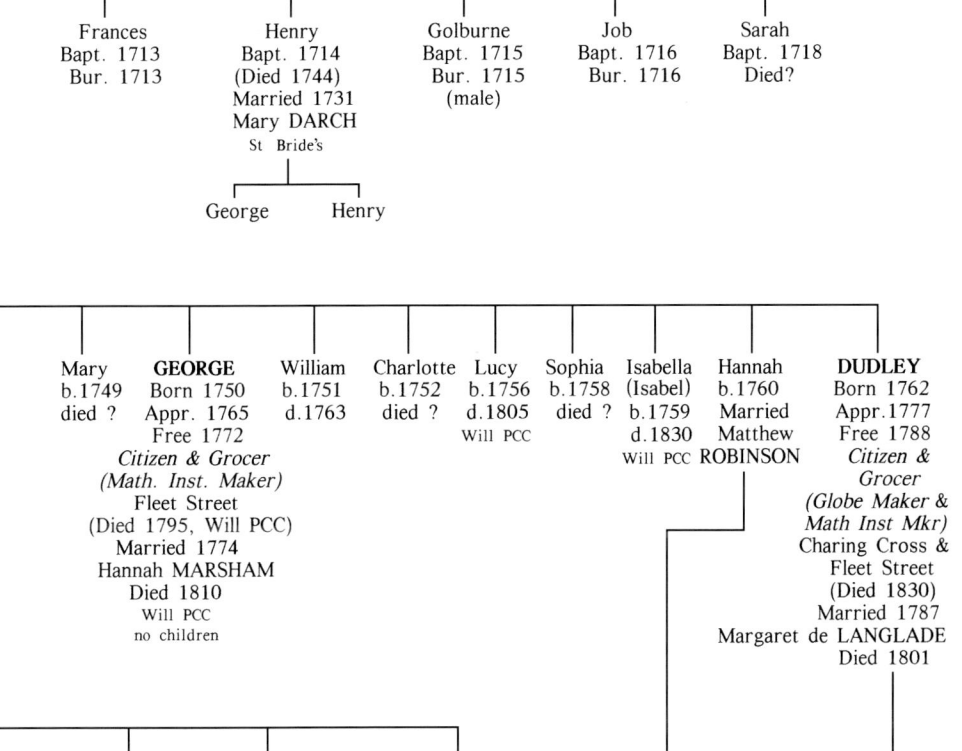

The ADAMS FAMILY
of FLEET STREET, LONDON
and their principal descendants
the BLUNT family of Charing Cross

Part I:

George Adams Senior

1 Family Background

George Adams senior, the first member of the family to become a mathematical instrument maker, was born in London in the spring of 1709, in the middle of Queen Anne's reign. His parents, Morris and Mary Adams, were then living in Globe Court, near the southern end of Shoe Lane, Fleet Street, a stone's-throw from the parish church of St Bride's where his baptism was registered on 17 April.[1]

George was the third in a family of nine children – five boys and four girls – born between 1706 and 1718; but the infant mortality rate was high. Of the boys, he was the only one to achieve a full lifespan and have a large family of his own. A younger brother, Henry, born about 1714, lived long enough to marry and have several children, but died in his early thirties. The other three boys all died within a few weeks of birth: Job in 1707, Golburne in 1715, and a second Job in 1716. At least three of the girls also died young. Consequently, George became the head of the family after his parents died, and by his late thirties was probably the sole survivor of his generation, with the possible exception of a younger sister whose death has not been traced.

When J-D. de Cassini, director of the Royal Observatory at Paris, conceived in 1788 a plan to send two French workmen to England to study under some of the leading instrument makers of the time, he referred to George Adams's eldest son (born in London in 1750) as being of French origin.[2] This statement has been repeated, and indeed elaborated, in some modern reference works. For example, Daumas asserted that George Adams senior and his contemporary John Dollond were both descended from French Protestant emigrants of the seventeenth century;[3] but while the Dollonds undoubtedly were Huguenots, 'Adams' is a common English surname with no particular French connotations. At the time of George senior's birth there were at least five separate Adams families living in the parish of St Bride's alone, and innumerable others in London as a whole. Coupled with the minimal information recorded in the early parish registers, this profusion of Adamses makes it difficult to trace family relationships with certainty, though one can all too readily draw tentative conclusions when names and dates appear to fit.[4]

George's father's first name, Morris (Maurice), might be thought to lend some weight to the story of his French origins. For convenience it is given here as Morris, except where direct quotations from manuscript records require otherwise, but in parish and guild archives it appears in a variety of phonetic spellings: Morris, Morres, Morrice, Maurice, or in Latin Mauritius. However, Morris (spelt thus) is a common English surname; there were at least six Morris families living in the parish of St Bride's in the first decade of the eighteenth century. As it was not unusual for maternal surnames to be perpetuated by using them as first names in later generations, it is quite possible that its occurrence in the Adams family commemorates a Morris forebear on the female side. The first name Golburne of one of Morris's short-lived sons may be similarly explained, and other instances of the use of maternal surnames will emerge as the Adams story unfolds.

Be that as it may, Morris Adams was certainly a citizen (freeman) of London, and those of his children and grandchildren who became instrument makers were undoubtedly Londoners by birth. This point is worth noting, as it is a curious fact that many of the well-known instrument makers who worked in London in the eighteenth century were not born there.[5] John Rowley, Master of Mechanics to King George I, came from Lichfield; George Graham, watchmaker and astronomical instrument maker, was born in Cumberland; John Bird was originally a weaver in Durham; James Short, the specialist maker of reflecting telescopes, was born and educated in Scotland; Jonathan Sisson came to London from Lincolnshire; Jesse Ramsden, famous for his accurate division of scales for instruments, was apprenticed to the textile trade in Halifax; and Benjamin Martin, one of George Adams's principal rivals in the 1760s, had a rural upbringing in Surrey and only entered the London instrument-making world (primarily as a retailer rather than maker) in his fifties.

In order to trade in the City of London one had to be a freeman,[6] usually by first becoming a freeman of one of the City's guild companies. In the manual trades this was normally achieved by serving an apprenticeship, typically for seven years, to a member of a guild, though there were other routes to the freedom such as redemption (purchase) and patrimony (the right of a child born to a freeman to claim the freedom himself).[7] Morris Adams was a 'citizen and loriner'; that is, a member of the Loriners' Company, founded in the fifteenth century and originally composed of 'bit-makers', bits being one of the principal metal parts of harnesses. It does not follow, however, that Morris Adams was a loriner by trade. Membership of some of the younger craft companies, such as the Clockmakers' (founded 1631), was confined largely to men practising that particular trade, but by the eighteenth century the membership of the medieval companies had become thoroughly mixed, and their original purpose (regulation of their own trade) largely lost. In the minute books of the Loriners' Company, which record – amongst other things – the binding of apprentices and granting of freedom to those who had completed their term, there are frequent marginal notes indicating that members of the company followed a variety of occupations including cutler, porter, engraver, cork-cutter, carpenter, and baker. There were also some members specifically described as bit-makers, or genuine loriners.

Morris Adams was, in fact, a cook, meaning in this instance a restaurateur with his own business rather than a domestic employee. St Bride's was a large parish, divided into three precincts, with a total of upwards of 1,000 separate properties, and its surviving archives are fairly extensive. Amongst them is a volume started about 1713, recording the names, addresses, and occupations of the principal ratepayers,[8] intended primarily to form an index to the entries in the vestry minutes showing when these men were chosen to serve in the various parish offices of questmen,[9] constables, scavengers, collectors (of the Poor Rate), sidesmen, and churchwardens. Once a year, at a general meeting of the principal inhabitants, the names of several people were proposed for each office in each precinct, and one or more (as appropriate) chosen by vote. In a large parish like St Bride's this could be a protracted process, as not only were there numerous offices to be filled (at least nine questmen, for example), but also each person voted into office could ask to be excused on payment of a fine. Sometimes as many as half a dozen persons in turn adopted this course for each office; indeed, this seems to have been one of the principal ways of raising money for the parish funds. The amount of the fine varied according to the importance of the office, the largest being around £12–£15 for churchwarden.

Sometimes an individual would ask to be allowed to pay a fine for several offices at once, thereby gaining exemption from serving for

1.1 The Fleet Street–Shoe Lane junction in the middle of the eighteenth century, showing the type of brick buildings erected in this area after the Great Fire of 1666. This engraving (from part of a view of St Bride's, drawn by John Donowell) shows the posts which divided the footway from the carriageway before the introduction of flagstone pavements. (*Guildhall Library, London*)

1.2 The Fleet Street–Shoe Lane junction photographed in 1986 before the exodus of newspapers from Fleet Street to the Isle of Dogs. The black glass *Daily Express* building, erected in 1931 and now only a preserved façade, covers the site of Adams's first shop. Racquet Court, the site of his second, was obliterated in 1970 by its eastwards extension. For 60 Fleet Street, his address from 1757, see figure 11.1. (*Photograph by the author*)

several years in advance. This is what Morris Adams did in April 1715, when he paid a combined fine of £8 to avoid serving as both scavenger and collector. It is the record of this transaction, when his son George was six years old, that provides the earliest evidence of his occupation of cook.[10] His address then was Shoe Lane. The ratebooks show that he moved from Globe Court, just off Shoe Lane, into the lane itself in about 1712, occupying the third house up from Fleet Street on the east side.[11] This remained the home of the Adams family for the next thirty years or so; it had a rateable value of £25. The whole of this region had been devastated by the Great Fire of 1666, and subsequently rebuilt in brick in accordance with the new standards laid down after the fire, so although no contemporary views of Shoe Lane itself have been located it is reasonable to suppose that Morris Adams's premises looked something like the buildings depicted in figure 1.1. The loading bay of the 1930s *Daily Express* building covers the site of the house where Morris lived and George Adams spent his childhood.

Some of the guild companies' records go right back to their medieval foundation, but unfortunately the earliest surviving minute book of the Loriners' Company starts in 1722. Indirect evidence suggests that Morris must have been a freeman by 1715, and probably somewhat earlier. By the 1720s he was not merely a citizen and loriner but also a liveryman of his company, one of the senior members who (having paid the fee of £10) was entitled to vote in City elections, and wear a gown in the company's colours on ceremonial occasions. There were about seventy liverymen in the Loriners' Company, out of a total membership in the 1720s of around 650.[12] The fact that Morris was a liveryman indicates a certain degree of standing in the business world.

During the years 1722–25 inclusive Morris Adams's name occurs several times in the court minutes, when he took or freed apprentices, and also when he was summoned to attend meetings at which stewards were chosen. On 7 August 1723, an apprentice of the late Job Adams, citizen and loriner, was presented for his freedom by Morris Adams, Job's executor. In view of the two attempts (both unsuccessful) by Morris to raise a son named Job, it is likely that he and Job were close relations, perhaps brothers. Like Morris, Job Adams was a cook by trade as well as a member of the Loriners' Company.[13]

The Loriners' court minutes show that in March 1724 Morris had at least two apprentices working for him. On 5 June that year he apprenticed his own eldest surviving son, George (then aged fifteen) to James Parker, a mathematical instrument maker and member of the Grocers'

Company, paying Parker a premium of £20.[14] What led Morris to adopt this course can only be conjectured: perhaps his own business was already overstaffed and he wanted his son to acquire skills that were more profitable than cooking. The choice of a member of the Grocers' Company could have been just chance; but Morris, a liveryman himself, could have taken into account the fact that the Grocers' was one of the 'top twelve', or 'great' companies, from whose ranks the lord mayor was normally chosen. (In the eighteenth century, the Grocers' provided the lord mayor on eleven occasions.) In the hierarchy of the City of London the Grocers' Company was second only to the Mercers'. The Loriners' Company, on the other hand, came fifty-seventh in the order of precedence and was relatively poor and insignificant. They did have their own hall, marked on Rocque's great map of 1746, but it must have been quite small as its rateable value was only £35, about the same as a single shop in a main street. Situated just outside the City wall, midway between Cripplegate and Moorgate, Loriners' Hall was demolished in the mid-eighteenth century when Basinghall Street was extended northwards to Fore Street. Redevelopment of this area following the total devastation of the Second World War has completely obliterated its site.

In September 1722, June 1723, and August 1724, Morris Adams was one of the members of the Loriners' Company summoned to be present at the meeting at Loriners' Hall when stewards for the forthcoming year were to be chosen. The office of steward, for which (in this company) one did not necessarily have to be a liveryman, was unpopular as it involved the holder in considerable personal expense. By the eighteenth century most of the ancient guild companies had largely become social and benevolent institutions, except for the formal business of binding and freeing apprentices. It was the duty of the stewards to 'make and provide a dinner for the Company as usual, upon the day my Lord Mayor goes to Westminster to be sworn'.[15] As with the parish offices, this obligation could be avoided by paying a fine, in this case £5, and many of those called upon to serve adopted this course. It was therefore usual to nominate a dozen or more people for the office, in the expectation that some of them would opt for paying the fine and thereby swell the company's funds. For example, in the year 1723, at the first meeting called on 26 June to choose stewards, five nominees paid the £5 fine, one pleaded poverty, and another claimed exemption on the dubious grounds that he was an excise officer. No fewer than eight more who had been nominated failed to attend the meeting: the beadle was instructed to order them to attend at the next court. From the entries in the minutes one gets the distinct impression that the affairs of the Loriners' Company were conducted on a somewhat casual basis, many members (having secured their freedom) simply ignoring their obligations, including the payment of quarterage. Barely half of the liverymen of the company kept their payments of sixpence per quarter up to date. Some of them were several years in arrears – one persistent offender had built up a backlog of forty-nine quarters unpaid by 1726.[16] It seems to have been common practice for arrears of quarterage to be paid only when the member concerned was obliged to attend the court for other matters, such as the binding or freeing of apprentices.[17]

Morris Adams was never more than four quarters behind with his payments, but he did his best to avoid serving as steward. As a professional cook, he would no doubt have been pleased to supply the annual dinner on a commercial basis, but not out of his own pocket. In 1722 he simply failed to attend the relevant meeting; fortunately two other members who were present accepted the office. (For anyone who had ambitions eventually to become a member of the company's governing body, the Court of Assistants, serving as steward was an essential preliminary.) He adopted the same tactics in 1723, though his escape on that occasion seems to have been due primarily to an oversight by the clerk, who omitted his name from the list of absentees summoned to appear at the next meeting. Precisely how he managed to avoid serving in 1724 is not clear, but presumably two volunteers came forward to accept the post. He would almost certainly have been obliged to serve as steward in 1725, had he not suddenly died in May that year.

The burial register of St Bride's shows that Morris Adams was interred in the parish churchyard on 30 May 1725. There is no record of his age, but if he married in his mid-twenties he was probably no more than forty-five or so at death. One might have expected Morris, a liveryman and employer, to have made a will, but he did not. Letters of Administration were granted to his widow Mary in the Commissary Court of London on 14 June 1725. The 'Admons' register entries are quite short and give no indication of the value of his estate.[18]

Lack of a will sometimes indicates sudden or accidental death, and that was indeed the case in this instance. Fatal accidents are news, so it is not surprising that Morris's demise was reported in one of the London newspapers. On Saturday 29 May (the day before Morris's funeral), the following paragraph appeared in the *Daily Post*:

On Tuesday last Mr. Adams, a noted Cook in Shoe Lane, having taken Horse at his own House in order to take the Air, was, by the starting of his Horse, thrown down near Poppings Alley in Fleet Street, and had the misfortune to break his Skull, so that he died soon after.

So, George Adams lost his father while still in the first year of his apprenticeship. His widowed mother continued to occupy the family home in Shoe Lane for another seven years, her younger son Henry and possibly a young daughter presumably living with her. If the normal procedure was followed, George would have lived with his master, James Parker, in White Fryars, who would have provided him with board and lodging as well as instruction in the art of mathematical instrument making.

Towards the end of the following year (1726) another upset occurred. George's master, James Parker, died: his will was signed on 11 November 1726 and proved in the Commissary Court of London in December.[19] On 26 January 1727 George Adams was turned over to a new master, Thomas Heath (another member of the Grocers' Company), whose premises were in the Strand.[20] Though it must have been upsetting at the time, in the long run this move undoubtedly proved beneficial to George's career. Thomas Heath had been trading for only about six years when George was turned over to him, but he was an ambitious businessman who had already taken four apprentices by then, and was keen to expand his trade through advertising and publishing as well as making and selling instruments. In his workshop George would have had the opportunity to learn not only the manual skills involved in the actual manufacture of mathematical instruments, but also the organization of a wholesale and retail business, and the art of promoting the firm's products in a competitive market. Heath also participated in the running of the Grocers' Company, eventually becoming its Master; he died during his year of office in 1773.

It should be noted that in the 1720s there was no such trade as 'scientific instrument maker'. Heath's principal products were mathematical instruments, defined in one of his advertisements in 1723 as instruments made of silver, brass, ivory, and wood, for geometry, surveying, navigation, gauging, measuring, dialling, geography, and astronomy, namely:

> Sectors, Scales, Compasses, Drawing-Pens, Land-Quadrants, Theodolites, Semicircles, Circumferentors, Plane-Tables, Water-Levels, Measuring-Wheels, Dials (Portable, or to be fix'd), Sea-Quadrants, Forestaffs, Nocturnals, Gunters, Sea-Compasses (Meridian, or Azimuth ones), Charts, Maps, Spheres,

Weather-Glasses, Burning-Glasses, Prisms, Sky-Opticks, Drawing-Boards, Sliding-Rules of all Sorts, Parallel-Rulers, &c., together with Books of their Uses.[21]

Globes, in 'wrack-work' mountings, were also mentioned, though the globes themselves would almost certainly have been bought in from specialist globe makers, only the brass mountings being made in Heath's workshop. Complex optical instruments such as microscopes and telescopes, and also spectacles, were generally made or retailed by 'opticians' and were apparently excluded from Heath's coverage at first; but by the early 1730s, when George Adams was nearing the end of his apprenticeship, Heath's advertisements included telescopes and reading-glasses, though not microscopes, in addition to the above.[22] Apparatus for 'Experimental Philosophy', such as air-pumps and electrical machines, was relatively rare at this time: it was not made in large quantities until the subject had been popularized by public lecturers in the mid-eighteenth century, especially after the invention of the Leyden jar about 1746. Instrument makers who ran large general businesses from the mid-century onwards, such as Benjamin Martin, Heath & Wing, and George Adams himself, usually called themselves 'Mathematical, Optical, and Philosophical' instrument makers, and their catalogues were divided into corresponding separate categories.

The Grocers' Company is fortunate in that its archives have survived largely intact from medieval times, enabling us to trace George Adams's progress through the successive stages of apprenticeship, freedom, liveryman, service as steward and later warden, and eventually election to the Court of Assistants, the company's governing body. But in the 1720s the latter distinction was still many years ahead. The first essential step was the completion of his apprenticeship, which if it followed the normal course would have been achieved by mid-1731. He did not apply for his freedom until two years later; it was quite usual for apprentices who had completed their formal term of servitude to continue working for their master for a few years, applying for the freedom only when they wished to move elsewhere or to take apprentices themselves.

About the end of April 1732, George's mother Mary Adams died, and was buried in St Bride's churchyard. George, who was then aged twenty-three, became the ratepayer at the family's Shoe Lane home. Mary's will, proved in the Commissary Court of London on 4 May, appointed George executor and left everything to him.[23] Her will also refers to a legacy of £60 left by her mother, Margaret Rogers, in a will dated 1716, to George and his brother Henry in equal shares, to be paid plus interest when they had both (or the survivor) reached the age of twenty-one. Morris and Mary had been joint trustees of this bequest, which by 1732 would have amounted to about £100 at the rates of interest then available. It may be relevant that a man named George Adams had an account with Hoare's Bank in Fleet Street for a few months from April 1732, which he used for the occasional deposit or encashment of notes totalling £170, but as no other details are given in the bank's ledger it is impossible to say whether this was 'our' man.[24]

On 10 October 1733 George Adams, then aged twenty-four, was granted his freedom and became a fully-qualified member of the Grocers' Company, ready to take his place, as an employer if need be, in the commercial instrument-making world. This cost a fee of 13s 6d plus one year's quarterage of 2s 4d, slightly dearer (at 7d per quarter) than the corresponding charges in the Loriners' Company. His address was given in a marginal note in the freedom register as 'at the Globe in Shooe Lane Eatinghouse'.[25] From this it would appear that Morris Adams's cookery business was continued by his widow after his death, and then by his son or sons, who presumably employed staff for that purpose. The premises remained in the occupation of the Adams family until the mid-1740s, but there is no indication that George attempted to run an instrument-making business from there: most probably he continued to work as a journeyman for Thomas Heath in the Strand (though living in Shoe Lane) until he felt sufficiently confident to set up as an instrument maker on his own. Just under a year after gaining his freedom, with the backing of some financial resources resulting from the deaths of his parents, he decided to take the plunge.[26]

2 Foundation and Development of the Fleet Street Business

At Michaelmas in 1734, George Adams started paying rates on some premises in Fleet Street four doors east of the Shoe Lane junction, just round the corner from the family home. This event marks the start of the Adams instrument-making business, which continued in Fleet Street (at various addresses) for eighty-three years. At about the same time, or earlier, he must have married his first wife Ann: their first child, Mary, was born in July 1735.[1] George remained the ratepayer at the Shoe Lane home as well for a few years, and as his address was given as Shoe Lane when his first two children were baptized, he probably lived there with his wife, younger brother, and possibly a sister. The Fleet Street premises had a rateable value of £25, the same as the Shoe Lane property but comparatively low for a shop in a main street, so they probably consisted of a small lock-up shop or workshop with perhaps a small yard at the back. They were located roughly on the site now occupied by the entrance to the former *Daily Express* building.

George's choice of this particular spot may have been decided by its convenient proximity to Shoe Lane, but in any case the land route from the City to Westminster (Cheapside–St Paul's–Ludgate Hill–Fleet Street–Strand) was a favourite location for opticians and instrument makers in the eighteenth century. In the 1730s, Adams's competitors in Fleet Street or its vicinity included John Cuff, facing Serjeants' Inn; Thomas Wright, mathematical instrument maker to King George II, on the corner of Peterborough Court; William Deane, mathematical instrument maker to the Office of Ordnance, just up Fetter Lane; and the globe maker John Senex, opposite St Dunstan's church. The Royal Society occupied a house in Crane Court, which opened off Fleet Street near Fetter Lane. Not far away in the Strand were Adams's former master, Thomas Heath, and also Jonathan Sisson, mathematical instrument maker to the Prince of Wales (that is, Frederick). These were all well-known men whose products are frequently found in the marketplace today. In addition, several lesser-known practitioners had their shops or workshops in or near Fleet Street in the 1730s, including George Bass, John Coggs, David Drakeford, George Hearne, Nathaniel Hill, James Lane, James Mann, and John Tracy.[2]

House and shop numbers were not allocated in this area until the mid-1760s. To identify his premises, Adams adopted the sign of 'Tycho Brahe's Head' – a rather surprising choice, which cannot have been of much assistance to illiterate messengers: only a person of some learning would have known who Tycho Brahe was. For the benefit of the uninitiated, Adams also described his location as being 'near the Castle Tavern' or 'over-against [that is, opposite] St.Bride's'.

The earliest newspaper advertisements mentioning George Adams that have been found are dated April 1735, about six months after he became a ratepayer in Fleet Street. Although London by this time had several daily and thrice-weekly newspapers, most of the commercial advertisements therein were for theatres, auction sales, books, and medicines. It was relatively unusual for traders such as instrument makers to market their products through newspaper advertising, possibly because instruments in general, unlike (say) books, were not standardized stock items sold at fixed prices. There were, admittedly, some exceptions: globes, for example, had been sold at fixed prices since the late seventeenth century, and occasionally an instrument maker would advertise some specific new product of his own design. Thus Jonathan Sisson advertised his 'new and curious' barometer, fitted with a telescope for observing small changes in the level of the mercury, several times from January to March 1736 in the *Daily Post*, the *Daily Journal*, and the *London Evening Post*; but this widespread advertising of a 'scientific' product was the exception rather than the rule.

There were, however, several indirect ways of bringing an instrument maker's name into the public eye through newspapers. For example, every time William Watts and James Stirling advertised a course of lectures at their academy in Tower Street, they announced that the astronomical part would be illustrated by 'The GREAT ORRERY, made by Thomas Wright of Fleet Street, Mathematical Instrument Maker to his Majesty'. (See figure 2.1.). This was first mentioned in 1730, and the same wording was still in use in 1735. Instrument

16 George Adams Senior

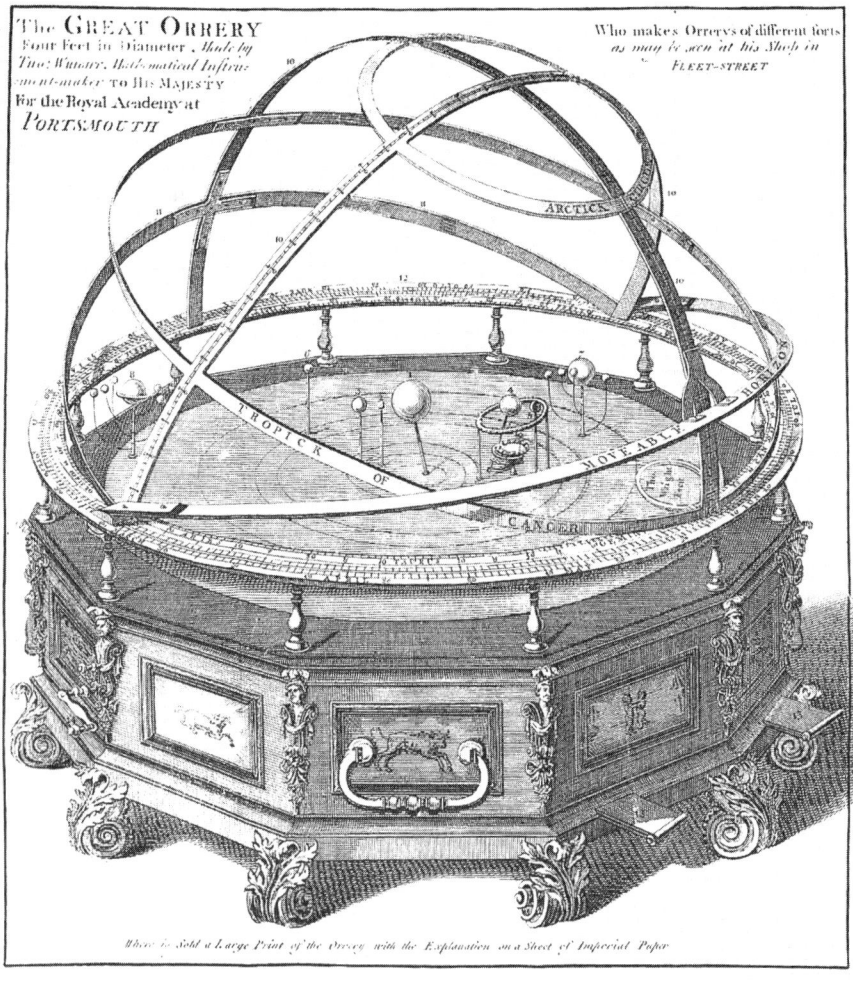

2.1 A typical grand orrery incorporating all the planets and their satellites, with the wheelwork housed in a twelve-sided wooden box surmounted by an armillary hemisphere, as made by Thomas Wright of Fleet Street in the 1730s and 1740s. This print, from Joseph Harris's *Description and Use of the Globes and Orrery*, which ran to twelve editions between 1731 and 1783, was widely copied by other engravers and publishers. (*Author's collection*)

2.2 A grand orrery made by George Adams senior before the change in the calendar in 1752. It is about 3 foot 6 inches wide, and is normally protected by a twelve-sided glass dome (not shown). Formerly at St John's College, Cambridge, this orrery was transferred to the Sedgwick Museum in 1904, and is now in the Whipple Museum. For details see the Museum's *Catalogue 4: Spheres, Globes & Orreries* item 29. (*Whipple Museum, Cambridge, Inv.Wh.1275*)

makers' shops were also mentioned occasionally as places where subscriptions to such courses were taken in: John Senex was associated in this way with J.T. Desaguliers, and later on (in the late 1740s) George Adams himself was similarly associated with the lecturer Joseph Sowerby.

Another way for instrument makers to obtain newspaper publicity was by dealing in books as well as instruments. Thomas Heath was one of those who mentioned on his trade cards that he sold mathematical books.[3] Between 1735 and 1740 he frequently advertised books such as Hammond's *Practical Surveyor* and Cunn's *Use of the Sector*, 'published by Thomas Heath, Mathematical Instrument Maker near the Fountain Tavern, in the Strand', mostly in the *London Evening Post* and sometimes in other papers as well. Conversely, the books themselves, of course, acted as advertisements for his instruments.

It was in association with Heath that George Adams's name first appeared in April 1735 in advertisements for 'A new Cata-Dioptrical Quadrant', in both the *London Evening Post* and the *Daily Journal*. The Longitude Prize offered by Act of Parliament in 1714 is generally remembered today for the stimulus it gave to the development of marine chronometers by John Harrison; but equally important at the time was its effect on the development of instruments for accurate measurement of angles, particularly for use on board ship for determining longitude by measurement of lunar distances, and also improving the measurement of latitude. Though Hadley's double-reflecting octant, in the form of its later development the sextant, eventually became the standard instrument for this purpose, several other designs were proposed and promoted by hopeful inventors before Hadley's device reigned supreme.[4] Compared with the backstaff or Davis's quadrant, to which seamen had become accustomed, Hadley's octant (devised *c.* 1731 and tested at sea in 1732) suffered from the disadvantages that it was not only more expensive but also needed careful manufacture and adjustment if its full potential was to be realized.

The catadioptric instrument devised by Caleb Smith about 1734 was so called because it had a mirror for observing the sun, and a separate internally-reflecting prism for observing the horizon. Caleb Smith and William Ward jointly submitted an application to the Commissioners of Longitude in January 1735, which included a description of a special telescope for observing Jupiter's satellites (for determining the longitude) as well as Smith's instrument for determining the latitude.[5] This was published (all submissions to the board had to be printed) by fourteen booksellers including John Senex and Thomas Heath. An advertisement on the last page stated that the 'astroscope' was made by the optician James Mann, while the 'astronomical instrument' was made by Thomas Heath. Adams was not mentioned, but on 12 April 1735 the advertisement transcribed below appeared in the *Daily Journal*, linking Adams with Heath as makers of the instrument in the preamble, though Heath alone is mentioned further down:

Made and sold,
By Thomas Heath, near the Fountain Tavern in the Strand, and G.Adams, at Tycho Brahe's Head, near the Castle Tavern in Fleet Street.
A NEW CATA-DIOPTRICAL QUADRANT, in which one Object is seen by Refraction, and another by one single Reflection; and likewise, A CAT-OPTRIC One, where each Object is view'd by Means of a single Reflection. Both contriv'd by Mr. Caleb Smith, for taking Altitudes from the visible Horizon, or any other Angles at Sea; whereby the Interruption and Incon-

venience occasion'd by the Ship's Motion, in observing with common Instruments, is remedied and prevented.

N.B. These Instruments have not only all the useful Properties of the Double Reflecting Quadrants, lately invented by Mr. Hadley, Mr. Godfrey, and Mr. Plank, but likewise excel them in many Particulars; as, 1st, The Construction of them is more simple. 2ndly, The Objects sought are found therewith more readily; and of consequence, Observations made thereby with great Ease and Expedition. 3dly, The Stars of lesser Magnitude, and other Objects, will be clearly and distinctly discern'd therein, which by two successive Reflections, lose so much of their Light, as to become very Obscure, or altogether Invisible. Lastly, they will be sold cheap.

At the Places above-mention'd may likewise be had, The Description and Use of A New ASTRONOMICAL INSTRUMENT, invented by Mr. Ward and Mr. Smith, and made by Mr. Heath, for taking Altitudes of the Sun and Stars at Sea without an Horizon; together with an easy and sure Method of observing the Eclipses of Jupiter's Satellites, or any other Phenomenon of the like kind on Ship-board, in order to determine the Difference of Meridians at Sea.

To which are added,
Tables for computing the Times when the Eclipses of the First Satellite of Jupiter happen under the Meridian of London.

By this time Smith had also produced a separate publication, describing his instrument alone, in which he said that a second mirror could be used instead of a prism if desired, to reduce the cost.[6] This was reviewed in the May 1735 number of *The Present State of the Republick of Letters*, ending with the comment: 'N.B. The curious may see this instrument at Mr. George Adams's, a Mathematical Instrument Maker, over-against St.Bride's Church in Fleet Street.' By the end of June, Smith's publication was being advertised in several newspapers, with a statement that the instruments described therein were 'made, and sold cheap, by G. Adams, at Tycho Brahe's Head, near the Castle Tavern in Fleetstreet'. Directions for its use were given with the instrument, and also sold by five booksellers in the City, plus Mann in Ludgate Street, Senex in Fleet Street, and Heath in the Strand, price 4d. These 'Directions' are rare today: only Cambridge University Library is recorded as having a copy of the original 1735 eight-page pamphlet. Only one extant example of Smith's quadrant itself signed by Adams has been reported (at Lisbon University), but even if only a few were sold, the numerous advertisements for it provided an excellent opportunity for him to publicize his own recently-established business.

According to some modern sources it was in 1735 or 1736 that Adams began making instruments 'for the East India Company'.[7] This story seems to have originated with the discovery, early in the twentieth century, of a letter written by George Adams and thought to be dated 7 January 1735/6. (The way the date is written is slightly ambiguous, but this seems the most likely interpretation.) It first came to light at a meeting of the Royal Microscopical Society on 17 January 1912, when an old microscope lent by Mr Alfred Hodgson was exhibited and described. The microscope itself, a 'Culpeper' or 'three-pillar' type, was unsigned, but in the drawer of the box foot was this letter, which the author of the subsequent article claimed 'leaves no doubt that it was made by this maker [George Adams] and about this date [January 1735/6]'.[8] Both the instrument and the letter are now in the Wellcome Museum for the History of Medicine at the Science Museum, South Kensington.[9] In the one-page letter, which is addressed to 'The Rev. Mr. Talbot' and is rather roughly written, Adams apologizes for some delay in sending the microscope, and blames it on being 'so prodigiously hurried by the East India Commanders to finish their works'. Unfortunately, the seal obliterated Adams's quoted price for the microscope, which is not described, so we have only the former owner's word for it that the letter and the existing unsigned microscope are indeed connected. One would have expected Adams, having just commenced trading, to have put his name on the instrument or its box somewhere, if only in manuscript, to let anyone seeing it know that he was in business.

This is the earliest document written by Adams that is known to have survived. Evidently he had succeeeded in obtaining orders for some navigational instruments, quite possibly quadrants of some sort, though this can only be conjectured. It should be noted, however, that the instruments concerned were for the East India commanders, not the East India Company itself, and were presumably ordered privately. According to R.S. Whipple it was usual for ships' officers at that time to purchase their own instruments, and the company only intervened if the instruments chosen were not of the required standard for safe navigation of their ships.[10] Consequently Adams's transactions with individual ships' officers are not recorded in the company's books.[11]

On 21 April 1736, Adams took his first apprentice, Erasmus Ham, whose father George Ham (deceased) had been a victualler in the Strand.[12] Adams received a premium of £10 for this, and at the same time paid to the Grocers' Company a binding fee of 18s 0d plus two years' quarterage of 4s 8d. The premiums charged by members of the Grocers' Company varied over a very wide range, from nil (which was usual if the apprentice was a close relative) to as much as £300.

Druggists, in particular, seem to have charged high fees, several hundred pounds being quite usual for members following this occupation, which had been brought within the scope of the Grocers' Company by a charter granted by William and Mary in the seventeenth century.[13]

As mentioned earlier, Adams's home address at this time was given in the baptism register of St Bride's as Shoe Lane. His first child, Mary, baptized in July 1735, died within a few days of her birth. A second daughter, Ann, was born in August 1736 but fared little better: she died at the age of three in 1740. The high infant mortality rate that prevailed in Morris's family seemed destined to recur in George's, and makes one wonder whether there was something particularly insanitary about the Shoe Lane premises.

Towards the end of this year (1736), Adams had his first taste of public office. At a parish meeting of St Bride's held on 1 December to choose the three constables for the following year, no fewer than eight people in turn were proposed but on being chosen opted to pay the fine of £8 instead. Then Adams (who was not present) was proposed, together with Joseph Eldershaw and William Webb, and by a show of hands the meeting chose Eldershaw to fill one of the three posts. Adams's name was again proposed, with Webb and Leonard Ashburner, and the last named was chosen to fill the second post. The minutes continue:

> This Vestry then ordered Mr. George Adams and Mr. William Webb should be again put up to chuse one of them for a Third Constable to serve for the year ensuing, by a great Majority of Hands Mr. George Adams the Mathematical Instrument Maker was chose, and a messenger being sent to acquaint him thereof brought word he would stand.[14]

To everyone's relief this enabled the meeting – which must have lasted several hours – to be brought to a satisfactory conclusion.

The office of parish constable was an ancient one, dating back to medieval times, which gave the holder power to apprehend offenders and hold them in the parish lock-up pending their appearance before a magistrate. It is not known how often Adams had to exercise his authority during his year of office. Five years elapsed before he was called upon again, and then in the mid-1740s he served in all the other offices in turn except scavenger, for which he opted (in 1745) to pay the fine of £3 instead.[15]

In midsummer 1738 Adams vacated his first premises in Fleet Street and moved a few doors eastwards, to the corner of the entrance to Racquet Court. At the same time he ceased to be the ratepayer in Shoe

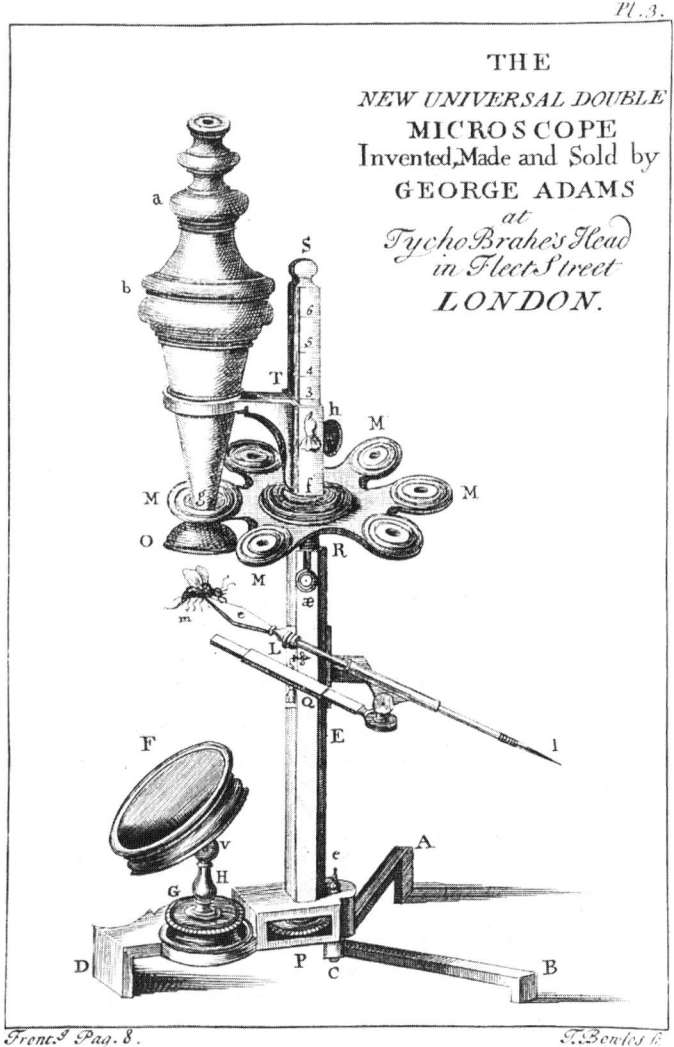

2.3 Adams's 'New Universal Double Microscope', as illustrated in the frontispiece of his *Micrographia Illustrata* (1746). The 'wheel' of eight objective lenses of different powers is mounted on a sleeve which can be slid to the appropriate height on the central pillar to suit the focal length of the objective in use. The compound (that is, 'double') body screws into a socket just above the 'wheel', as shown; alternatively it can be replaced by an eye shield if the instrument is to be used as a simple microscope. For fine adjustment the stage or forceps can be raised and lowered on a vertical screw by means of the knurled wheel in the base, conveniently situated so that it can be turned with the observer's hand resting on the table. No complete extant instruments exactly matching this illustration are known. (*Whipple Museum Library, Cambridge*)

2.4 An Adams compound microscope, '1746' pattern, similar to the design shown in figure 2.3 but with a different form of body mounting. In this version the body and its associated wheel of objectives is fixed to the top of a column which slides inside the central pillar, and can be clamped at the appropriate height. The fine adjustment is by a vertical screw as in figure 2.3. In both designs the mirror and its support can be removed, exposing a threaded hole in line with the optical axis, which can then be screwed on to a scioptic ball to covert the whole instrument into a solar microscope. (*Whipple Museum, Cambridge, Inv.Wh.197*)

Lane, where his younger brother Henry Adams took his place. It may be presumed, therefore, that henceforth George lived with his wife and children in Fleet Street 'over the shop'. His 'Corner of Racquet Court' premises had a rateable value of £50, which was, if anything, a little above the average for a single shop in a main thoroughfare. His family at this time consisted of a daughter Ann, aged two, and another daughter, Sarah, baptized in February 1738. He may also have had a son, George: a child named George Adams, whose baptism and hence parentage has not been traced, was buried in St Bride's aged two in January 1740/1. Sarah, born in 1738, was the first of George's children to live to maturity and have a family; one of her sons, Robert Blunt (George's grandson) trained as a mathematical instrument maker towards the end of the eighteenth century and became a member of the Grocers' Company.

George Adams remained at the corner of Racquet Court for almost twenty years, and would probably have stayed there indefinitely had he not been forced to move by an event outside his control, as will be seen later. The earliest of his trade cards has this address, which became 114 Fleet Street when numbers were allocated (after he had moved) in 1766.[16] Racquet Court itself survived until the mid-1970s, when it was obliterated by extensions to the *Daily Express* building (see figure 1.2), so neither of Adams's first two locations in Fleet Street is exactly traceable today. The last occupiers of number 114 before its demolition were the globe makers Messrs Geographia Ltd.

Adams's activities during his first ten years as a businessman are not well documented: there is only fragmentary evidence of his interests from a few newspaper advertisements and extant manuscripts, insufficient to reveal whether he had to struggle to survive or managed to secure a succession of lucrative contracts. Unlike his later rival, Benjamin Martin,[17] George Adams (senior) did not produce a continuous stream of books, tracts, and pamphlets on a variety of subjects. In the course of his working life of just under forty years his literary output comprised only two major works, on microscopes in 1746 and on globes twenty years later, plus a number of minor descriptive tracts issued between those dates. His first official appointment (to the Office of Ordnance) was not obtained until he had been in business for thirteen years. No dated shop bills or ledger entries prior to 1744 are currently known, and although a large number of instruments of all types bearing Adams's name are in existence today, few can be dated with certainty to better than the nearest decade or so.

From one surviving manuscript, now in the Science Museum Library, it seems that around 1740 Adams's thoughts turned to the construction of orreries – mechanical models representing the motions of the planets and their satellites.[18] This was a totally different field from that of navigational instruments such as Caleb Smith's or Hadley's quadrants. While the latter were produced in considerable quantities, orreries in the 1740s were essentially one-off products costing several hundred pounds, bought mainly by wealthy patrons for their prestige value, or occasionally by high-class educational establishments such as Watt's Academy. Thomas Wright, the principal maker of 'grand orreries', had attempted to introduce cheaper ones for schools, but surviving examples of these are extremely rare and there is no evidence to suggest that Adams was interested in this end of the market – at any rate, not until Benjamin Martin had popularized cheaper astronomical models of a different design twenty years later.

The manuscript concerned is dated 1741 and entitled 'The Description and Use of a Terrestrial Planitarium, Including the Orbits of Mercury and Venus. Commonly called an Orrery. Made by George Adams who makes all sorts of Mathematicall, Phylosophical, and Optical Instruments, at Tycho Brahe's Head, the corner of Racquet Court in Fleet Street, London. MDCCXLI.'[19] It comprises thirty-two octavo-size leaves, written on both sides, plus the title-page; the latter is laid out like the title-page of a book, with 'ORRERY' in Gothic characters, suggesting that Adams intended it to be printed though apparently this never took place. The first fifteen pages are concerned with arguments in favour of the Copernican system (with a reference to Newton's *Principia*), Kepler's laws of motion, and proofs of the Earth's motion round a stationary sun. A fold-out diagram of the solar system, 'Mr. Whiston's Scheme', taken from a printed book, is accompanied by several pages of explanatory text and tables. Then follows a caption title, 'The Description and Use of the Orrery', and the rest of the manuscript is devoted to this.

Adams's orrery, according to this description, consisted of a mahogany frame 2 feet in diameter and 6 inches high, standing on twelve brass ball feet; twelve brass ribs, with shields carrying the signs of the zodiac in between, supported twelve pillars which in turn supported a large flat circle representing the ecliptic. Metal plates covering the top of the box were painted blue, 'because the blew colour of the sky depends wholly upon the Earth's atmosphere, and not on a blew arch or canopy at an immense distance'. Clockwork (by which Adams meant wheelwork) housed within the mahogany box gave motion to the sun (rotation on its inclined axis), Mercury, Venus, the Earth, and the moon. One turn of the handle represented one day, and for convenience in demonstrating the motions of the Earth–moon system alone the annual motion of the Earth could be disconnected. Several interchangeable Earth-balls were provided for different demonstrations, including a small one for showing eclipses (with a lamp in place of the sun) and a large one with a map on its surface and an adjustable horizon. For demonstrating the changing elongations of the inner planets as seen from the Earth, a two-armed steel index rather like a pair of compasses could be fitted to the tops of the axles supporting the sun, Earth, and planet. The Venus ball was provided with a black cap like the moon's to illustrate its phases.

The description implies that Adams had made only one such model when this manuscript was written, and it is not clear whether it was a speculative venture or was made in response to a specific order. In comparison with the grand orreries made by Thomas Wright, William Deane, and Benjamin Cole (a former employee of Wright's), it was somewhat smaller and lacked the outer planets with their satellites. By this time (1741) Wright was advertising in Harris's *Description and Use of the Globes and Orrery* that he had made 'large ORRERYS', which by implication were all 4 feet in diameter, for the new Royal Academy at Portsmouth, Watt's Academy in Tower Street, and for the king at Kensington, as well as 'several other large ones for Noblemen and Gentlemen'.[20] Deane had published in 1738 his own description of a similar orrery which he said was 5 feet in diameter.[21] Cole, in his trade cards of this period, used an illustration of an orrery which was copied directly from the plate of Wright's orrery in Harris's book.[22] Desaguliers was using in his lecture courses two outwardly similar models each 3 feet in diameter, one being a planetarium and the other a true orrery incorporating the moon's motions.[23] All of this publicity for orreries may have led Adams to consider there was sufficient demand to justify an attempt to secure some of the business himself.

At least seven large orreries by Adams which may be loosely described as grand orrery type, with twelve-sided or circular wooden boxes housing the mechanism, are in existence today, but as the basic design remained much the same for half a century they are difficult to date.[24] Most have an overall diameter of about 2½ feet, but one at the Whipple Museum, Cambridge, which is definitely pre-1752, has an overall diameter of about 3½ feet. This model, illustrated in figure 2.2, incorporates the outer planets and may be regarded as being a slightly later development of Adams's 1741 design.[25]

It is possible that Adams was induced to enter the orrery market by a chance encounter with an agent for the third Earl of Bute. Though no direct contact between Adams and the earl at this time has emerged, a letter in the Bute archives indicates that in 1739 or 1740 a man named Adams, presumed to be him, was consulted about a wheel-cutting engine. The writer, Archibald Bothwell (1699–1756), made purchases of books and so on on behalf of the earl, before the latter moved permanently to London in 1746. On 3 March 1739 (probably old style, that is, 1740) Bothwell wrote from Forrest's Coffee House, Charing Cross, to say that some time ago he had asked 'Adams in Fleet Street' about the price of a machine for cutting clock (that is, wheel) teeth, but he could not tell him then because it was usual for clockmakers to make the greater part of such machines themselves. Adams had since told him that the lowest price with all the associated apparatus would be 10 guineas. Bothwell said that if the earl decided to order one, Adams would want to know how many divisions and circles he should provide on the dividing plate.[26]

Why the earl should have wanted a wheel-cutting engine is not clear, but it could be that he was thinking of having an orrery made by one of his workmen in Scotland. If so, no doubt Adams would have offered to do the whole job himself. Be that as it may, Adams certainly had more than a passing interest in the subject, for when he published his *Micrographia Illustrata* a few years later (1746) he appended a list of instruments which included descriptions of not only a variety of planetariums up to 3½ feet in diameter, but also a 'COSMOTHEORION, of a new Invention, which at present is without a Parallel'.[27] According to his lengthy (about 400 words) description of this device, it was 4½ feet in diameter and stood on 'a pedestal of curious workmanship'; it incorporated all the known members of the solar system – satellites as well as planets – including the inclinations and eccentricities of their orbits.

No evidence has come to light to suggest that the 'Cosmotheorion' was ever made. From Adams's description it sounds suspiciously like the Leyden sphere, thought to have been made *c.* 1670 by Steven Tracy of Rotterdam, and located since the early eighteenth century at Leiden University.[28] Nearly half a century after George Adams senior published his description of the Cosmotheorion, his son George junior made use of an engraving of the Leyden sphere printed *c.* 1711 for an encyclopaedia plate purporting to show one of his own products,[29] so

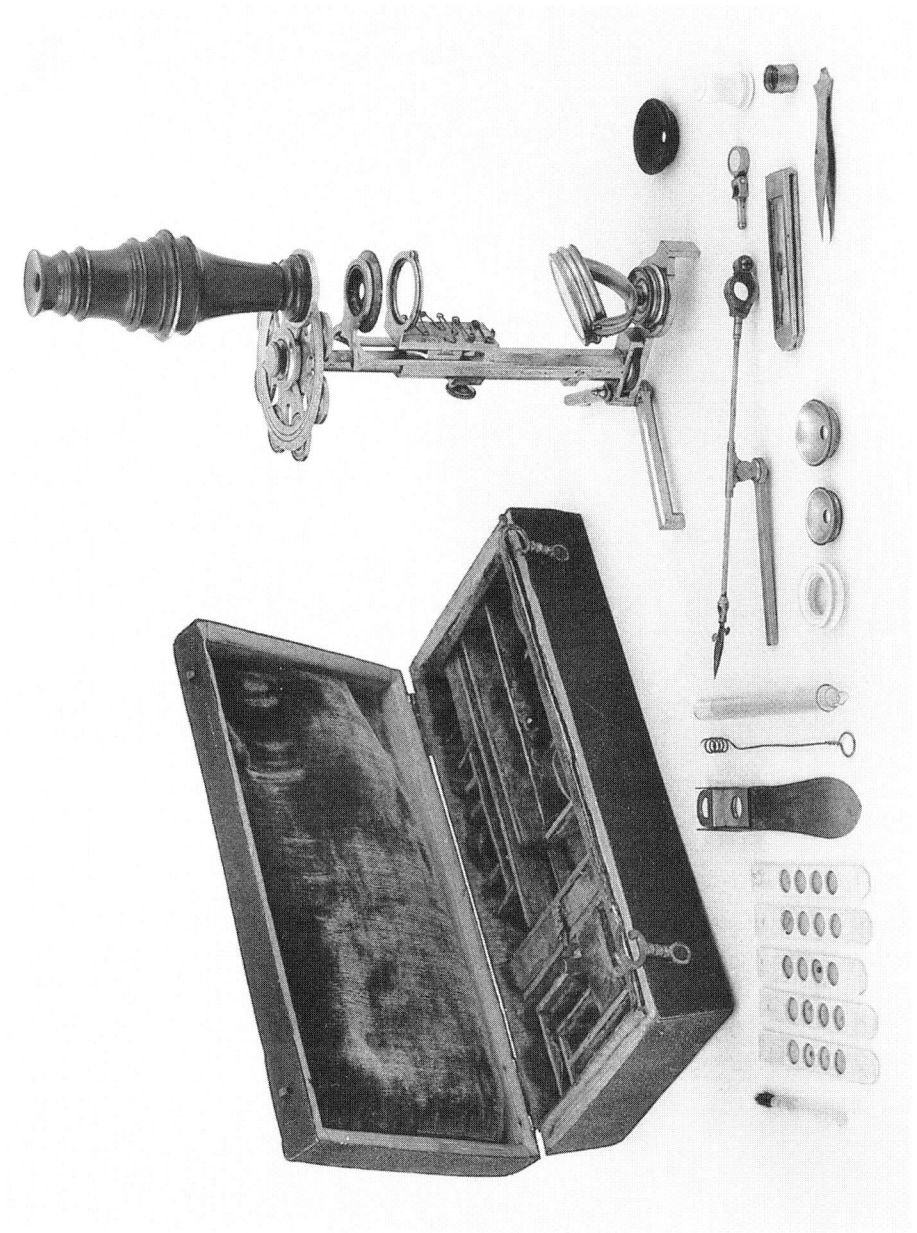

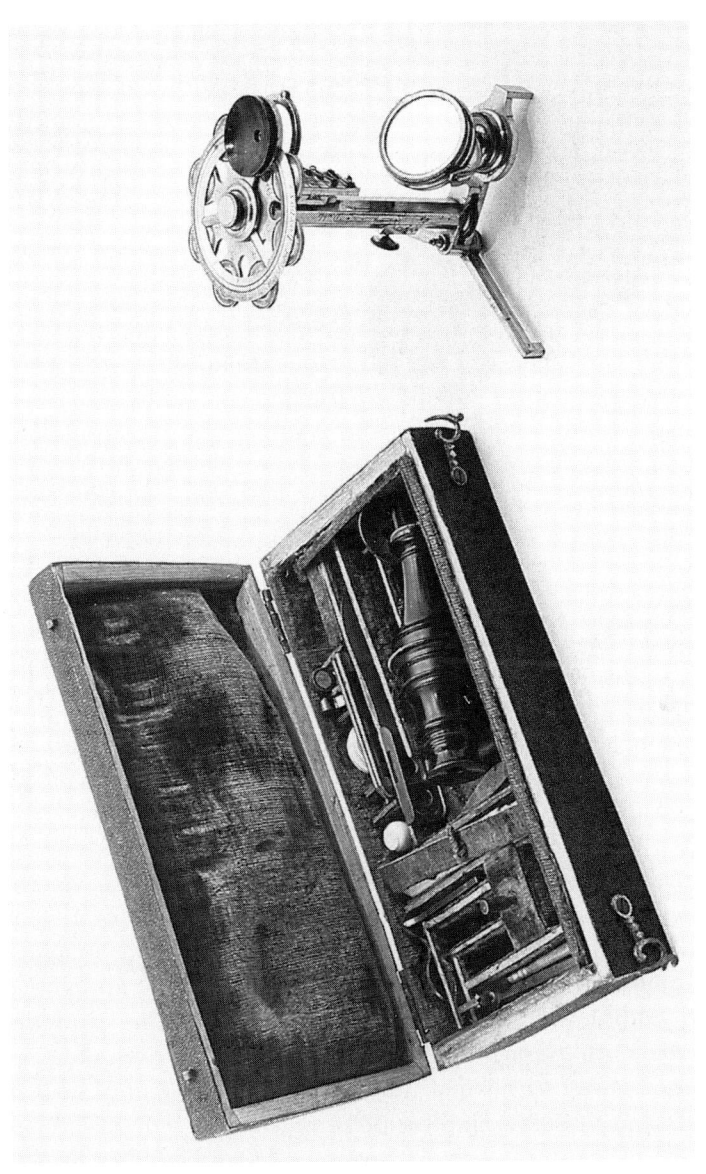

2.5 Another microscope of Adams's '1746' pattern differing from that shown in figure 2.4 only in minor details, such as the form of mirror mounting and the ornamental turning of the wooden body tube. (a): assembled as a compound microscope. (b): assembled as a simple microscope, with the rest of the apparatus and accessories in the fitted wooden case. (*Museum of the History of Science, Oxford, inventory no. 54497; G.L'E.Turner cat. no. 238. Photographs by G.L'E.Turner*)

it is likely that George senior possessed a copy of this now rare engraving in the 1740s and considered making a similar sphere himself. He may even have seen the actual sphere at Leiden, for we do not know whether he travelled abroad in the course of his education. In a catalogue appended to his *Description and use of the Trigonometrical Octant*, published in 1754, the Cosmotheorion is mentioned briefly as a machine 'by which all the celestial Phenomena are plainly and clearly exhibited', but after that nothing more is heard of it.

Adams did make one fairly elaborate grand orrery in the 1750s, possibly commissioned by Princess Anne, daughter of King George II, though this model did not have all the features attributed to the Cosmotheorion. In 1757 it was amongst the effects of the late Samuel Koenig (1712–57), librarian to the Stadholder William IV, when it was described in a sale catalogue as 'Le grand planetaire inventé par Mylord Orrery & executé par George Adams in Fleetstreet à Londres en 1754. Ce superbe Instrument a environ 3 pied de diamètre & il est posé sur une table de bois de Mahogny, garni d'un tiroir conteneant tout l'apparat propre au planetaire'. It sold for 1,310 Dutch florins, equivalent to about £110, including a manuscript treatise by Adams describing its functions.[30]

Returning to the early 1740s, in February 1741/2 Adams's family was increased by the birth of another daughter, named Ann after her mother, the first to bear that name having died at the age of three in July 1740. This second Ann certainly lived into her fifties but apparently never married (she was a spinster when she had some communication with the Grocers' Company in 1793).[31] Ann Adams being a very common name, her residence as an adult has not been identified, nor is it known what happened to her after 1793.

A month after Ann's baptism on 7 February 1741/2, on 8 March, Adams took another apprentice, Charles Baynham. The son of a 'yeoman' of St Giles-in-the-Fields, his apprenticeship attracted a premium of £30, three times the sum that Adams charged for his first apprentice in 1736.[32] In addition to the binding fee of 18s, Adams paid the company six years' quarterage of 14s, so he seems to have adopted the common practice of letting his quarterage payments lapse until he was obliged to attend at Grocers' Hall for other reasons. Baynham did not complete his seven-year term with Adams: on 27 October 1744 he was turned over to William Collier, a member of the Goldsmiths' Company.

In 1742 Adams was chosen for his second period of service in a parish office, as collector of Poor Rates for the Salisbury Court precinct of St Bride's.[33] The churchwardens' accounts show that he collected around £110 for each of the four quarters.[34] In the following year (1743) he served as both a questman and a sidesman, so quite a substantial proportion of his time must have been devoted to parish matters at this period: the vestry minutes show that he attended at least four vestry meetings in 1742–43 as well as carrying out his duties as collector, questman, and sidesman in the parish.

It is worth noting here that one of Adams's fellow-members of the Grocers' Company, who also had premises in Fleet Street in St Bride's parish, was a druggist named Shute Adams. Shute, whose first name probably commemorated a forebear on the female side, was a son of George Adams of Bristol, and may have been a distant relation of the instrument maker, though this has not been established. George Adams of Bristol was a goldsmith and banker, and evidently a man of some means, as a premium of £300 was paid to apprentice Shute to a druggist and member of the Grocers' Company in 1734. Furthermore, having served five years of his seven-year term, Shute Adams purchased his freedom in the company in September 1739 by redemption.[35] Like George Adams the instrument maker, Shute Adams the druggist was called upon to serve in all the parish offices of St Bride's, and also like George, he eventually served as a warden of the Grocers' Company. Casual references to 'Mr. Adams of Fleet Street' in the mid-eighteenth century need to be scrutinized for subsidiary information before they can be positively assigned to one or the other.[36] Shute's shop was on the south side of the street, almost facing Shoe Lane.

About 1743, George Adams's interests switched to the field of microscopes and their uses, perhaps prompted by the almost simultaneous appearance of two books on this subject towards the end of 1742.[37] *Micrographia Nova*, by Benjamin Martin, first advertised in November 1742,[38] did not pose any direct threat to Adams's business as Martin at this time was an itinerant lecturer with no shop of his own. The microscopes he designed had some novel features, but were fairly crude in comparison with those produced by London makers; his 'universal' model (so called because the body could be tilted to any angle) incorporated an unobscured stage supported by a brass side pillar, which was one of the principal features of the slightly later 'Cuff' design, but it had no provision for fine focusing, and apart from the pillar the construction was largely of wood and cardboard. This model and a smaller 'portable' microscope, based on those devised by him at Chichester in 1738–39, were illustrated by full-size engravings in Martin's

1742 book. A cartouche informed readers that they could be obtained from the publisher John Newberry in Reading, Berkshire, and a few other agents elsewhere. No complete extant examples of either type matching his plates are known, and probably few were made, as they were rapidly superseded by the more professional models in brass produced by London opticians and instrument makers from the mid-1740s onwards.

The other publication issued in late 1742, Henry Baker's *The Microscope made Easy*, was of more direct interest to Adams as it contained references to his rival, John Cuff, whose shop was on the same side of Fleet Street near Fetter Lane.[39] Both Martin and Baker devoted most of their texts to describing a great variety of microscopic objects; but while Martin introduced his work with a description of his own two microscope designs, Baker surveyed the various types then available on the London market, with specific examples drawn from the products of John Cuff. Baker's book, which bore the imprimatur of the Royal Society dated 28 October 1742, included an account of Cuff's improvement of the solar microscope shown by Liberkhun to the Royal Society about 1740. (Martin also dealt with solar microscopes, but only in the scioptic ball form.)

The Microscope made Easy proved to be a popular work. A first edition of 1,000 copies is said to have been sold out by March 1743 and an enlarged second edition appeared in April that year;[40] for this, Baker added a plate showing Cuff's improved solar microscope (which had only been described in the first edition). Martin's book, which evidently did not sell so well as Baker's, was re-advertised in the London papers at about the same time. In June 1743, Adams started advertising a product of his own, in the following terms:

> A New Portable MICROSCOPE adapted for viewing all Kinds of Minute Objects, as well Opake as Transparent, in a more conspicuous and concise Manner than any other heretofore made. It is also a very curious Solar Microscope when apply'd to a proper Apparatus.
>
> Invented, made, and sold by G.Adams, at Tycho Brahe's Head, the Corner of Racquet Court in Fleet-Street; Who also makes and sells all sorts of Mathematical and Optical Instruments, viz. Choice of Pocket Cases of Drawing Instruments, curious Reflecting Telescopes, together with an Azimuth Compass of a new Contrivance, for finding the Variation of the Magnetical Needle at Sea, with great Ease and Certainty, without any Astronomical Calculation, and by the new invented Reflecting Sea Quadrants of either sort, &c.[41]

Adams does not appear to have produced a printed descriptive leaflet or tract dealing with this microscope at the time of its introduction, but according to Clay and Court (1932) there was then in the Court Collection a manuscript description by Adams, with illustrations, of an instrument closely resembling the 'new universal' microscope depicted in his *Micrographia Illustrata* (1746) except that it was called a 'new portable' microscope.[42] Unfortunately this manuscript has since disappeared. As it used the same term as the 1743 newspaper advertisements, the latter most probably referred to what is nowadays always known as Adams's 'new universal' model, generally assumed to have been introduced three years later in 1746, when his book was published.

Meanwhile, in November 1743 Baker's *The Natural History of the Polype* appeared, inspired by the work of Abraham Trembley;[43] this was followed in 1744 by a virtually unchanged third edition of *The Microscope made Easy*. In the same year (1744) Baker also wrote the descriptive text to accompany a reprint of the plates from Hooke's *Micrographia* of 1665, then just rediscovered, published in March 1745 as *Micrographia Restaurata*. All this activity must have generated (or perhaps reflected) an upsurge of interest in microscopy amongst the educated public, as well as in purely scientific circles, and persuaded Adams that a full-scale book on the subject, drawing attention to his own microscope designs, would be worthwhile.

Advance publicity for Adams's *Micrographia Illustrata: or, the knowledge of the microscope explain'd, together with an account of a new invented universal single or double microscope* began early in January 1746 and indicated that the book would be published on Saturday 11 January.[44] It would be a quarto volume, containing a description of the nature, uses, and magnifying powers of microscopes in general; directions for the preparation of specimens; an account of the principal microscopical discoveries mentioned by the most celebrated authors; a great variety of new experiments and observations; a translation of Joblot's observations on animalculae; and an account of the freshwater polyp, translated from the French treatise of Mr Trembley. It had been compiled, said Adams, 'for the Assistance of those, who are desirous of surveying the extensive Beauties of the minute Creation', and would be illustrated with sixty-five copperplates containing upwards of 560 pictures of microscopic objects.

The book was duly published on 11 January 1746, price 16s bound; it was 'Printed for and sold by the Author', and the initial newspaper advertisements indicated that it was sold also by Staples & Barlow at

York, though their name did not appear in the imprint. It was technically a quarto, being printed in four-leaf sections, though the 242 pages measured only about 9 × 6½ inches overall. (Martin's *Micrographia Nova* had the more usual quarto format of 10 × 8 inches; Baker's *The Microscope made Easy* was an octavo.) A further twenty-one pages at the end contained a 'Catalogue' of instruments. It was illustrated by sixty-five copperplates, as promised.

Its appearance immediately produced a strong protest from Baker, who complained that it was a blatant plagiarism of his own book. On 17 January he wrote to six of his friends, trying to enlist their aid in exposing Adams's underhand behaviour.[45] The third edition of *The Microscope made Easy*, which had been published in 1744, was re-advertised (by Baker's bookseller, Dodsley) several times in February 1746 in the midst of Adams's publicity campaign,[46] the rival advertisements often appearing side by side. Those for Baker's book stressed that it had the approval of the Royal Society, of which he was a fellow. But his efforts to denigrate Adams's book apparently had little effect: from the beginning of April 1746 the advertisements for *Micrographia Illustrata* indicated that it was sold not only by the author and Staples & Barlow, but also by Fletcher at Oxford, Thurlbourne at Cambridge, and Wickes at Norwich. If these advertisements are taken at face value, it is evident that the book achieved distribution in all the important centres of learning; however, there is a suspicion that Adams cited the names of these booksellers without their knowledge or approval, hoping to give the impression that his book was widely available. Be that as it may, a second edition appeared in the following year,[47] bearing the name of a London bookseller, S. Birt of Ave Maria Lane, jointly with Adams's in the imprint. Even if the provincial distribution claimed for the first edition was fictitious, Adams could hardly have used the name of a London bookseller in this way without his approval. This 'second edition' seems to consist simply of the original sheets with a new title-page; since it is not known who actually printed the book, no information on the total number of copies printed is available.

If *Micrographia Illustrata* had been straightforward plagiarism of *The Microscope made Easy*, Baker's failure to undermine its sale might have been hailed as an example of the superior power of media advertising over the efforts of an individual; but in reality Baker's complaints had little substance. It is undoubtedly true that the two books had much in common – when both were quoting the works of previous authors, such as Leeuwenhoek, that is hardly surprising – but Adams's

2.6 Adams's '1746' microscope could be converted into a solar microscope by attaching the foot to a scioptic ball, but the fixed horizontal tube design made by Cuff and shown in the second edition of Henry Baker's *The Microscope made Easy* (1743) was much more convenient to use. Cuff's instrument had a wooden base and a wheel rotated by a cord and pulley. The one shown here is a more advanced version, of all-brass construction with geared axial rotation; it is signed 'Made by GEO ADAMS at Tycho Brahe's Head in Fleet Street LONDON', and probably dates from c. 1750. (Museum of the History of Science, Oxford, inventory no. 71754; G.L'E.Turner cat. no. 239. *Photograph by G.L'E.Turner*)

35

not only covered a considerably wider field (with due acknowledgements to his sources), it also cost over three times as much. Nobody in their senses would have bought Adams's book at 16s if they could have obtained the same information from Baker's at 5s. Moreover, Baker himself had been guilty of, if not plagiarism, at least bad manners, in rushing into print with his own work on the *Polype* in 1743 before the original discoverer (Trembley) had published anything himself. This action of Baker's had particularly upset Martin Folkes, president of the Royal Society, to whom Trembley had sent the first specimens of polyps early in 1743.[48]

In his preface, Adams admitted that his book was 'rather a faithful collection of everything that has hitherto been mention'd by the best writers upon microscopes, than matters of my own invention'. For this purpose, he said, he had been 'allowed free access to one of the finest libraries in England, belonging to a noble personage, whose name I am not permitted to mention'. (Possibly either the Duke of Argyll or the Earl of Bute.) But he had, he said, enlarged on other authors where necessary, and (as stated in the newspaper advertisements) had attempted a translation of 'two very valuable pieces: the one containing Mr. Joblot's observations upon the animalcula, that are found in many different sorts of infusions; the other Mr. Trembley's account of the fresh water polype; neither of which, I believe, has hitherto appeared in our language'. If Adams translated these himself, as is implied, it indicates that he was familiar with the French language. However, this does not necessarily mean that he had French origins. When Francis Watkins of Charing Cross decided to write a book on microscopy in 1754 he did so in French, because (he said) that language was more widely understood amongst philosophers than English.[49] Nothing has presently been discovered about George Adams's education prior to his apprenticeship at the age of fifteen. London abounded with private schools and academies, and his father Morris Adams was no pauper, so George may well have had quite an extensive education before being handed over to James Parker to learn a trade.

As for the hundreds of illustrations in Adams's book (far more than in Baker's), while some can be identified as copies of well-known plates by Hooke and other authors, Adams claimed that he had produced large-scale drawings by first placing objects in a solar projection microscope, and then reducing them to a size suitable for book illustration by means of a camera obscura, before sending them to the engravers. At least two engravers were employed: T. Bowles and J. Wigley; some other plates, mainly of apparatus, are unsigned. The large number of copperplates (sixty-five, some with numerous figures), Adams said, accounted for the high price of the book.

In practice the market proved large enough for both Adams's and Baker's books to be viable. (Martin's *Micrographia Nova*, on the other hand, rapidly faded into oblivion, probably because the instruments depicted therein were relatively crude and soon became obsolete.) Baker's certainly suffered no permanent damage from the publication of Adams's, as it went through two more English editions and was translated into Dutch (three editions) and French. Adams, being concerned with instruments in general, turned his attention to other matters after 1747, though he did produce an updated edition of his microscope book towards the end of his life, in 1771.[50] Baker's comments, in his correspondence with friends in 1746, strongly suggest that his indignation at Adams's effrontery in publishing *Micrographia Illustrata* was really prompted by the fact that Adams was a mere tradesman, while Baker considered himself a philosopher and a gentleman.[51]

The importance of *Micrographia Illustrata* in this account of the Adams family and their activities is twofold. In the first place, the instruments described and depicted therein secured for George Adams senior a permanent place in the history of the microscope. This aspect has been discussed in the literature,[52] so it will suffice here to say that Adams's 'new universal' microscope (see figures 2.3 to 2.5), though incorporating some novel and useful features, was eclipsed by the more practical Cuff design introduced at Baker's instigation at about the same time.[53] This was copied by numerous other opticians and instrument makers in London and abroad, including Adams himself. The second point is that Adams appended to his book a twenty-one-page 'catalogue' of instruments said to be made and sold by him. As this 'catalogue' has given rise to some misconceptions by present-day writers, it merits careful examination.

In his newspaper advertisements for *Micrographia Illustrata*, Adams claimed that such an extensive and explanatory catalogue of instruments had never been printed, and he was undoubtedly correct; but none of the 335 items listed is priced, and in a lengthy preamble he said that he had numbered each one 'so that if a Gentleman is desirous of any one or more of them, and is at any Distance from London, he need only send me the Numbers adjoining to those that he intends to purchase, and he shall be served with Fidelity, and at the lowest Prices'. In other words, what Adams had done was to list every instrument that he could think of, in sufficient detail for intending purchasers to identify their requirements by simply quoting the reference numbers. The 'catalogue' does not indicate the range of goods that he manufactured, much less kept in stock, nor does it necessarily have any bearing on the size of his workshop. Some of the items listed, such as the Cosmotheorion mentioned earlier, were most probably never made at all. Others, such as item 333, 'Astronomical Clocks, or Regulators carefully performed', are unlikely to have been made on Adams's own premises, though he would have known where to obtain one quickly if a customer placed an order. Some specific items were simply taken from published books, such as 'The famous Glass Sphere of the Reverend and Learned Dr.Long's Invention' and 'The Uranium invented by the Reverend Dr.Long'. Entries such as these indicate Adams's willingness to supply virtually anything that his customers wanted: they do not imply that he kept such items in stock.

That Adams certainly employed subcontractors in the course of his business is revealed by one of his policy statements at the end of his preamble. This (number 2 below) has frequently been quoted, but to obtain the full picture one needs to read the whole set:

> In the Construction of all the Machines I have ever made, my first and greatest Care hath been to procure good Models and Drawings, several of them I have imitated from the best Authors, as well Foreigners, as those of our own Country; I have alter'd and improved others, and have added new ones of my own Invention. And,
>
> 1. In all my Performances I endeavour not to augment the Instruments and Machines with superfluous Ornaments, both that they may be of frequent Use to those of middling Fortunes, and that their Neatness may render them not unworthy of a Place in the Cabinets of the Curious.
>
> 2. That their Exactness may be particularly attended to, I always inspect and direct the several Pieces myself, see them all combined in my own House, and finish the most curious Parts thereof with my own Hands.
>
> 3. To the End that their Construction may be as simple and substantial as the Uses of the Instruments will admit; it is my constant Study to contrive them in such a Manner that they may be managed with the greatest Ease.
>
> 4. I also have Respect to their being made applicable to several Operations, especially when the Extent of their Use does not prejudice their Simplicity, to the End that Instruments may not be multiplied without Necessity.[54]

At this time (1746) Adams had no registered apprentices, Charles Baynham having been turned over to another master in October 1744.

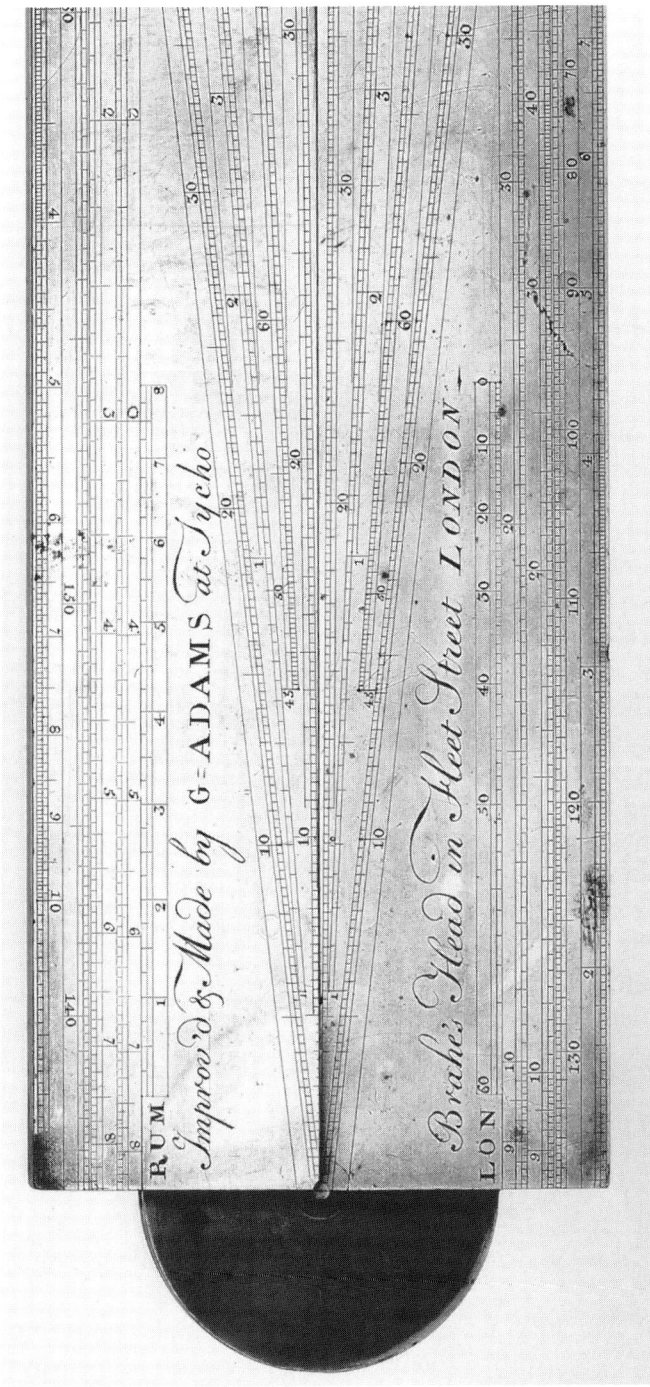

2.7 Detail of the signature on an early sector by George Adams citing his shop sign, 'Tycho Brahe's Head'. (*Christie's, London, 27 September 1990, lot 251*)

Unfortunately we have no way of determining directly the number of journeymen, or trained workmen, that he employed. The interpretation sometimes made of his second point, that he was one of the first instrument makers to employ mass-production methods,[55] can almost certainly be discounted, however, except in the limited sense that he used division of labour to make the various component parts of an instrument (as did watchmakers, for example). His remarks about finishing the 'most curious' pieces himself implies that each product despatched from the Adams workshop was individual, in which the components had been carefully 'matched' to make them fit; this is the very opposite of mass production.

In November 1746, Adams was chosen to be junior churchwarden of St Bride's, and for the next year and a half parish matters must have dominated his life. Churchwardens were normally appointed each Easter, to serve until the following Easter. In the parish of St Bride's it was the custom for both wardens to be chosen by the parishioners; it was also customary for the junior warden, having spent a year effectively learning the job, to be elected senior warden, the most responsible of all the parish offices, in the following year. In Adams's case what happened was that the senior warden, John Simmons, died in November 1746 and both offices became vacant.[56] William Merryfield was chosen to serve as senior warden for the rest of the year, and three names, including Adams's, were proposed for the junior post. As in 1736 when he was voted into the office of constable, Adams was not actually present at this meeting, and on being chosen had to be informed by messenger.[57] The minutes show that he was present as junior warden at meetings of the vestry on 4 and 18 December 1746, having agreed to serve.

Three months later, in March 1746/47, a major domestic upset occurred: Adams's first wife, Ann, died at the age of forty-four, and was buried in St Bride's churchyard.[58] George, then aged not quite thirty-eight, was left with a family of two surviving daughters, aged nine and five respectively. It was natural that he should consider marrying again, but as he expected to be heavily involved in parish matters for the next year or so he did not take this step at once. Following precedent, at Easter 1747 he was elected senior churchwarden of St Bride's, and served in that capacity until the next annual meeting on 14 April 1748.[59]

The senior churchwarden was one of the most important men in the parish (much more so than today), having charge of all the parochial income and expenditure and generally taking the chair at vestry meetings. It must have been an excellent way for a businessman to make himself known; but it also involved a great deal of work, so it is not surprising that some people, on being chosen, opted instead for payment of the fairly substantial fine of around £14 (equivalent to three months' wages for a journeyman).[60]

At a time when each parish, not the state, looked after its own sick and poor and elderly, the day-to-day management of parochial affairs involved quite large sums of money. During his year of office George Adams made about 400 entries in the churchwarden's accounts, with income and expenditure totals for the year each approaching £2,000, equivalent to perhaps £½ million in today's money.[61] A substantial part of this was the expense of running the parish workhouse, though this activity did produce a small amount of income by way of sale of the paupers' labour, chiefly in sewing shirts. The major source of parochial income was, of course, the Poor Rate, which amounted to about two-thirds of the total every year, but there were several other sources as well, summarized in Adams's accounts for 1746–47 as:

	£	s	d
Poor rate	1,242	19	9
Fines	144	0	0
Manufactures in the Workhouse	51	14	1
Casual receipts	196	7	6
Pitts & knolls[62]	37	14	2
Rents & benefactions	218	1	4
	£1,890	16	10

This income was used to pay the following expenses:

	£	s	d
Staffman's bills[63]	124	3	6½
Mr. Jones's weekly bills[64]	262	11	3½
Provisions and apparel for the Workhouse	729	16	1¼
Gifts to charities etc.	54	1	0
Apprentice bindings[65]	32	0	0
Officers' salaries and necessaries for the Church[66]	145	3	1½
Expenses at Vestries[67]	163	0	10
Divers [misc.] expenses[68]	276	15	3½
Casual and other poor	84	4	3
For Mr. Green	5	0	0
	£1,876	15	5¼

From these totals it will be seen that Adams completed his term of office with a balance in hand of £14 1s 4¾d.

The parish church of St Bride's in Adams's day was a fairly new structure, built to Wren's design in the early 1670s to replace the previous church destroyed in the Great Fire of 1666. It was the seventh to stand on this site since Roman times. The well-known 'wedding cake' steeple, originally 234 feet high, was added in 1703. Adams was fortunate in having no major repairs to contend with during his term of office. A successor seven years later was not so lucky – in 1754 the steeple was struck by lightning and lost about 8 feet in height when it was rebuilt. This event gave rise to discussion in the contemporary journals about the merits of the new Franklin-type lightning conductors, but the building remained unprotected for many years.[69] Unlike the old church of St Dunstan's in the West, which in Adams's time abutted Fleet Street (except for some small single-storey shops which were eventually removed in the 1770s), the church of St Bride's was completely surrounded by high buildings, and apart from the towering steeple could not be seen from the street; the perspective view shown in figure 2.8 is an artist's impression, from an imaginary viewpoint to the north-west. After a fire destroyed several buildings on the Fleet Street side in 1824,[70] a public subscription enabled a small part of their site to be left open when the shops were rebuilt, so that Wren's tower and steeple, though not the body of the church, remained permanently visible. The interior of the church itself was almost completely destroyed by firebombs in December 1940 but has since been restored. So far as is known, the pre-1940 structure did not contain any memorials or other references to the Adams family.[71]

Three months after handing over to the next churchwarden at Easter 1748, on 5 July that year George Adams obtained a Bishop of London's licence to marry Ann Dudley, spinster, aged twenty-six, thirteen years younger than himself.[72] The wedding took place not at St Bride's but at St Martin in the Fields, Westminster, on the following day. During the next fourteen years they were to have nine children – three boys and six girls. The overall survival rate was rather better than in the previous generation, at least five of the nine (including two of the boys) living to maturity, in addition to George's two surviving daughters, Sarah and Ann, from his first marriage.

Curiously, Adams does not appear to have become involved in the scramble to produce and demonstrate electrical apparatus and phenomena in the late 1740s, following the invention of the Leyden jar in 1746.

The Fleet Street Business   43

2.8   The parish church of St Bridget alias St Bride, Fleet Street, as rebuilt to Wren's design after the Great Fire of 1666, with its distinctive 'wedding cake' spire. George Adams was senior churchwarden here in 1747/48. This mid-eighteenth century print is a conjectural view from the north-west, impossible to achieve in practice as the church has always been closely surrounded by buildings. (*Guildhall Library, London*)

The newspapers of this period contain numerous advertisements by people of various trades and professions offering to give demonstrations of electrical phenomena,[73] and there was great activity in this field by members of the Royal Society, especially William Watson. Possibly Adams was too preoccupied with first parish matters, and then his second marriage, to consider entering this new and highly competitive field. Instead, he reverted to the more familiar field of navigation, with the launch in the autumn of 1748 (soon after his second marriage) of a 'new sea quadrant' purporting to be of his own design. This immediately produced a storm of protest from one of his rivals in Fleet Street, Benjamin Cole (senior), reminiscent of his brush with Henry Baker in 1746 over the publication of *Micrographia Illustrata*.

Some years earlier Benjamin Cole had been an employee of Thomas Wright's in Fleet Street, mathematical instrument maker to King George II, but he had been in business as a mathematical instrument maker on his own account since the late 1730s, first in Poppings Alley, Fleet Street, and then in Ball Alley, Lombard Street. When Wright retired in 1748, Cole took over his shop 'At the Sign of the Orrery' in Fleet Street on the corner of Peterborough Court, but not the royal appointment. He was a member of the Merchant Taylors' Company, and had apprenticed his own son, Benjamin Cole junior (1725–1813), to himself in 1739.[74] His son-in-law, John Sudlow, who married his daughter Elizabeth, was also a mathematical instrument maker, though not in the same company: he was apprenticed to Joseph Jackson in the Grocers' Company in 1736 and obtained his freedom in 1750.[75] It is not at present known whether Sudlow worked for Cole senior after qualifying, but the latter certainly employed some staff as well as his own son, for in February 1749 he advertised for 'a journeyman Clockmaker, that can be well recommended'.[76] As the elder Cole carried on making grand orreries after Wright had left this field, he would obviously have needed to employ some workmen with experience of wheelwork.

On 14 June, 1748, not long after taking over Wright's shop, Cole advertised a new quadrant:

To the CURIOUS
BENJAMIN COLE, Mathematical Instrument-Maker, at the Orrery in Fleet-Street, London, has invented A New Quadrant for taking of the Sun's Altitude at Sea, in a manner more easy and correct than by any of the other Instruments in Common Use. Price 18s.

The Quadrant, and a Book of its Use, may be seen at the Jerusalem and Portugal Coffee-Houses, in Sweeting's Alley; the Pennsylvania Coffee-House, in Bell-Yard; and at the Inventor's in Fleet-Street: But as the Sign of the

Orrery is taken down to repair, Gentlemen are desir'd to take Notice, that there is a Hadley's Quadrant, and this new one, in its Stead.

Note, The Construction of this Instrument is so simple, that a Person unacquainted with the Manner of taking of the Sun's Altitude, may at first Trial become a perfect Master of its Use; it not being so perplexed, or liable to the common Inconveniences, as is Davis's Quadrant. And with a small additional Charge, it can be fitted to take Altitudes without an Horizon; and serve all the Purposes of a Gunner's Quadrant, for setting of Artillery to any Degree of Elevation.[77]

Three days later, on 17 June, Cole advertised under the heading '*Just publish'd, Price 1s.*' his *Description and Use* of this new quadrant. The wording of the rest of the advertisement is similar to the above. It illustrates the point made earlier, that while advertisements for publications were common, advertisements for instruments were relatively novel: they had to be disguised as, and located amongst, book advertisements to encourage people to read them. (On 17 June, when this second version of Cole's advertisement first appeared, the *Daily Advertiser* had three columns of 'book' advertisements for over thirty titles, including Cole's.)

Cole's advertisement was repeated several times during the next few months. His new quadrant was more like a backstaff than a Hadley's quadrant: the index arm with a sighting vane was held stationary in a horizontal position from the observer's eye, while a pivoted frame carrying the sun vane and a 90-degree arc was moved against it (figures 2.9 and 2.10).

At the end of September 1748, Adams published his *Description and Use of a New Sea Quadrant*, 'invented, made and sold by George Adams'. It described an instrument almost identical to Cole's, and incorporated a folding frontispiece (dated 30 September) similar to Cole's, depicting a mariner using the instrument (see figures 2.11 and 2.12). This publication was advertised no fewer than nine times in the *Daily Advertiser* between 1 and 14 October, almost every weekday. The quadrants then used at sea, said Adams, were either incapable of taking measurements with sufficient accuracy, or were too expensive for ordinary use; in the first category were Davis's quadrants, and in the second Hadley's. 'In order therefore to remove the Imperfections of the former, and Expence of the latter, I have constructed a Quadrant, which is freed from all the Incorrectness of Davis's; and have also added to it, one of the most useful Properties, for which the Reflecting Quadrants have been so much esteemed.' He explained that he had based his design on a description of the 'Mariner's Bow' invented by

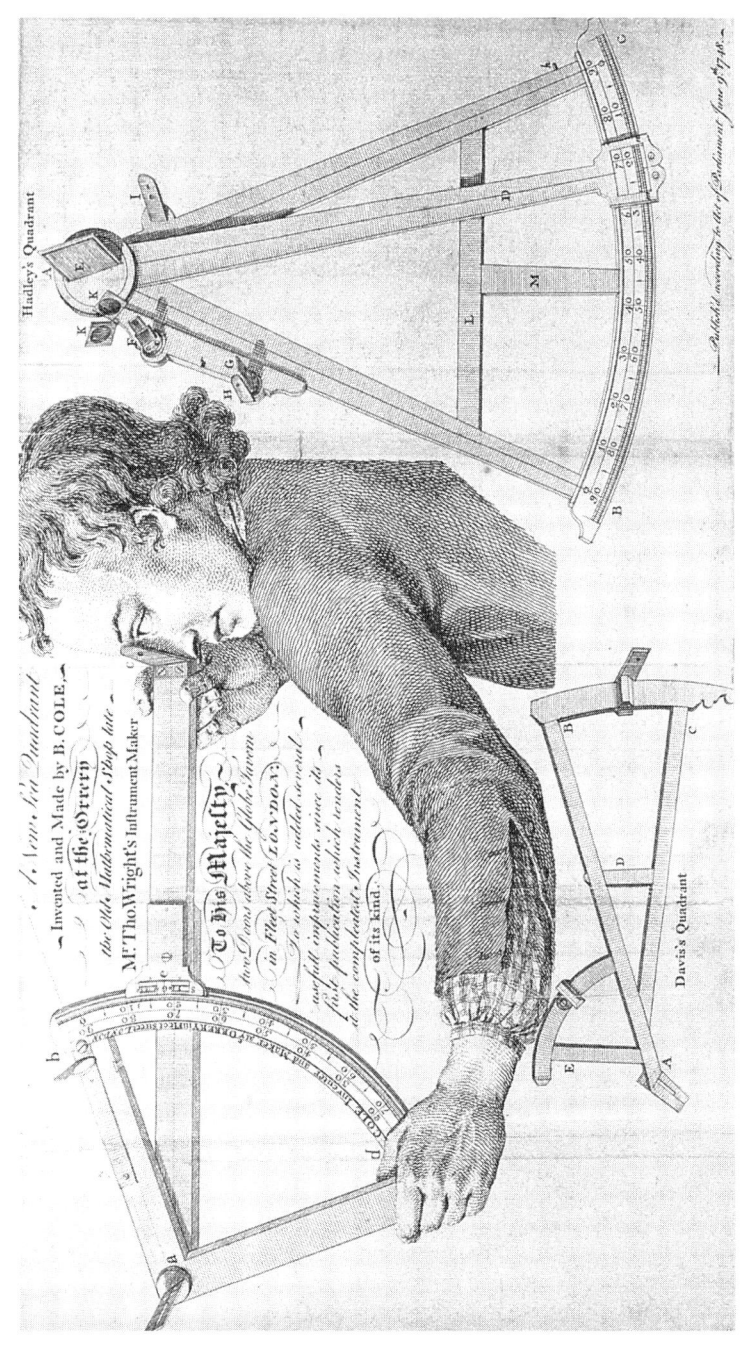

2.9 Frontispiece of Benjamin Cole's tract describing his 'New Sea Quadrant', published in June 1748. As Cole held no official appointment himself, he gave prominence in the inscription to the fact that his shop was formerly occupied by Thomas Wright, mathematical instrument maker to the king, that is, George II. (*Museum of the History of Science, Oxford*)

2.10 A quadrant signed 'COLE Inventer and Maker at Y^e ORRERY in Fleet Street LONDON No.200', broadly similar to that shown in the frontispiece of his tract. The radius of the divided arc is 9 inches, and the brass pointer on the movable arm has a vernier reading to 10 minutes. (*Museum of the History of Science, Oxford, inventory no. 40241*) Another Cole quadrant, differing only in minor details, is located at the Science Museum, London.

Thomas Godfrey of Philadelphia in the early 1730s, at almost the same time as Hadley was independently devising his double-reflecting quadrant in England. Godfrey's instrument, as described in the *Philosophical Transactions*,[78] had a radius of about 2 feet; Adams considered this was too large, as it was liable to lead to errors through distortion of the frame, so his instrument had a radius of only 10 inches, but was provided with a vernier to enable readings to be taken to 1 minute of arc. He claimed that because the observer looked directly towards the horizon, keeping the central arm of the instrument horizontal, observations were not materially affected by the pitching and rolling of the ship: apparently this was what he meant by his instrument having 'one of the most useful properties' of reflecting quadrants like Hadley's. For use when the horizon was not visible, Adams's instrument was fitted with a bubble level to assist in keeping the sighting arm horizontal. He recognized, however, that observations taken by this means were not very accurate, so he included in his tract descriptions of two 'artificial horizons', namely Searson's 'whirling speculum' and Ellis's floating mirror.

Cole did not respond to Adams's claim immediately. Over two months elapsed after Adams's run of advertisements in the first half of October (1748), and then both Adams and Cole placed advertisements for their respective quadrants in the *Daily Advertiser* on the same day, 19 December.[79] By this time Cole had produced a second edition of his descriptive tract. As well as mentioning this, his advertisement had an extra paragraph:

> Note, As the judicious are not to be puffed into a Belief, which shall in Prejudice of one Man favour another's Claim to an Invention, therefore the Publick are left to determine, Whether another modern Quadrant, publish'd the latter end of September last, is not a manifest Copy of that of Mr. Cole's, publish'd the Beginning of June last, with only the Addition of Mr. Godfrey's Horizon Vane.

In a slightly longer advertisement on the same lines in the *General Advertiser* four days later, Cole added a reference to Godfrey's paper in the *Philosophical Transactions* of 1734 'which a vain Boaster would persuade the World was never before made Publick but by him', and went on:

> As Mr. Cole is the sole Inventor of the above Quadrant, and first published it many months since, he thinks it is his Duty (to prevent the Publick from being deceived) to inform them that Cole, Inventor and Maker is stampt thereon, to distinguish it from any other pretended Invention whatsoever.[80]

The second edition of Cole's tract was considerably longer than the first, and included a sixteen-page section at the end directly attacking Adams, as well as a note at the front which claimed that 'several Gentlemen seeing it [Adams's instrument] at a Shop, somewhat nearer than mine to the Royal Exchange, have thereby mistaken that Shop for mine'. Since the publication of the first edition, he said,

> ... there has appeared a Pamphlet shewing the Use &c. of a New Sea Quadrant, which the Publisher, Mr.A— deceives the Public as one of his own invention: As this Action of his does very much interfere with my affairs, I hope the unprejudiced Reader will excuse me for examining the right Mr.A— has to the invention which he thus ascribes to himself.

Outlining the sequence of events, Cole said that he had made gunner's quadrants similar to this design five or six years ago, and had published it for general use last June (1748). Comparing Adams's instrument with his, they were seen to be alike in almost every particular: the same radius (10 inches), same scale (100 degrees divided into thirds), same length of sighting arm (18 inches), same nonius divisions to show minutes of arc, and so on; in fact the only differences were in the details of the horizon vane and glass vane. Adams's pretended new device, Cole said, was only his (Cole's) with the addition of Godfrey's horizon vane, taken from the latter's paper in the *Philosophical Transactions* of 1734. Moreover, Adams's instrument and tract had not been published until September 1748, three months after Cole's. Referring to the episode of the microscope book in 1746, Cole said that Adams had 'cunningly informed himself' of the fact that Baker had omitted to register *The Microscope made Easy* at Stationers' Hall, and then proceeded to copy it with impunity.

On the face of it, then, it was a clear case of plagiarism by Adams, who had been accused of the same practice before. But matters were not really so simple. Why did Cole wait nearly three months before attacking Adams in public, and why did he need sixteen pages plus a character assassination to prove his point? Basically, the reason was that neither Cole's nor Adams's quadrant was really new. Both were based on ideas that had been around for at least a decade, and, as Cole admitted, he had made similar devices himself several years earlier, though for a specific purpose (gunnery) rather than general navigation.

This reference to gunnery may be the key to Cole's annoyance at the appearance of Adams's instrument so soon after his own, for Adams's tract describing it informed readers on its title-page and frontispiece that he (Adams) was 'Mathematical Instrument Maker to His Majesty's

2.11 Frontispiece of Adams's *The Description and Use of a New Sea Quadrant*, published in 1748 a few months after Cole's. As well as showing how to hold the instrument, examples of different forms of vernier divisions are given. Note that Adams drew attention in the inscription to his recently-acquired Ordnance appointment. (*British Library, London, 716.f. 5–3*)

THE

Description *and* Use

Of a NEW

Sea Quadrant,

FOR

Taking the ALTITUDE of the SUN
from the visible HORIZON;

Which is so contrived,

That the Observer will be liable to no Interruption, or Inaccuracy from the Ship's Motion.

AND

The LATITUDE at SEA,

May be obtained with greater Certainty, and more frequently than by any Instrument of the like Kind hitherto made publick.

Invented, Made, and Sold by

GEORGE ADAMS,

Mathematical Instrument Maker to His MAJESTY'S Office of ORDNANCE, at *Tycho Brahe's Head*, the Corner of *Racquet Court*, in *Fleetstreet*, London.

To which is added,

The DESCRIPTION and USE of two Curious METHODS
for procuring an

ARTIFICIAL HORIZON at SEA,

To be used with *Hadley*'s Quadrant; one by Means of a *Whirling Speculum*; the other by a Floating Mirrour.

LONDON:
Printed by JOHN HART, MDCCXLVIII.

2.12 Title-page of Adams's *New Sea Quadrant* tract, which also mentions his Ordnance appointment. (*British Library, London*)

52 *George Adams Senior*

Office of Ordnance', an appointment that Cole clearly thought should have been his. Indeed, Adams may well have heard of Cole's quadrant in the first place through the Office of Ordnance, before it was made public.

For more than twenty years prior to 1748 this appointment had been held by William Deane, who had been apprenticed to John Rowley.[81] Deane's last bills for instruments supplied to the office were dated in the middle of 1747. The records show that on 15 September 1747 another instrument maker, Joseph Jackson, supplied a measuring chain and theodolite for the use of the Royal Military Academy at Woolwich,[82] and a few weeks later Thomas Wright (who, like Deane, had been apprenticed to Rowley) supplied various drawing instruments for the use of Captain Goodyer, an artillery officer who was going to the Far East with the fleet under Admiral Boscawen.[83] (The French were trying to gain possession of British territory in India.) Wright had occasionally supplied instruments to the office ever since Rowley gave up in the mid-1720s, although Deane was nominally the appointed instrument maker to the office in general, and Jackson to the Royal Military Academy in particular. In late 1747 and early 1748 Wright fulfilled two more Ordnance orders, and it is possible that, had he not decided to retire shortly afterwards, the office would have appointed him to succeed Deane as their instrument maker in general; if so, it must have been a double blow to Cole when he failed to obtain not only Wright's royal appointment but also his Ordnance work. Instead, on 22 April 1748, it was Adams who supplied some surveying equipment for the use of Ordnance engineers in Scotland.[84] From then until his death in 1772 Adams was the office's appointed mathematical instrument maker, and the records show that he supplied hundreds, if not thousands, of instruments (and also carried out repairs) during those twenty-four years. For example, between April and October 1748, while Cole was busy promoting his new quadrants, Adams submitted seven bills, namely (abbreviated):

Date	Description	£	s	d
22 April	A large plain table with ball & socket, brass ruler sights etc. for the use of the Works in North Britain	£7	17	6
3 May	Repairing the ball & socket on a plain table for Flanders	£0	6	0
24 May	3 pairs of drawing compasses, dividers, drawing pens, rulers, for the Officers at Portsmouth	£1	7	6

24 May	Small [magnetic] compass with brass box, for Sheerness, and 6 steel pens for Minorca	£2	6	0
24 May	A set of brass shot gauges for the Stores in the Tower	£1	15	0
26 Sept.	Brass [magnetic] compasses for H.M.'s sloop 'Montague'	£3	5	6
2 Oct.	Plain tables, drawing compasses and pens etc. for the Drawing Room in the Tower	£18	1	0

At this time (1748) the War of the Austrian Succession was drawing to a close with the Treaty of Aix-la-Chapelle, so Ordnance requirements were fairly small. The situation changed drastically when the Seven Years' War with France in America and Europe broke out in 1756. Ordnance orders then became a significant part of Adams's business: between 1756 and 1763 he supplied instruments worth over £1,700, detailed in seventy-eight bills (see Table 5.1 in chapter 5).

Although the records of the Office of Ordnance held at Kew are extensive – over 10,000 manuscript volumes or boxes of papers – no board minutes have survived for the period around 1747–48 when the appointment of mathematical instrument maker passed from Deane to Adams, via Jackson and Wright. It is not possible, therefore, to determine precisely why the board chose Adams rather than (say) Cole for this appointment. However, a long run of surveyor-general's minutes starts in January 1749, and entries in the first year of this series provide some retrospective information on this point. At a meeting on 27 January 1749 a letter from Joseph Jackson was considered, complaining that a warrant for repairs to a theodolite and chain belonging to the Royal (Military) Academy at Woolwich had been issued to Adams, although he (Jackson) had a 'Signification' from the master-general appointing him instrument maker to the academy. This was correct: for funding and administrative purposes the academy was an Ordnance institution, but it did not automatically come within Deane's mandate for the supply of instruments. On its foundation in 1741 the board initially gave an order for gunnery instruments to Deane, and one for drawing instruments to Jackson. All subsequent academy orders in the 1740s were placed with Jackson, who is described as 'Mathematical Instrument Maker to the Royal Academy' in the bill books, while Deane was 'Mathematical Instrument Maker to this Office'. Hence, when Jackson heard about an order relating to the academy being given to the newcomer, Adams, he naturally protested. He asked the board to cancel the warrant and redirect this repair job to him. In response, the board decided to seek guidance from the master-general himself over this conflict of interest. The board pointed out that after Deane had been dismissed, some instruments were ordered from Jackson, but he 'being gone to Sea and no Person left to attend to his business', and Adams in the meantime having been appointed mathematical instrument maker to the office in place of Deane, they had given the academy order to him. The master-general (the Duke of Montagu) replied that he still wanted orders for the academy to go to Jackson. The warrant was accordingly recalled and redirected to Jackson, as he had requested.

Attempts to find a date of death for Deane have not, so far, been successful. The above passage indicates that he did not die in office but was dismissed, perhaps (but this is just conjecture) due to inability to meet orders through ill health. Though the precise date of this event has not been discovered (due to the absence of surviving minutes), from 1749 onwards Adams is positively identified in some bill book entries as 'Mathematical Instrument Maker to this Office'.

Adams was not entirely unknown to the board when he succeeded Deane. Four years earlier, in August 1744, he had supplied a plane table and protractor for the use of Archibald Bontein, the Ordnance engineer at Jamaica. (This order, worth £8 and recorded in bill book WO51/154, is the earliest connection between Adams and the Office of Ordnance that has been discovered.) For the same application the board had obtained from Deane in July a theodolite with a vertical arch and telescopic sight, mounted on a conical ferrule on a brass-headed staff, and also two chains (100-foot and 50-foot). It is not clear why the plane table was ordered – exceptionally – from Adams in this instance, but possibly these surveying instruments were wanted urgently and Deane could not cope with the whole order himself. Adams's bill (that is, the entry in the bill book) is dated only a few days after the date of the relevant warrant, suggesting that he fortunately had a suitable plane table and protractor in stock at the time.

A year or so after Adams became the official instrument maker to the Office of Ordnance, Jackson incurred the board's displeasure by repeatedly ignoring requests for information about his prices, which were appreciably higher than those of his competitors; and when the Duke of Montagu died in July 1749, leaving the post of master-general vacant, the board apparently decided that they were no longer bound by his personal preference for Jackson. Henceforth, almost all orders for drawing, surveying, gunnery, and other mathematical instruments,

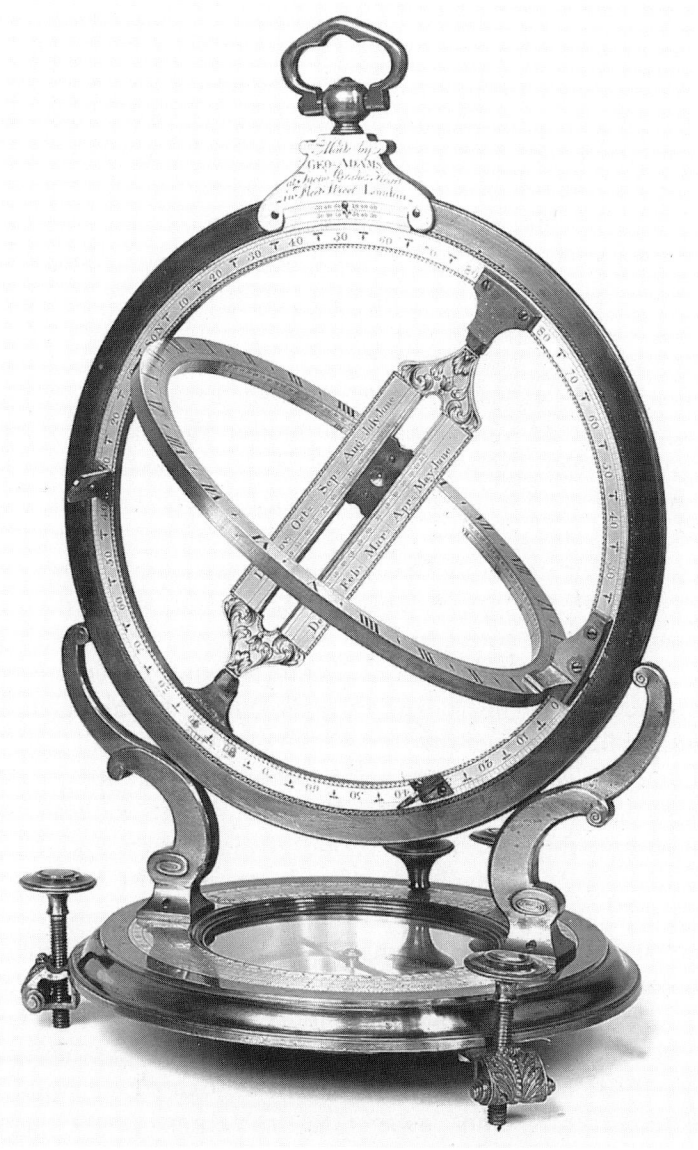

2.13 An early standing universal ring dial signed 'Made by GEO. ADAMS at Tycho Brahe's Head in Fleet Street London', similar to instruments produced by his former master Thomas Heath. (*Science Museum, London, Inv.1916–318; Neg.944*) An Adams ring dial similar to this but with different ornamentation, at Columbia University, New York, is said to have been presented to the college by King George III. For three examples by Heath see *Bulletin of the Scientific Instrument Society* 21 (1988) p. 26.

including those for the academy, went to Adams. By the time he died in 1772 he had submitted at least 146 bills, totalling nearly £2,400.[85]

Deane had also described himself as 'Mathematical Instrument Maker to the Royal Mathematical School', which was part of Christ's Hospital in Newgate Street. The archives of that institution show that he supplied mathematical instruments to the school annually from the death of John Rowley in 1728 up to and including 1747.[86] Benjamin Cole had a marginal connection with the school, for in 1739 (the same year that he apprenticed his own son) he took a boy from there, Thomas Paine, as an apprentice.[87] However, Cole does not appear to have supplied any instruments to the school. Instead, the next instrument maker's name to be mentioned in the ledgers is George Adams (in 1750). Though he did not actually call himself 'Instrument Maker to the School', as Deane had done, the records show that Adams was the official supplier of instruments to the school for the rest of his life. Furthermore, when he died in 1772 the appointment continued into the next generation, and only lapsed when George's son, George junior, died in 1795. The annual value of this business was fairly small, varying between £20 and £50 from year to year, but it must have provided useful publicity for Adams. Detailed invoices have apparently not survived, but the instruments supplied for boys going into sea service would have been mostly quadrants of some sort, together with instructional books. The ledgers also record some small payments for instruments (and books) supplied for boys going on to university: these would most probably have been sets of drawing instruments.

Thus, by 1750 Adams had a double advantage over Cole, being the official supplier of mathematical instruments to both the Office of Ordnance and the Royal Mathematical School. (The appointment of mathematical instrument maker to his majesty, which Cole also failed to obtain, apparently remained with Thomas Wright despite the latter's retirement: nobody else used this title during the remainder of the reign of George II.) It is understandable, therefore, that Cole should have expressed considerable annoyance at Adams's entry into the quadrant market with an instrument so closely resembling his own. In the late 1740s, during the controversy with Cole, Adams also achieved some newspaper publicity through advertisements by the lecturer Joseph Sowerby, who announced on 29 September 1747 that he would give public lectures in London on the use of the globes and orrery: subscriptions for these were taken in by George Adams and Mary Senex in Fleet Street, Thomas Ribright in the Poultry, and a bookseller.[88] In the previous year (1746), and again in the spring of 1747, Adams's name had been similarly mentioned in connection with advertisements by an anonymous teacher of mathematics who was looking for pupils;[89] this may also have been Sowerby, though it could equally well have been Richard Jack, with whom Adams was associated in 1750 in a patent application (discussed below).

It was about this time that a number of advertisements appeared in the London newspapers which, though probably having no direct connection with George Adams the instrument maker, indicate just how common the surname Adams was in London. The series began with the following announcement in the *Daily Advertiser* on 23 April 1748:

> By particular Desire
> Of several eminent Persons of the Name of ADAMS,
> (Out of a particular Regard for their Forefather)
> An Invitation is hereby given to all their Brethren of that Name, to meet at the House of GEORGE ADAMS, the Royal Swan in Kingsland-Road, on Tuesday the 26th Instant, at Four o'Clock in the Afternoon, in order to establish a Brotherly Meeting and Loving Society, to be held once a Month at his House, for the Welfare and Prosperity of each Brother, and to support and commemorate the eminent Dignity and Antiquity of our original, most antient in the World, most venerable and paternal Name, from which all Monarchs, Kings, and States are descended, and are Sons of; which, if encouraged, will be a most laudable, and commendable and most antient Foundation and Society in being, and may claim the Precedency of Title before all others, and will be the most lasting to the World's End, while there is a Son of Adam living, at the last great Day, who must be a Brother of ours.

This probably did not attract much support immediately, as six months elapsed before, on 1 October 1748, a similar advertisement appeared proposing a meeting at the Royal Swan on Monday 3 October, to establish a society which would meet on the first Monday in every month. 'We will let the World know, it is not in their Powers, / To say so much of their Names, as we can of ours.' To this advertisement George Adams (presumably an innkeeper) appended a note saying that at his address could be seen 'the greatest Number of Curiosities and Rarities of any House in the whole World'. Evidently this particular George Adams was an enterprising showman. Similar advertisements just before the first Monday in each month during 1749 indicate that the society was duly formed, and flourished for a while, but as it does not appear to have left any written records it is not known whether 'our' George Adams was a member.

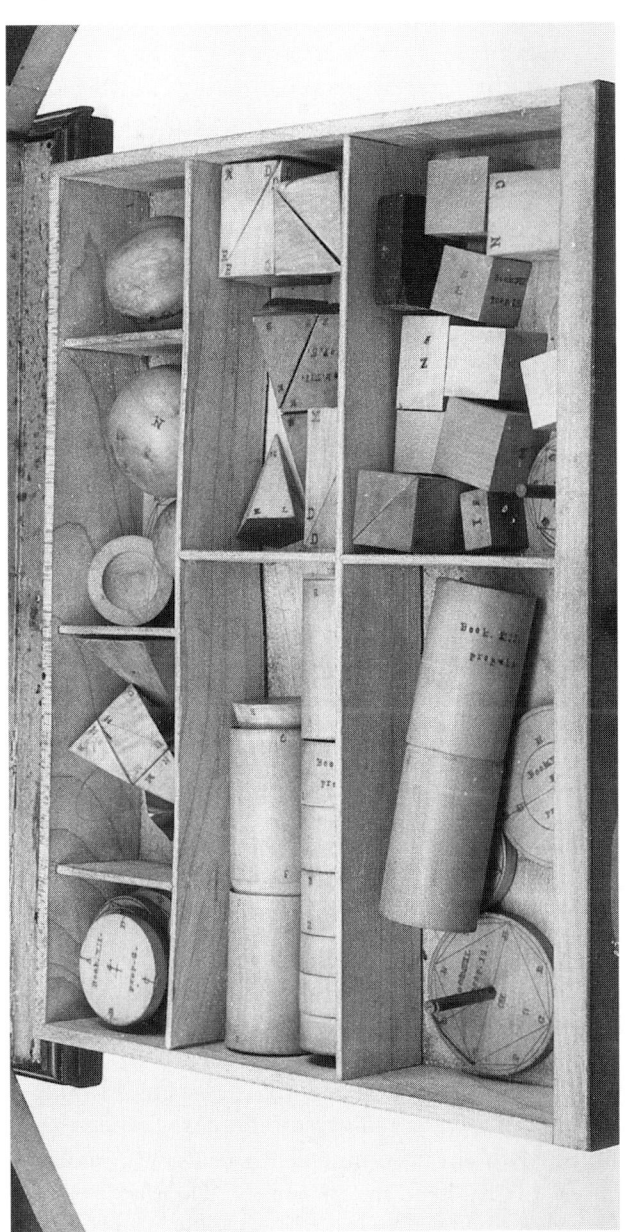

2.14 A set of geometrical solids contained in a three-drawer wooden cabinet 6¼ inches high. Each of the solids, which are made of boxwood, some having wire attachments, is stamped with the number of Euclid's proposition to which it refers. The two doors of the cabinet have scroll-shaped ivory inserts stamped (left) '*GEOMETRICAL*/*Made by Geo.Adams at Tycho*/*Brahes Head*' and (right) '*SOLID*BODIES*/*the Corner of Raquet Court*/*in Fleet Street London*'. (*Museum of the History of Science, Oxford, inventory no. 49936*). Geometrical solids are mentioned in Adams's pre-1757 trade card and (unpriced) in the list of products at the end of his *Description ... Trigonometrical Octant* (1753). This set was evidently made before the 1757 fire. On a similar but larger set, with four drawers instead of three (*Private collection*), the address is given as 'Tycho Brahe's Head in Fleet Street', omitting 'Corner of Raquet Court'; the latter set was probably made after Adams's move in 1757 but before his appointment to his majesty in 1760.

In August 1749, George Adams the instrument maker took the first step on the ladder of promotion in the hierarchy of the Grocers' Company, by becoming a member of the livery. He was then aged forty, and had been running his own business in Fleet Street for fifteen years. His election to the livery was not a purely personal matter, however – more likely it arose through the company being short of money. At a meeting of the Court of Assistants on 17 August 1749, no fewer than forty freemen were nominated for election to the livery, amongst them Thomas Heath, George Adams, and John Gilbert.[90] They were summoned to appear at the next meeting on 31 August. On that occasion, fifteen members who were present, including Thomas Heath, were elected, paid the fee of £20 (twice as much as in the Loriners' Company), and were 'cloathed' with livery gowns. Two more elected members who were present, one of them being George Adams, 'gave a Note of their Hands' for payment of the £20, and were similarly 'cloathed'. (Presumably the other fifteen came to the meeting prepared for election, and paid in cash.) The remainder of those nominated asked to be excused. This request was granted, but the fees from the seventeen elected members boosted the company's funds by a useful £340 at a stroke.

The date of Adams's and Cole's rival publications on their respective quadrants (1748) was just fourteen years after John Hadley was granted a patent for his reflecting quadrant in 1734. Perhaps this was just coincidence, but as fourteen years was the normal life of a patent, both Adams and Cole may have been seeking, through the widespread publicity for their own designs, to acquire customers for the admittedly superior Hadley pattern once the patent had expired. In June 1749, John Gilbert & Son felt it necessary to advertise several times that only they sold quadrants of Hadley's own make.[91] Both John Gilbert senior and his son John junior were, like Adams, members of the Grocers' Company: it was the elder man who was nominated for election to the livery at the same time as Heath and Adams in 1749.

In the following year (1750) George Adams applied jointly with Richard Jack for his only patent, perhaps thinking that it might take the place of Hadley's. Its title, in the printed version prepared when the Patent Office was established in the mid-nineteenth century, runs to about 200 words, reduced in the heading to simply 'Quadrant and Telescope'.[92] Richard Jack described himself in the preamble as a 'Gentleman'; he was a teacher of mathematics and author of several works in this field, who during the 1740s and 1750s gave courses of public lectures on experimental philosophy, fortification, and gunnery, so his activities would naturally have brought him into contact with mathematical instrument makers.[93] When the patent application was submitted he was living in Golden Square, in the parish of St James's, Westminster. Though primarily a pure mathematician, he was evidently a practical man, for when his library was sold by auction in 1760 after his death he was described as 'Teacher of the Mathematicks, and Assistant Engineer in the late Expedition against Guadaloupe'.[94] His effects then included an air-pump, microscope, telescopes, and other instruments used in experimental philosophy.

Adams's and Jack's patent specification unfortunately did not include diagrams, and the description of their invention is far from clear. Apparently the main feature of the quadrant was the use of lenses to throw the images of the sun and horizon on to the observation point, and also (perhaps more important) the use of a telescopic sight instead of a vane with a sighting hole. The telescope part of the specification, which really had no connection with the quadrant part, described what amounted to a refracting telescope with both a double objective and a compound eyepiece, giving much higher magnification than usual for a given physical length. The aberrations with this arrangement must have been very large, but according to newspaper advertisements published during the next year or two, some quadrants and telescopes as described in the patent specification were certainly made. No extant examples of either have been identified. The earliest relevant advertisements that have been noted began in November 1751 and continued into December, in at least three papers. On 18 November a short advertisement appeared in the *General Advertiser*, stating that 'A new SEA-QUADRANT, and a REFRACTING TELESCOPE', invented by Richard Jack and George Adams, were made and sold by the latter in Fleet Street. This was repeated on the following day, and twice in the following week; it also appeared in the *London Evening Post* on 16/19 and 19/21 November, in the *London Gazetteer* on 28 November, and possibly in other papers too (the runs of some titles in the British Library are defective).[95] A longer version was published in the *General Advertiser* on 12 December:

> By his Majesty's Royal Patent,
> a New REFRACTING TELESCOPE, and a SEA QUADRANT,
> invented by
> RICHARD JACK, Teacher of the Mathematicks, and
> GEORGE ADAMS, Mathematical Instrument-Maker

at Tycho Brahe's Head, the corner of Racquet Court, in
Fleet Street, at which place they are made and sold.
N.B. The Magnifying Power of this Telescope greatly exceeds that of any other of the common Sort, although they are considerably longer than this; also its particular Property, of various magnifying Powers without changing the Glasses (which no other Construction hath) renders it extremely serviceable for viewing distant Objects.

They are prepared for Use at SEA in Mahogany Tubes two feet in Length. The Quadrant also hath several peculiar Advantages in taking Altitudes at Sea, either in a forward or backward Observation, and both by Refraction, whereby the Coincidence of the Sun and Horizon are preserved with the greatest Ease, although under all the Disadvantage of a rough Sea.

Magna est Veritas & prevalebit.

This advertisement appeared five times in the *General Advertiser* from 12 to 17 December 1751 – every day except Sunday. It was also placed in the *General Evening Post* at least once, in the number for 21/24 December.[96] On Saturday 28 December, Francis Watkins, optician, of Charing Cross, felt obliged to protest with a lengthy statement published in the *General Advertiser* for that day. Too long to quote in full here, Watkins said that he had carefully compared the telescope of Jack and Adams with others of the 'common construction' and found 'all its boasted Excellencies to be intirely false and groundless'. Directing his remarks at Jack, and ignoring Adams completely, he said that the inventor appeared to be 'totally ignorant of the true Nature of Refractions', with the result that his patent telescope not only 'colours the Image of the Object more than any other of the ordinary Construction', it also covered a field only one-third that of the latter. He claimed that the proportions of the eyeglasses in the patent telescope had been considered earlier by 'P.Cherabim D'Orleans, in his Dioptrique Oculaire, Page 187', but rejected by him. To demonstrate the inferior performance of the patent telescope in comparison with those of the common construction, Watkins had taken the trouble, he said, to construct two telescopes of the same length as the patentee's, with different eyeglass arrangements, one of which magnified to the same extent as his, and the other more, and with a larger field. He concluded by challenging 'the said Mr.Jack, or Mr.Adams' to a fair comparison before impartial judges.

Jack and Adams do not appear to have replied to this challenge directly. On 9 January 1752 they inserted their standard long advertisement in the *General Advertiser* again, repeated on the following two days; these advertisements have a footnote stating that Adams also made ordinary Hadley's quadrants, with truly parallel mirrors, for 2 guineas. After this, neither Watkins nor Jack and Adams seem to have made any public announcements on the subject. (One cannot be certain of this without examining every issue of every paper, which is impracticable as some of the surviving runs are incomplete.) However, evidently during this period some form of practical trial took place, for on 18 March 1752, repeated on 20 and 23 March, the *General Advertiser* carried the following statement by the patentees defending their product:

> TO THE PUBLIC
> We are authorized and concerned to disabuse the Public, in the Relation which was published of the Trial of the Patent Telescope, and those of the common Construction; which, in most of the material Circumstances is false or absurd, impertinent to the Question, defective in the Particulars, and injurious to the Honour of Personages of the most distinguished Rank and Abilities. And that such Gentlemen as choose to take unquestionable Authority for the Measure of their Judgement in this Affair, may be fully satisfied of the Merits of the Competition, and of the superior Properties and Excellencies of the Patent Telescope, by condescending to call at Mr. ADAMS's House, in Fleet Street, they will see the same determined and authenticated under the Hands of the most Noble and Honourable Personages and Gentlemen, who were present at the Trial; whom they will find to be of so superior Rank, Honour, and Abilities, as to be above the mean and base Suspicion insinuated, of being partial in their Decision; and to be incapable of mis-applying their Authority, to give Sanction to any Design to mislead or impose on the Public; and where such Gentlemen, as in scientifical and experimental Researches, would choose to receive Conviction from the evidence of their own Senses, may have an Opportunity of making the Experiment. And if the great Demand for this Instrument can be of any Weight to determine the Preference, we can assure, and prove to the Public, that the Sale thereof, within the Compass of a few Days after the Trial, to Gentlemen who were present, or who waited for the Issue and Report thereof from Persons, on whose Judgement and Veracity they could depend, amounted to above the sum of £100. We are satisfied to rest the Merits of our Pretensions to the future Favour and Patronage of the Public, as these Facts shall appear to them to be established, or conclusive.
>
> The Patentees.   RICHARD JACK
> London, March 16, 1752.   GEORGE ADAMS.

It is conjectured from the above that Jack and Adams may have persuaded the Lords of the Admiralty, or the Board of Ordnance, to carry out a trial of their patent telescope in comparison with one of the common construction. No record of such a trial has yet been discovered, but it would appear from the above statement that the patent

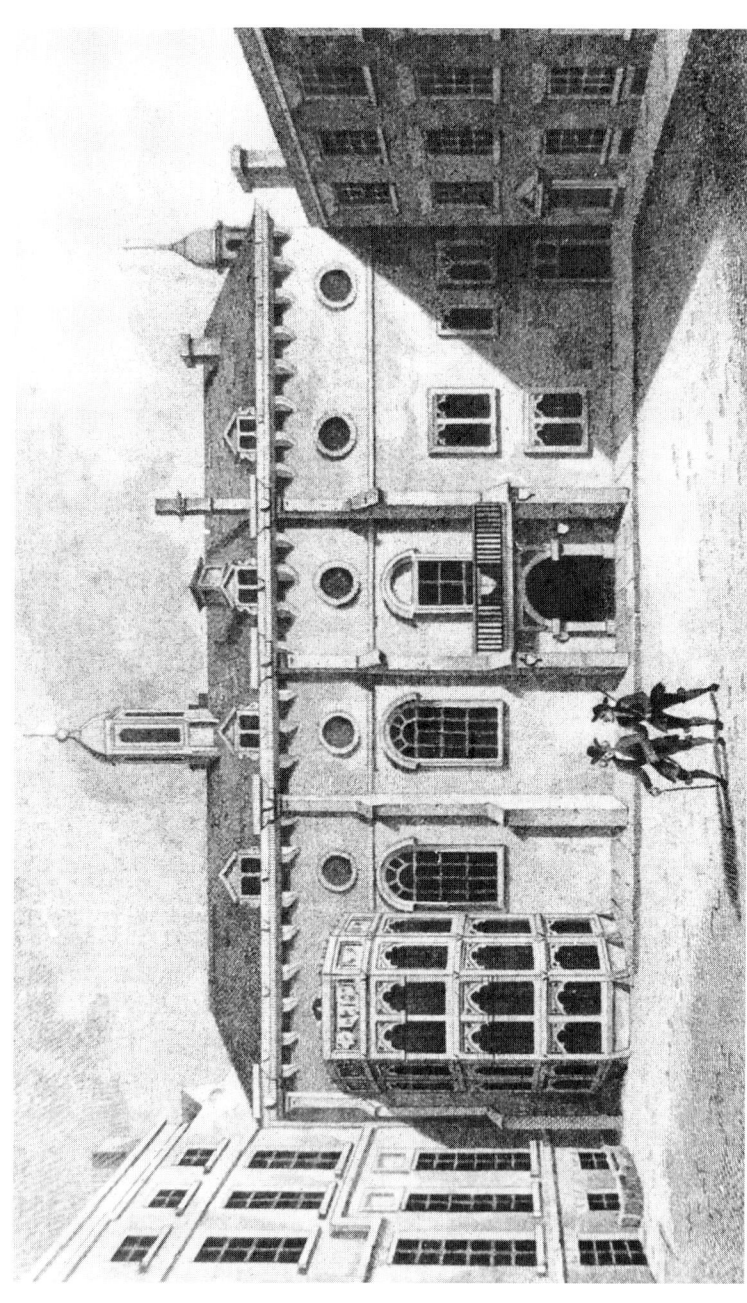

2.15 'South View of Grocers' Hall, as it was restored after the Fire of London 1666'. From a plate in J.B. Heath's *Some Account of the Worshipful Company of Grocers* (1854), facing p. 1. (*Author's collection*)

telescope found favour with some customers at least. It should be borne in mind that at this time (1752) the achromatic lens had not been introduced (Dollond's patent was granted in 1758), so the false colours and distortion that the patent telescope undoubtedly produced may not have been considered important by these purchasers, in comparison with the increased magnification provided in a given physical length. For certain applications, such as detecting a small static object at a distance (for example, a ship on the horizon), the small field, false colours, and distortion were irrelevant: the most important factor was magnification. It is sometimes thought today that the introduction of the achromatic lens made other telescopes obsolete overnight, but that was far from being the case.

Nevertheless, correction of chromatic aberration in refracting telescopes remained a desirable objective, if it could be achieved at reasonable cost, if only because an achromatic telescope would have more general uses than one in which the design concentrated on one feature (such as magnification) to the detriment of others. The apparent success of the Jack and Adams patent telescope, despite its defects, may well have been the spur that prompted Dollond to investigate and develop the achromatic lens. When Dollond's own patent (that is, achromatic) telescopes were first advertised in the newspapers in 1758–59, his name was always coupled with that of Francis Watkins of Charing Cross, who had so vehemently objected to Jack and Adams's patent a few years earlier.[97] It is known that Watkins provided Dollond with some financial assistance in the development of the achromatic lens, but the reason why he, rather than another optician, should have done so does not appear to have been fully investigated by historians. From the Jack and Adams episode it is clear that Watkins was an experimenter in his own right, not just a working optician, so his part in the development of the achromatic lens may have been greater than is generally realized.

Returning to the early 1750s, it often happens that when one tradesman persistently advertises a product, others in the same line of business soon follow suit. In the middle of Jack and Adams's advertising campaign for their patent telescope, on 21 November 1751 John Cuff thought it desirable to remind potential customers of his existence, with an advertisement in the *General Advertiser* (repeated the following day) stating that 'he continues to carry on Business in the same Shop as formerly, where every Thing will be performed in the best and most expeditious Manner, at reasonable Prices'. Cuff at this time was in severe financial trouble. About a year earlier, he had been declared bankrupt,[98] but had apparently managed to raise enough money to prevent his stock-in-trade being seized. He was not so fortunate with his household effects: these were removed from his Fleet Street premises and sold by auction on 20 December 1751.[99] On 4 March 1752, just before Jack and Adams inserted their long rebuttal of Watkins's criticism of their patent telescope, Cuff advertised his optical instruments in the *General Advertiser* with a reminder that it was he who had 'chiefly contributed to bring Microscopes, both single and double, to the Perfection they are in at present'. The opticians James Ayscough and George Sterrop also advertised frequently in late 1751 and early 1752. From the middle of November 1751, Heath & Wing, who had been advertising the third edition of Hammond's *Practical Surveyor* (published by themselves) for some time, always added a footnote pointing out that they 'make and sell all Sorts of Mathematical and Philosophical Instruments, accurately finished, according to the best Improvements of the most eminent Professors'.

Of more concern to Adams were advertisements inserted in the *General Advertiser* from 21 January 1752 onwards by his old adversary Benjamin Cole, who, leaving Watkins to deal with the patent telescope, concentrated on quadrants instead. Commenting that many Hadley's quadrants had been 'badly made with false Glasses, to the Prejudice and Disreputation of that valuable Instrument', he said that his instruments were 'carefully made in a perfect Workman-like Manner, with true Ground-Glass parallel Plates', at 2 guineas. He also mentioned that he sold the seventh edition, just published, of Joseph Harris's *Description and Use of the Globes and Orrery*.[100]

Cole's advertisement was repeated at least thirteen times during February and March 1752. Together with those by Jack and Adams, Watkins, Cuff, Ayscough, Sterrop, and Heath & Wing, the winter of 1751–52 saw an unusual concentration of advertising by scientific instrument makers. On 10 February 1752 Edward Nairne, 'successor to the late Mr. Matthew Loft', joined the fray, but only to announce a change of address from Bartholomew Lane to Cornhill.

This activity may have arisen partly through the death of Frederick, Prince of Wales, in March 1751. Though this event did not affect Adams immediately, it meant that appointments such as Optician to the Prince of Wales (held by Francis Watkins) lapsed, and tradesmen who had expectations of eventually holding the same appointments to the king could no longer count on that happening. Frederick's eldest

2.16 A folding universal equinoctial sundial signed 'Made by G:Adams in Fleet Street LONDON'. Dials of this type differ from 'horizontal inclining dials' (see figure 10.16) in having equal divisions and a gnomon at right angles to the plane of the dial. In this example the gnomon incorporates a pinhole, adjustable for date, as in a universal ring dial (figure 2.13). (*Museum of the History of Science, Oxford, inventory no. 50638*)

son, George (later King George III), was then aged twelve; he was created Prince of Wales shortly after his father's death, but did not have his own household until he attained his majority at the age of eighteen in 1756. As will be seen later, this interval of five years enabled Adams to secure a royal appointment himself, which carried over to the next reign. If Frederick had not died before his father (George II), the relative importance of several instrument-making businesses in the second half of the eighteenth century might well have been different.

In the earliest of Adams's trade cards currently known, probably published about 1750, his 'new Sea Quadrant by Refraction' is listed first, followed by 'Mr.Jack's New Refracting Telescope'. Francis Watkins, in his attack on the patent telescope, also referred to the latter as Jack's, virtually ignoring Adams. As there was really no connection between the telescope and the quadrant, it is likely that Adams, having made Jack's design of telescope for him, applied jointly with him for the 'quadrant and telescope' patent simply to reduce the cost of doing so. The publicity provided by the patent telescope must have helped Adams's business, despite the unfavourable comments by Watkins. Meanwhile, Adams was continuing to supply goods and services to the Office of Ordnance, and to the Royal Mathematical School. He also had some export orders: Peter Collinson, writing to Benjamin Franklin on 3 July 1750, mentioned that he had obtained some (unspecified) instruments for him from Adams,[101] and Thomas Penn is known to have presented a copy of *Micrographia Illustrata* to the Library Company of Philadelphia, together with various instruments, though precisely when is not clear.[102]

Early in 1753, Adams took advantage of an Act passed in November 1750 by the Common Council of the City of London, which enabled masters to employ 'foreign' journeymen (that is, not Londoners) if insufficient London-trained freemen were available for hire. On 27 February 1753 he registered four men, to be employed by him under licence from the court for three months:[103]

Thomas Ray, at Mr Gretton's, lock founder, Long Acre;
Thomas Hinton, New Change Court, Strand;
William Andrews, Saffron Court, Saffron Hill; and
Samuel Tarry, at Mr Evers's, Lambeth Street, Goodman's Fields.

In practice it seems that temporary licences of this nature were often renewed if there was a continuing shortage of skilled labour. On 2 October, Adams reregistered Thomas Ray (who was then living in Shoe

Lane), Samuel Tarry (then in High Holborn), and William Andrews, and also registered a new man, William Sharyer of Dean Street, Holborn. Benjamin Cole also registered several 'foreigners' during the next few years, as did other instrument makers such as Edward Nairne. These records indicate that the demand for trained men in the instrument-making field was outstripping the supply in London. Evidently Adams was extremely busy at this time, for on 20 March 1753 Aaron Burr wrote to Benjamin Franklin saying that he had asked Adams to quote for supplying philosophical apparatus but could get no reply.[104]

It is likely that a large proportion of Adams's trade was in navigational instruments, especially quadrants of various types, for although he is better known today for his microscopes and globes, the subject to which he returned again and again was the improvement of angle-measuring instruments. In January 1754 he published *The description and use of the Universal Trigonometrical Octant*, 'invented and applied to Hadley's Quadrant, by George Adams'. This tract of 135 pages, 'given only with the instrument', has a title-page dated 1753 but one of the plates is signed 'Geo. Adams Jan:24 1754'. The instrument described and illustrated therein was basically a standard Hadley's quadrant (octant) with two additional features. In the first place, the scale of degrees had adjacent to it a sliding scale, like the slider of a slide rule but curved to match the angular scale, bearing calendar data which gave the sun's declination on every day of the year. This enabled the latitude to be read directly from meridional observations of the sun, without recourse to tables. The other innovation consisted of linear scales along one edge of the frame and index arm, plus a separate T-square (also bearing a scale) which could be placed against the scale on the edge of the frame. This enabled trigonometrical functions corresponding to the angular setting of the index arm to be read directly from the scales: hence the name, 'Trigonometrical Octant'. In effect this was using the octant as an analogue device for solving the basic navigational problem, 'What is the course run when the latitude change and ship's heading are known?'

With this device a navigator should have been able to find his position directly without the aid of tables or laborious calculation. However, in practice the errors introduced through reading the numerous scales must have been a serious disadvantage. The practical trend at this time was away from complicated multi-purpose instruments and towards the improvement of accuracy of a single reading – the angle measured by the octant. As the various sources of instrument error were recognized and steps taken to overcome them (for example, non-parallelism of the mirrors, and irregularity of scale division), so the accuracy of measurement gradually improved to the point where analogue devices for carrying out the necessary computations were not good enough, and tables calculated to several places of decimals became essential.

Nevertheless, instruments like this must have generated a certain amount of business for Adams. In addition to employing 'foreigners', on 17 December 1754 he increased his workforce by taking on another apprentice (the first since 1744), Nathaniel Kettle, for a premium of £30.[105] As well as quadrants, another navigational instrument with which Adams was closely concerned, and which probably contributed to the need to increase his staff at this time, was the mariner's compass. His first batch of Ordnance orders in mid-1748 included several for magnetic compasses. These were specifically for use in Ordnance, rather than naval, vessels, and did not amount to much financially; but a few years later a much more extensive market opened up, when Adams managed to obtain orders to make compasses of Gowin Knight's pattern for the Admiralty. Dr Gowin Knight (1713–72), who had been elected FRS in 1745 for his researches on magnetism, devised an improved form of compass for use at sea which so impressed the Admiralty that from 1752 only compasses of Knight's pattern were approved for use in naval vessels on foreign service.[106] Knight's compasses were initially made commercially by Smeaton, but by the mid-1750s Adams was their chief supplier. In a tract on the magnetic variation published in 1758, Mountaine and Dodson wrote that compasses 'not only for steering but also for taking the sun's amplitude and azimuth ... are made by George Adams ... and before they pass out of his hands, are examined and attested by Dr.Knight, whose certificate is fixed to the cover of the box'.[107]

Magnetic compasses were also used, of course, for other purposes besides navigation at sea. Equipped with some form of sights for taking bearings, they provided the simplest means of surveying land, especially if a high degree of accuracy was not required. Surveying instruments are considered in more detail in chapter 7, in association with George Adams junior's *Geometrical and Graphical Essays* (1791); George senior did not write anything on this subject. In the Ordnance context, 'miners' compasses' feature in several of Adams's bills. Mining (that is, excavating tunnels to place explosives under an enemy's fortifications) was an important part of siege operations, often employed

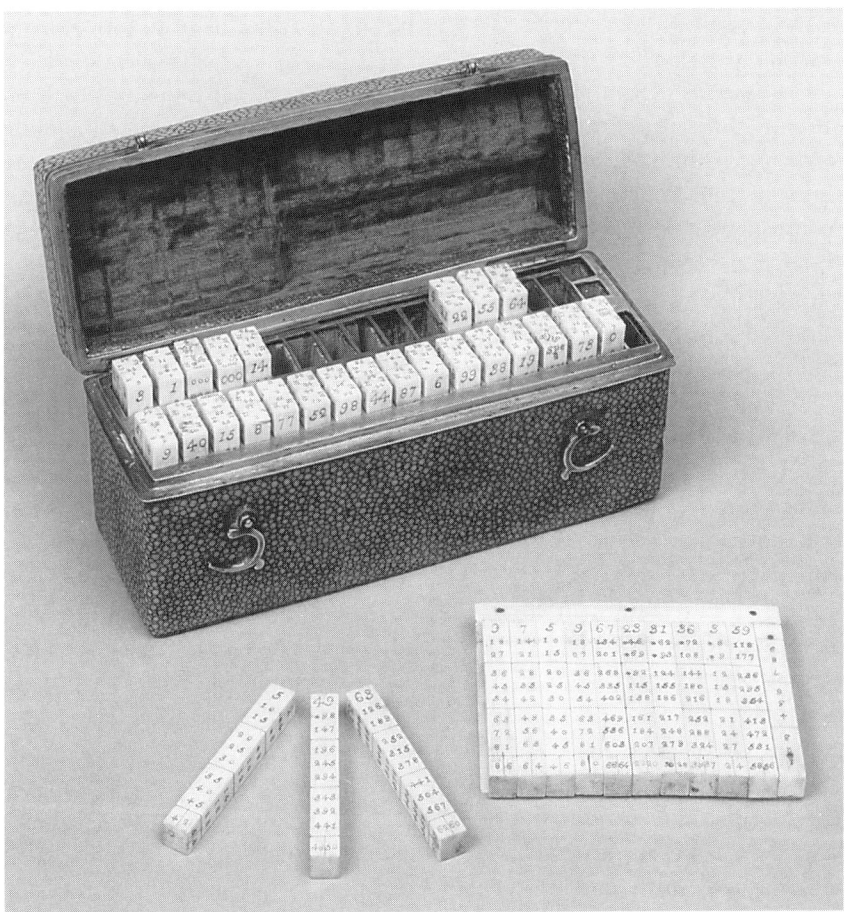

2.17 A part-set of ivory Napier's Bones, in a silver-mounted shagreen case signed 'Made by GEO=ADAMS in Fleet Street LONDON'. The square-section rods are stamped on the top and sides with columns of numbers for mathematical calculations. A rectangular ivory board incorporating a single index rod forms part of the set. Napier's Bones are not mentioned in any of George Adams senior's catalogues or trade cards, so this item was probably made to a customer's special order, quite early in his career. (*Christie's, London, March 29 1990, lot 145*)

when circumstances rendered battering by heavy guns impracticable or ineffective.

Compasses also formed an integral part of several types of portable sundial. Like most other instrument makers, Adams produced a wide variety of sundials, from elaborate 'standing universal ring dials' (see figure 2.13) designed more for ostentatious display than practical use, to folding or pocket dials for travellers (see figure 2.16). He also made fixed dials (not incorporating a compass) for use as garden ornaments or on buildings. At least one of the latter type, costing 10 guineas, was supplied to the Office of Ordnance in 1766 for installation at Woolwich, where the turret clock was continually giving trouble.

In 1755 Adams almost had his first taste of office in the Grocers' Company, when at a meeting of the company on 14 May he was one of eight members nominated for the post of steward; but like his father before him in the Loriners' Company, he managed to avoid serving on this occasion.[108] In 1756 he was not so lucky: he was again nominated, and this time was one of the four members chosen to serve.[109] The Grocers' 'Entertainment for the Livery' in the eighteenth century was always held on 29 May, the day appointed by Act of Parliament for celebrating the restoration of the monarchy in 1660. King Charles II had been enrolled as a member of the company, and their Sovereign Master, in that year. The entertainment was apparently on a more lavish scale than the corresponding feast in the Loriners' Company, as the fine payable on refusing the office was £25, against the Loriners' £5. Moreover, as an extra inducement to serve, each steward was allowed £15 towards his expenses, so there were relatively few refusals in the Grocers' Company. Thomas Heath, Adams's former master, served his turn in 1754.

The Grocers' Hall that Adams came to know well was not the present edifice standing in a courtyard off the western side of Princes Street, which was originally erected in 1893 and partially rebuilt after fire damage in 1965, but a much older building on the same site (see figure 2.15). Dating in part back to medieval times, it was set in extensive grounds between the Poultry and Lothbury, with a garden on the northern side and a courtyard to the south, approached by a narrow carriageway from the Poultry. (The realignment of Princes Street in the 1840s, when Moorgate [the street] was constructed, opened up an access from the east instead.) The Great Fire of 1666 left the original hall a blackened ruin and melted all the company's plate; but fortunately all the company's papers had been placed for safe keeping

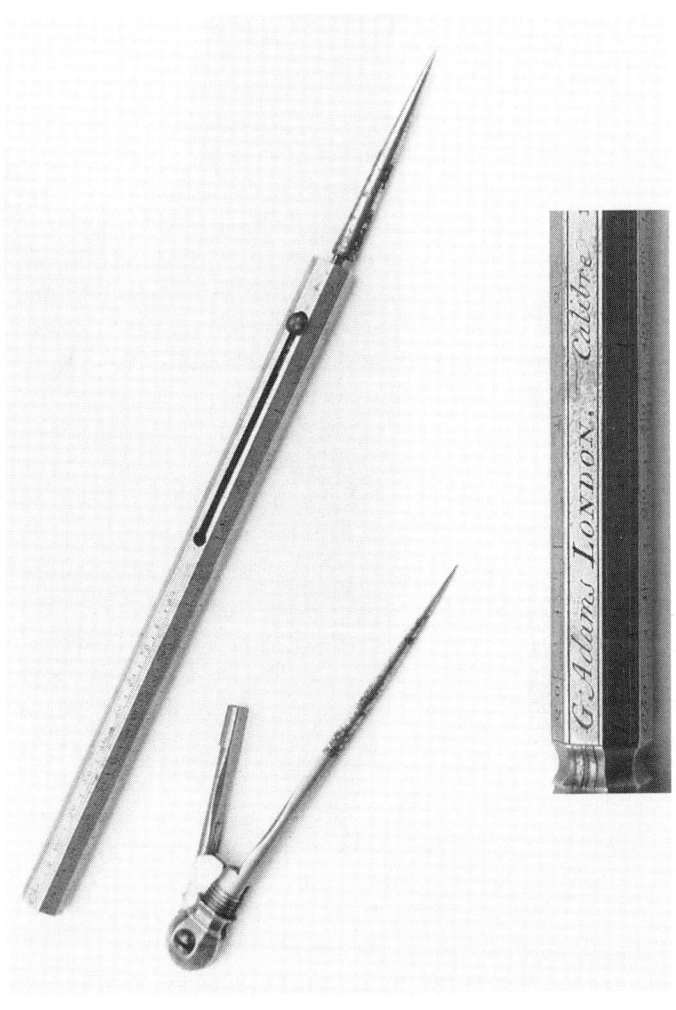

2.18 A draughtsman's tool consisting of a hollow octagonal prism engraved on each face with scales of inches, trigonometrical functions, and calibres of guns, with a pair of dividers housed in one end. One leg of the dividers can be removed and fitted into the other end of the prism to form a scriber, as shown. Signed simply 'G.Adams, LONDON', its date is uncertain. As it seems to be based on an illustration in Edmund Stone's *Construction ... Mathematical Instruments*, first published (in the English version) in 1723, it could be quite early despite the short signature, which is probably the result of space limitations. *(Tesseract, Cat.16 [1987] item 44)*

in a stone tower in the garden – a remnant of the medieval building – and this proved sufficiently strong and isolated to withstand the flames, so the archives were saved and are available for consultation today.[110] However, the fire destroyed all the properties owned by the company in the City, the rents from which provided much of its income, so for several decades financial problems were severe. The buildings constituting the hall were partially restored after the fire, but were not fully rebuilt until 1682 when Sir John Moore (a member of the Grocers) was lord mayor. After this, the company derived a useful income and helped to discharge its debts by letting the hall to successive lord mayors, and then from 1694 to 1734 to the newly-formed Bank of England.[111] By the time Adams became a regular frequentor of the hall, the company had managed to regain financial stability and was once again the prosperous City institution that it had been before the disaster of 1666. In the mid-eighteenth century the rateable value of Grocers' Hall was £300, over eight times as much as Loriners' Hall, reflecting the greatly superior status of the Grocers over the Loriners.[112]

3 Royal Connections

On 4 June 1756, the Prince of Wales attained the age of eighteen, and soon began to form his own household.[1] During the next few months the names of various appointees were announced in the newspapers, especially in October after the prince had returned from Kew to Leicester House, London, for the winter. Both the prince and his younger brother, Prince Edward (later Duke of York), born in 1739, still had much education to undergo, and Dr Stephen Demainbray still called himself 'Tutor to their Royal Highnesses the Prince of Wales and Prince Edward' in advertisements for his courses of public lectures on natural history and experimental philosophy in November and December 1756.

The precise date of Adams's appointment as Mathematical Instrument Maker to the Prince of Wales has not been discovered, but newspaper announcements indicate that it was towards the end of 1756. James Rice, a teacher of mathematics, inserted a long advertisement in *Jackson's Oxford Journal* on 1 January 1757 for an academy that he was proposing to open in Gloucester, where the subjects taught would include navigation and astronomy; and in a comment about the instruments that would be available, Rice said that they had been supplied by Adams of London, 'now Mathematical Instrument Maker to his Royal Highness the Prince of Wales'. The earliest advertisment by Adams himself mentioning this appointment that has been noticed was in the *London Evening Post* on 18/20 January 1757 (repeated in the next few issues), where he announced that he would shortly be publishing some new globes in various sizes.

With his official Ordnance appointment, regular orders from Christ's Hospital, and now this personal appointment to the Prince of Wales, at Christmas 1756 Adams must have been congratulating himself on his success. Admittedly some of the Ordnance orders were of little more than nuisance value (such as repairs to drawing instruments costing only a shilling or two), but the surveying and military instruments that Adams supplied for the use of engineers and artillerymen were sent to many different parts of the world, and must have helped to publicize his business overseas. The frequently-repeated orders for new drawing instruments for use at the Tower of London, where the Drawing Room, established as a separate entity early in the century, was continually being expanded to cope with new demands, must have provided a useful income and helped to keep his staff fully employed. With the outbreak of the Seven Years' War in 1756, Ordnance orders increased substantially in frequency and value. Between May and December 1756 Adams submitted thirteen bills totalling almost £300, for a variety of products which included 3 theodolites, 3 plane tables, 4 100-foot chains, 20 miners' compasses, 12 brass gunner's quadrants, 16 ink/pencil drawing compasses, 12 bow compasses, 23 parallel rulers, 7 pairs of gunner's calipers, 6 gunner's levels or 'perpendiculars', and 12 sets of shot gauges.[2] Also in the calendar year 1756, Adams is mentioned three times in the ledgers of Christ's Hospital, when he received payments totalling just over £100, though some of these may have been postponed from previous years as the annual average was normally less than this.

However, just as Adams was beginning to feel that his future looked secure, an event outside his control nearly brought his career to a premature and sudden end. A news item in the *London Evening Post* for 3/5 February 1757 gives the bare facts of the story:

> Yesterday morning, about one o'clock, a fire began at Mrs Binfield's, a milliner, near Racquet Court in Fleet Street, and burnt about an hour and a half with great violence. Mrs Binfield's house was entirely consumed, and she and her family with great difficulty escaped the flames (with scarce cloathing to cover them) by getting in at the top of the next house, an oilman's; which, with Mr. Adams's a mathematical instrument maker, and Mr. Rutter's house backwards, were much damaged, and with great difficulty preserved. A porter belonging to the Union Fire Office brought down 30 lb. of gunpowder, out of the garret belonging to the oil shop, while the said room was on fire.

The churchwarden's accounts of St Bride's for this period record payments to three people for bringing the fire engines of the parishes of St Dunstan's, St Bride's, and Bridewell. Mr Atwood of St Dunstan's, who was evidently quick off the mark, got £1 10s 'for bringing the first

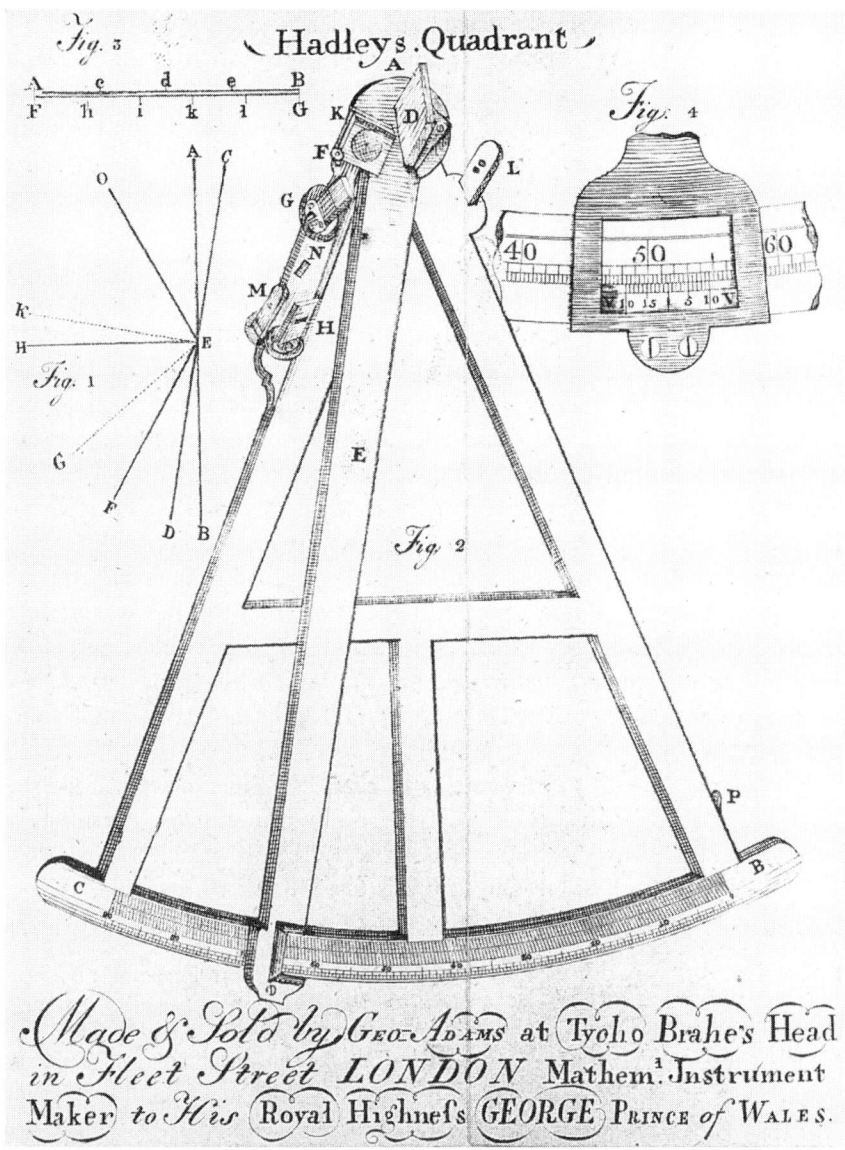

3.1 Frontispiece of George Adams senior's tract describing Hadley's quadrant, probably published early in 1757 shortly after he had been appointed instrument maker to the Prince of Wales (George). The main illustration has a scale with diagonal divisions, while the inset shows the alternative of a vernier on the index arm. (*Whipple Museum Library, Cambridge*)

engine', Mr Graves of St Bride's £1 for the second engine, and Mr Dell 10s for the Bridewell engine.[3] As there was evidently an engine from the Union Fire Office also, it must have been quite a blaze. A further payment of 10s was made to 'the Turncock' for opening the cocks to supply the engines, the narrow-bore pipes to individual houses being much too small for this purpose. Altogether this incident cost the parish funds of St Bride's £3 10s in unforeseen expenses.

The cost to Adams was the virtual loss of both his business and his home. On Saturday 5 February (the same day that the thrice-weekly *London Evening Post* was published), he inserted an announcement in the *Daily Advertiser* saying that the injury to his house and goods 'renders him incapable of carrying on the said Business in the same House for the present'. As soon as he had found new accommodation, he said, he would give public notice.

Adams's family at this time consisted of his (second) wife Ann, two daughters aged nineteen and fifteen from his first marriage, and two somewhat younger sons and two daughters from his second marriage. The sons, George junior and William, were then aged six and five respectively. (William subsequently died at the age of twelve, before he was old enough to become involved in the family business.) The eldest surviving daughter from his second marriage, Charlotte, was aged four, and the youngest child, Lucy, a baby aged three months. How the family coped with this emergency can only be conjectured, as no personal letters or diaries describing the event are known; presumably they found temporary accommodation with friends or relatives nearby.

The City land tax books show that by the middle of that year (1757) Adams was paying rates and taxes on premises farther to the west in Fleet Street, five doors from Bolt-and-Tun Court on the southern side of the street.[4] This was in the parish of St Dunstan's in the West, so the fire resulted in not only a change of address but also a break with the parish that had been the Adams family's home for half a century at least. Henceforth the baptisms of further children were registered in St Dunstan's instead of St Bride's, though Adams himself eventually returned to St Bride's, where his parents and first wife were buried, on his own death in 1772.

An incidental effect of his change of address was that it made him liable for service in the parish offices all over again, this time in St Dunstan's. Although there were precedents for exempting newcomers to a parish who had already served elsewhere,[5] either this did not apply in St Dunstan's or else Adams chose not to avail himself of this privilege: in 1762 he paid a fine of £10 to avoid serving as overseer, and in 1765 £16 to avoid serving as junior churchwarden.[6] This additional expense of £26 (equivalent to at least £5,000 today) was the direct result of moving from one parish to another.

Adams's new premises had a rateable value of £40, which was £10 lower than his 'corner of Racquet Court' shop. On the other hand, his will signed in 1772 shows that the premises were freehold and were (then) owned by him. When house numbers were allocated in 1766 this address became 60 Fleet Street; despite extensive redevelopment in the Fleet Street area in modern times it is still a separate address today, though the actual building on the site is not that occupied by the Adams family in 1757.

Re-establishing his business must have been a slow process. The earliest newspaper advertisements by Adams after the February fire that have been found were in November of that year (1757). On Friday 18 November, and the next day, the following announcement appeared in the *Daily Advertiser*:

> GEORGE ADAMS, Mathematical Instrument Maker to his Royal Highness the Prince of Wales, is removed to a commodious Shop, nearly opposite to the Horn Tavern in Fleet-Street, at his old Sign, The Tycho Brahe's Head and Hadley's Quadrant, the Prince's Arms over the Door; where the Nobility, Gentry, Merchants, &c. may have great Variety of all Sorts of Mathematical, Philosophical, and Optical Instruments, of the newest Invention, and at the lowest Prices, he being the Maker.
> Note, Three second-hand Azimuth Compasses, as good as new.

It will be observed that the shop sign quoted above had 'and Hadley's Quadrant' added to it, reinforcing the suggestion made in the previous chapter that navigational instruments, rather than microscopes, formed a major part of his business. Just before the fire he had produced another publication dealing with the subject, namely *Instructions for the use of Hadley's Quadrant*. This rare twelve-page tract (only three extant copies are known) is not dated on the title-page but must have been printed before the fire, as it cites his 'corner of Racquet Court' address; and also before he received his royal appointment, as it cites his Ordnance appointment instead. The three known copies,[7] however, are each attached to a similar but separate eight-page tract entitled *Instructions for the Use of Hadley's Quadrant ... as made and sold by GEORGE ADAMS, Mathematical Instrument-Maker to his Royal Highness George Prince of Wales, at Tycho Brahe's Head in Fleet Street, between Serjeant's Inn and Water Lane*. This must have been printed

80 *George Adams Senior*

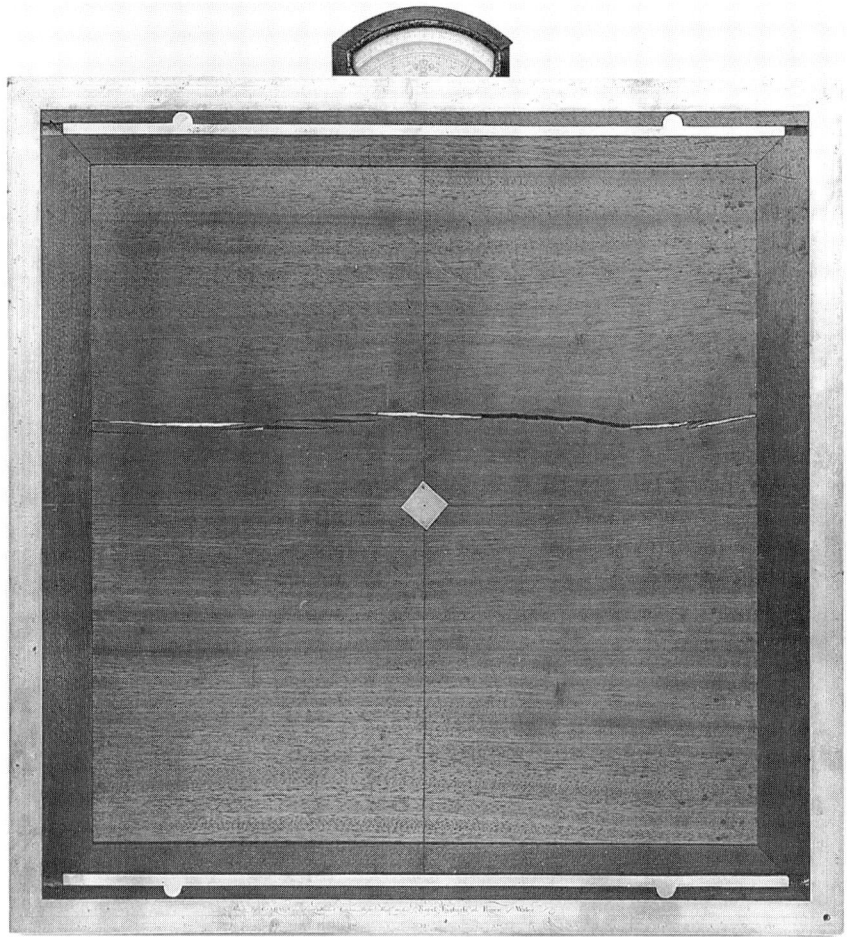

3.2 A plane (or plain) table of unusually complex construction, signed along one face of the brass edging 'Made by G*ADAMS in Fleet Street London, Inst: Mak^r to his Royal Highness the Prince of Wales'. Brass rules at two opposite sides can be adjusted in height, or withdrawn flush with the table top, by milled-head screws underneath (see figure 3.3). A 90-degree compass can be inserted in grooves in the middle of any of the four sides. The table itself, excluding the compass, is 18¾ inches square. All four sides have engraved scales on the upper face of the brass edging. (*Museum of the History of Science, Oxford, inventory no. 37998*)

Royal Connections 81

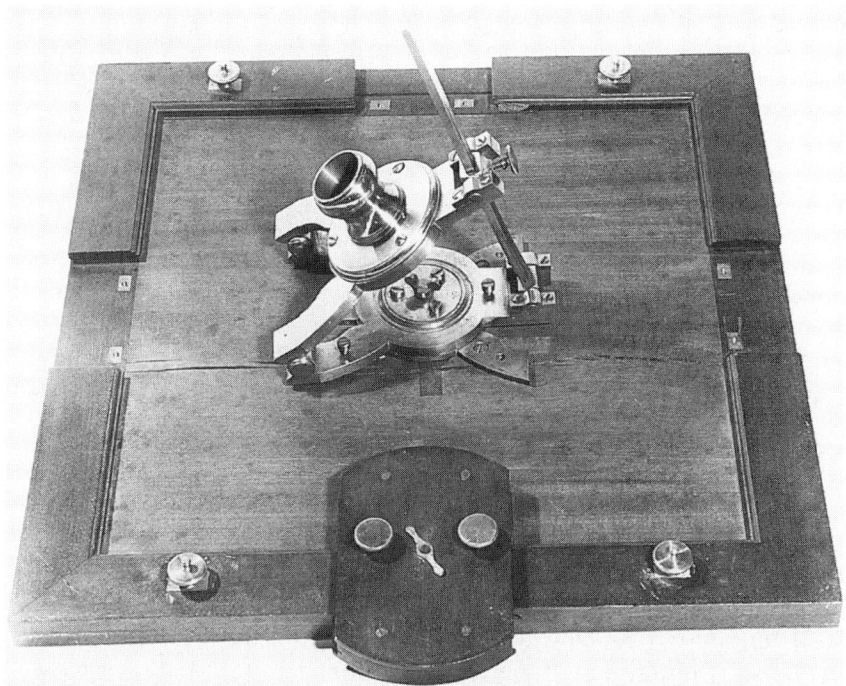

3.3 The underside of the table illustrated in figure 3.2, showing the special tilting arrangement which enables the table to be set and held at any inclination to the supporting staff or tripod. R.T. Gunther called this 'the finest plane table which we have ever seen' (*Handbook of the Museum of the History of Science in the Old Ashmolean Building*, Oxford, 1935, p. 81). In the mid-eighteenth century it was the property of the surveyor John Thompson of Witherly (1722–83), and is said to have been made to his own design. (*Museum of the History of Science, Oxford, inventory no. 37998*)

after the fire, but was nevertheless intended to be sold with the earlier tract appended to it, as the text ends with the catchword 'IN-' at the foot of page 8. What seems to have happened is that the earlier twelve-page tract was printed, but perhaps not distributed, just before Adams received his royal appointment; and then the fire a few weeks later, necessitating a change of address, provided an opportunity for Adams to have an additional eight-page section printed, quoting both his royal appointment and his new address.[8]

The earlier, twelve-page, portion was said in a footnote to be 'Extracted from a Description of a new Instrument invented by John Hadley Esq.', namely Hadley's paper in *Philosophical Transactions* of over twenty years earlier. It includes Flamsteed's table of refraction, and a table of the angle of dip of the visible horizon at heights up to 40 feet. The later, eight-page, portion is more theoretical than Hadley's own paper, and includes a description of the nonius or vernier; this is illustrated by a folding plate showing both the whole instrument and the vernier in detail. The legend on the plate cites Adams's royal appointment, confirming that it belongs to the later rather than the earlier portion.[9]

Over thirty years later, George Adams junior published yet another tract on Hadley's quadrant, which has sometimes been confused with his father's; but although there is some duplication of the text, George junior's *Description, use and method of adjusting Hadley's Quadrant and Sextant* runs to seventy pages and is positively dated 1789 on the title-page, so it can easily be distinguished from his father's publications.

About a year after establishing himself at his new shop, on 18 July 1758 Adams was chosen to be a junior warden of the Grocers' Company.[10] This must have further enhanced his status in the business world. In accordance with a charter dated 1640, the Grocers' Company had four wardens, the first (most senior) being called the Master. (The number of officers, and their titles, varied from company to company.) Each was elected to serve for one year, and although in the mid-eighteenth century they were not actually called 'second' or 'third' (etc.) warden, their names were always entered in order of seniority in the minute books, with the Master's at the top. It was usual, though not invariable, for a man to progress through the ranks from the most junior (fourth) warden until he became Master, though in some cases long gaps occurred between successive periods of service, and many wardens died before reaching the highest office. The principal function of the wardens was to supervise the day-to-day running of the company, such as binding of apprentices and granting freedoms. The real seat of power, and the ultimate governing body of the company, was the self-perpetuating Court of Assistants, whose members (about fifty) were elected for the remainder of their lives, usually after many years on the livery. According to a contemporary writer, their privileges included 'a share in profits known only to themselves'.[11] Members of the Court of Assistants met privately to determine policy, but generally left the conduct of ordinary meetings to the wardens (some of whom were members of the court). The minute books show that Adams was present in his capacity as the most junior (fourth) warden at seven ordinary meetings during his year of office,[12] the last occasion being on 20 September 1759. Ten years were to elapse before he served as a warden (third) again, by which time he was a member of the court himself.

A domestic event that occurred during Adams's first period as a warden is worthy of note, as its consequences will be mentioned when we come to consider the next generation. On 21 June 1759, Robert Blunt, a linen-draper of Charing Cross (number 64 when numbers were allocated), obtained a Bishop of London's licence to marry George Adams's elder surviving daughter by his first marriage, Sarah, then aged twenty-one.[13] The wedding took place two days later at St Martin in the Fields, where George had married his second wife. By this time the registers are a little more detailed than earlier in the century: George Adams signed the register himself as a witness, as did a member of the Blunt family (William), as well as, of course, the bride and groom.[14]

Thus, one of Adams's daughters left the family home in the City and went to live in Westminster, where the Blunt linen-drapery business continued into the nineteenth century;[15] but his family remained predominantly female, as two more daughters had been born since the fire. As well as his two sons, George (9) and William (8), he now had five daughters at home: Ann (17), Charlotte (6), Lucy (2½), Sophia (1½), and Isabella (later Isabel) (3 months). It would be interesting to know how many domestic servants the household included, and whether Adams employed a governess for the older children or sent them to school somewhere in the vicinity, but no information on this aspect of the Adams story has been discovered.

Adams's appointment as Mathematical Instrument Maker to the Prince of Wales lasted just under four years, from the end of 1756 to

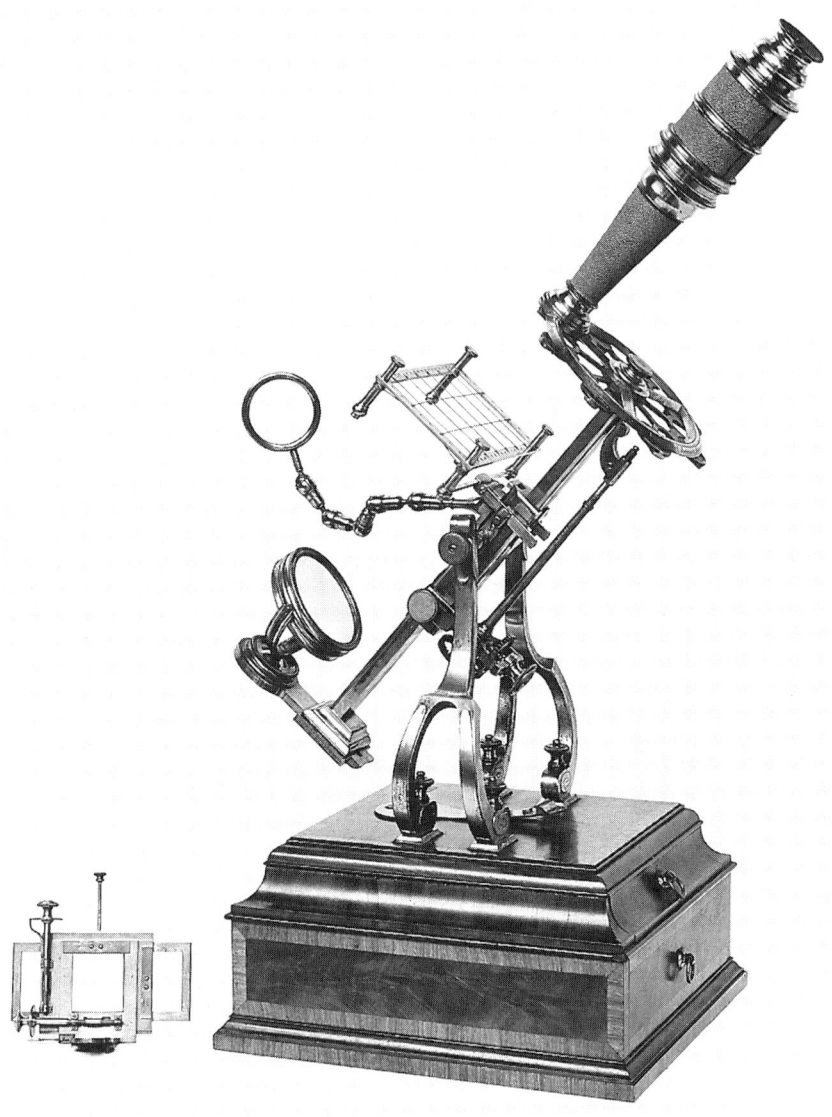

3.4 The 'Prince of Wales' microscope, so called by Clay and Court in 1932 because it is inscribed round the edge of the wheel of objectives: 'Invented and made by George Adams in Fleet Street, Mathematical Instrument Maker to His Royal Highness the Prince of Wales'. At that time it was thought to be the only microscope by Adams with trunnion mounting, but several others have since come to light. (*Science Museum, London, Inv.1925–136, Neg.2891*)

the prince's accession to the throne as King George III in October 1760. Any of his instruments that cite this title (and a surprisingly large number do) can therefore be firmly assigned to the period 1757–60. One example is a sector at the Science Museum, London, signed in full 'Improved and made by GEO. ADAMS Math^l Instrum^t Maker to his Royal Highness GEORGE Prince of Wales, London'. Another, rather better known, is the so-called 'Prince of Wales' microscope at the same museum (see figure 3.4). The distinctive feature of the latter is that the limb carrying the body and stage is supported on trunnions near its centre of gravity, so that it can be tilted to any convenient angle. It is known as the 'Prince of Wales' pattern because an instrument of this type, inscribed 'Invented and made by George Adams in Fleet Street, Mathematical Instrument Maker to his Royal Highness the Prince of Wales', was originally in the King George III Collection. By some mischance the microscopes from this collection became separated from the rest of the apparatus and passed into private hands. Several, including this one, turned up in the sale of the late Sir Francis Crisp's collection in 1925, when this particular instrument was purchased for the Science Museum. It was illustrated and described by Clay and Court shortly afterwards in a journal paper, and again in 1932 in their book.[16] It was also subsequently illustrated and described by J.A. Chaldecott in his handbook of the King George III Collection,[17] and by Palmer and Sahiar in a Science Museum booklet.[18] However, none of these authors was aware that there are three extant microscopes by Adams with the trunnion mounting which is characteristic of the 'Prince of Wales' design. Details of the other two were not published until 1986, when the Whipple Museum's catalogue of microscopes appeared. Both are currently at the Whipple Museum, Cambridge, one having been presented by R.S. Whipple while the other is on loan from Gonville and Caius College.

One of the instruments at the Whipple Museum is similar to that at the Science Museum but is signed 'Invented and made by GEO.ADAMS at Tycho Brahe's Head in Fleet Street LONDON', without any reference to the Prince of Wales.[19] The other, signed simply 'Made by G.Adams Fleet Street LONDON', is broadly similar but has a different form of construction for the coarse and fine focusing arrangements (figure 3.5).[20] In this respect the latter instrument bears a strong resemblance to Adams's 'New Universal' microscope of 1746; indeed, it might almost be a 'New Universal' removed from its folding brass base and supported on trunnions on the wooden box base of a standard

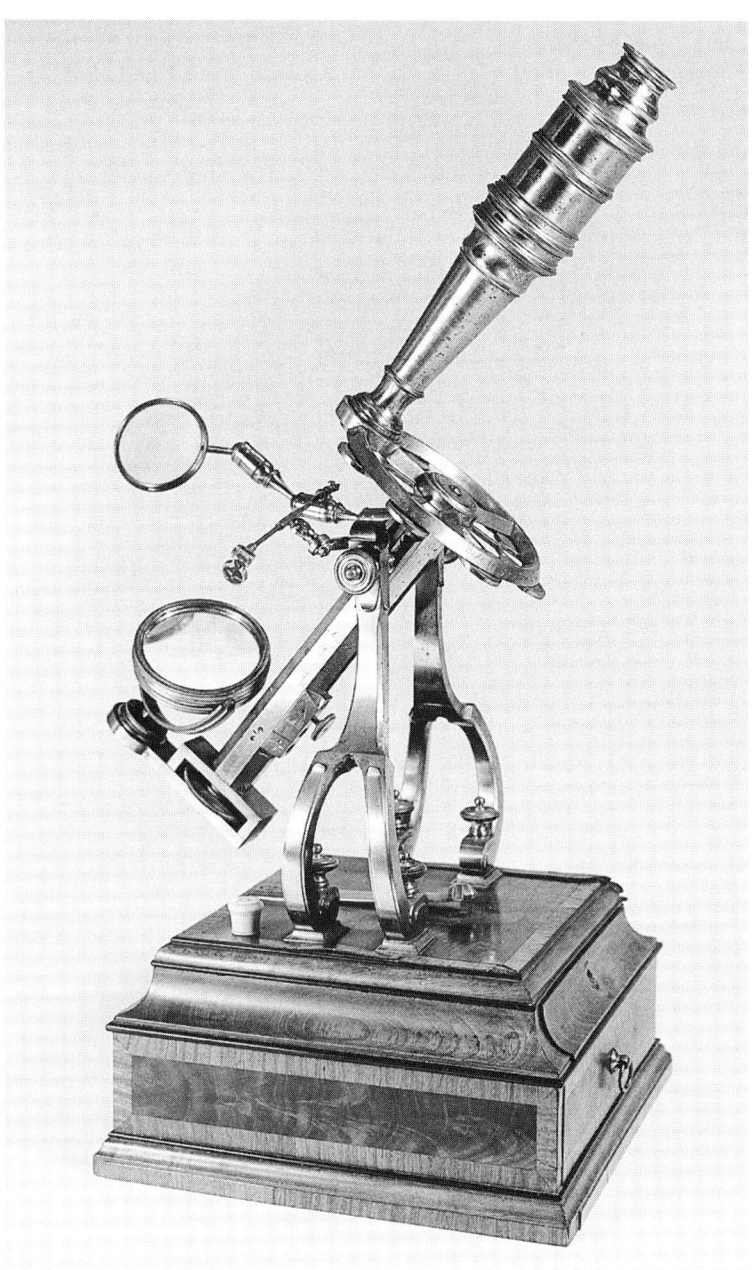

3.5 A compound microscope by Adams with the 'Prince of Wales' trunnion mounting but with a limb similar to the central column of his '1746' model. (*Whipple Museum, Cambridge, Inv.Wh.841*)

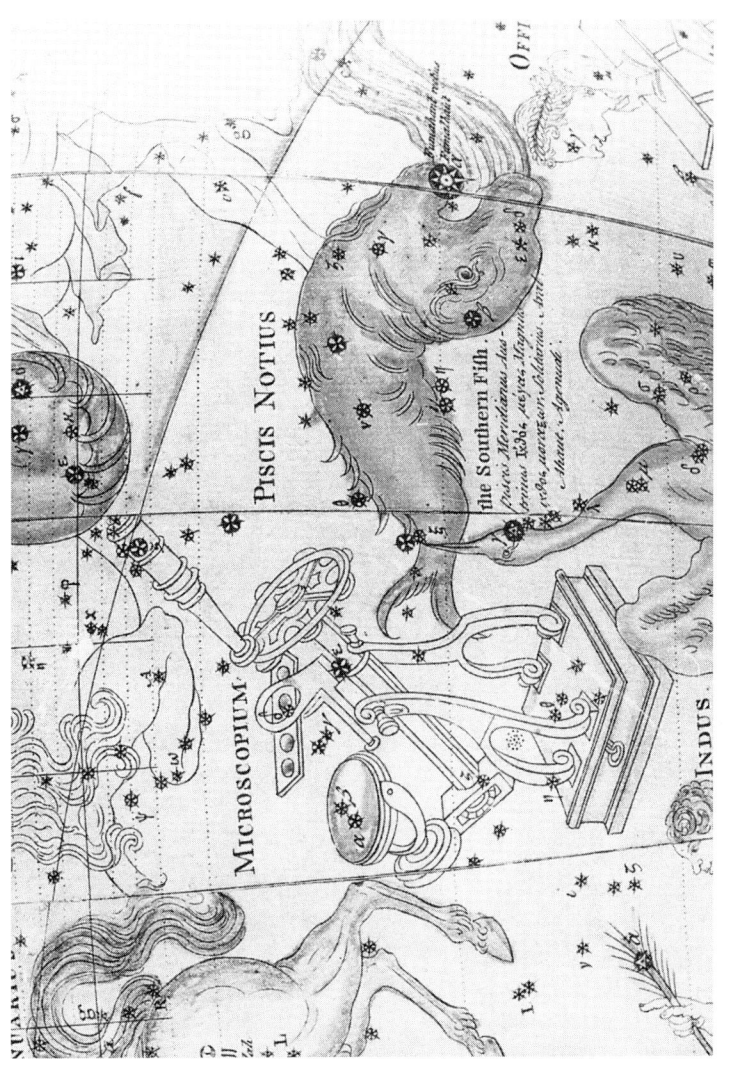

3.6 When Adams launched new terrestrial and celestial globes in 1766 he inserted several new constellations amongst the ancient ones, including 'Microscopium', depicted as a trunnion-mounted instrument like that shown in figure 3.5. (*Photograph by G.L'E. Turner from the 18-inch Adams celestial globe at Teyler's Museum, Haarlem*)

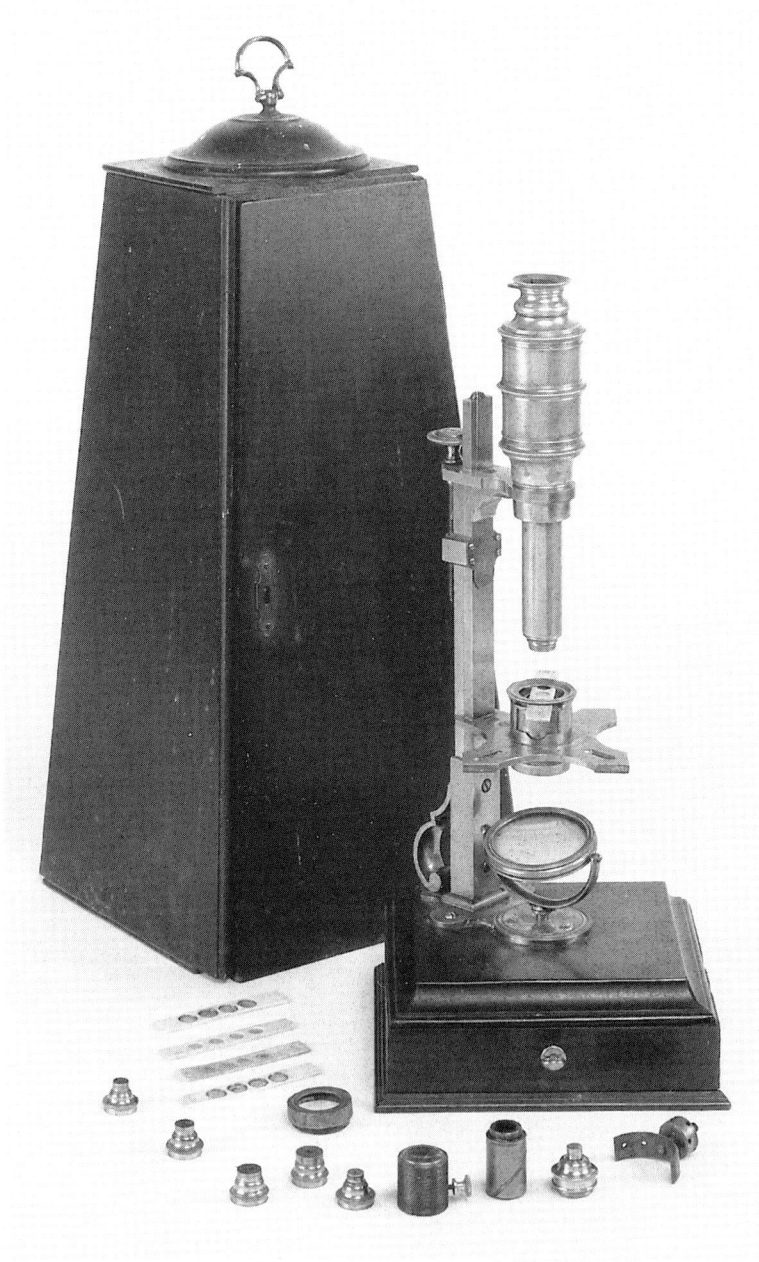

3.7 A Cuff-type microscope signed on the cruciform stage 'G. ADAMS Inst. Maker to HIS MAJESTY', mounted on a mahogany plinth base with a drawer containing five objectives. (*Christie's, London, 29 September 1988, lot 268*)

Cuff-pattern microscope (figure 3.7). The other two instruments, on the other hand, have a 'limb' consisting of two rectangular bars, one sliding on the other, as in the (vertical) pillar of Cuff-pattern microscopes. Adams certainly made – or at least sold – microscopes of the Cuff pattern in the late 1750s, for there is one signed 'Made by Geo.Adams Inst. Maker to the Prince of Wales' in the Billings Collection.[21] For fine adjustment these two trunnion-mounted instruments have the stage coupled to a long screw running parallel with the limb, actuated by a contrate wheel and pinion connected to a knurled handwheel. This indirect arrangement is necessary because otherwise the handwheel would be inconveniently located between the trunnions.

Apart from its unusual mounting, the 'Prince of Wales' microscope at the Science Museum and its counterpart at the Whipple Museum have another interesting feature in common, a rectangular mechanical stage which can be fitted in place of the usual forceps or frog plate. This device incorporates two micrometer screws mounted at right angles, each having 100 threads per inch, with milled heads divided into 100 parts, so that in theory the movement of the stage across the field of view can be controlled and read to 1/10,000th of an inch. In practice, of course, this accuracy would be achieved only if the errors in the screw threads, and backlash in the mechanism, were negligible in comparison with the scale divisions.

Clay and Court, in their comments on the 'Prince of Wales' microscope at the Science Museum, wondered what prompted Adams to produce the trunnion form of mounting and why he did not persist with this clearly advantageous design. As can be seen, it is now apparent that he did persist to the extent of making at least three instruments with the trunnion mounting, though the chronological order in which the three were made is not clear. A possible explanation for the initial trial of this mounting is that the instrument concerned was made for a child. The Cuff-pattern, the 'New Universal', and the popular 'three-pillar' or 'Culpeper' microscopes were all designed to be used in a vertical position, which must have been difficult for a child even when standing up. Mounting the whole instrument on trunnions at its centre of gravity enabled the eyepiece to be brought to any level without danger of the instrument tipping over. If this was the real reason for the construction of the first model, the one most likely to have been made first is that shown in figure 3.5, in which only the trunnions are basically new, the rest of the instrument being constructed largely from components of standard design. The provenance

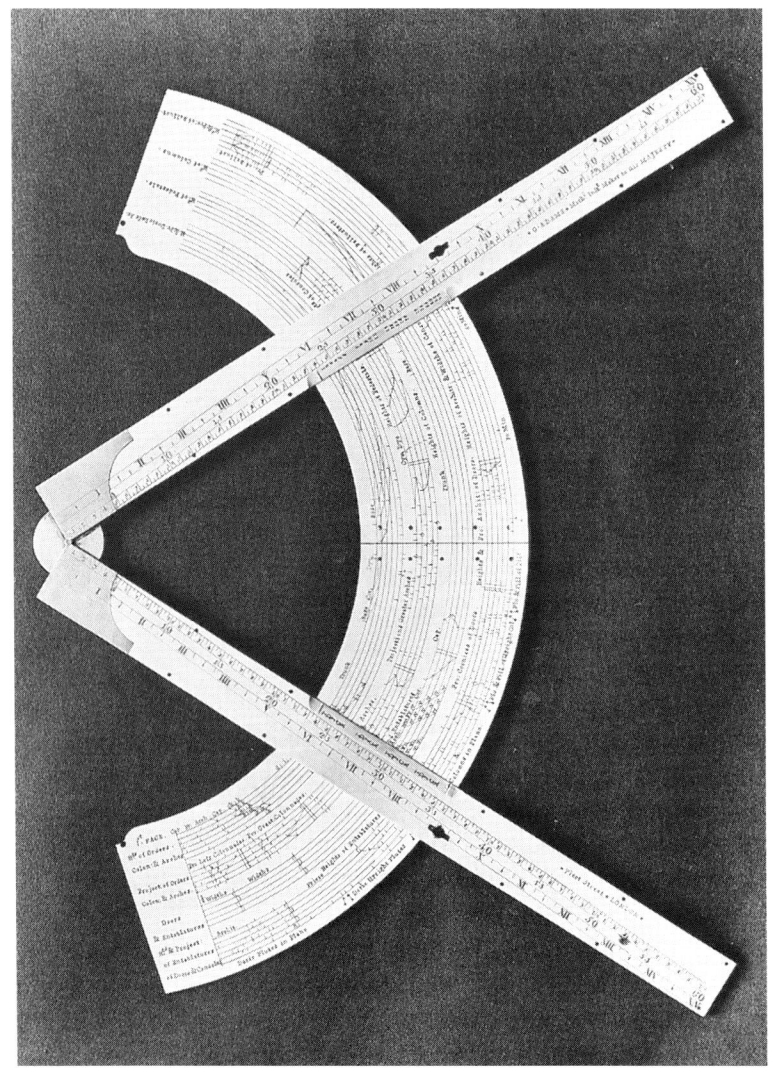

3.8 One side of an architectonic sector made of ivory with a silver hinge, inscribed 'G.Adams Math¹ Inst¹ Maker to His Majesty' on one arm and 'Fleet Street London' on the other. (*Science Museum, London, Inv.1927–1010; Neg.5243*)

of this instrument has not been traced back to the eighteenth century, so its original owner is not currently known; it is conceivable that Adams made it in the late 1750s for his own sons, George and William, and that he then made the other two, with some improvements such as a more robust 'limb' and a mechanical stage, for the Prince of Wales and Prince Edward. The absence of any further instruments of this pattern (so far as is currently known) can be explained on grounds of higher cost and lack of portability in comparison with later all-brass designs.

Another instrument that belongs to the late 1750s is the 'architectonic sector', a development of the usual two-armed sector specifically intended to facilitate architectural drawing. Adams would have heard about an earlier version while undergoing his training under Thomas Heath, for in 1723 the latter published a small book by T. Carwitham of Twickenham describing such a device. Classical architecture was based on strict rules of proportion, the relative dimensions of the various components being expressed in terms of the diameter of a column but differing from one order to another. The instrument initially described by Carwitham was specifically intended for use in designing fluted columns. In a second edition, also published by Heath (in 1733), he extended the text to show how his sector could be used to facilitate drawing plans and elevations (which he called 'uprights') of whole buildings. A more elaborate instrument, similar to that shown in figure 3.8, was described in English by Thomas Malie in 1737, 'from the original Italian of Ottavio Revesi Bruti'. The curved plates in this version bear separate scales on each face for each of the five orders of classical architecture, those on one face giving the relative heights of arches, doors, bases, and entablatures, while those on the other face give the corresponding widths. Malie's book, which was published jointly by Heath and a bookseller, incorporated full-size drawings of a sector with arms about 11 inches long, the outer radius of the sliding plates being about 7 inches. These were intended to be either cut out and stuck on to cardboard to make a working sector, or copied by engraving on to a metal instrument.

Similar full-size drawings were subsequently included in a large-format book on architecture by Joshua Kirby, which, though not published until 1761, was commissioned by the Prince of Wales a few years earlier. In his introductory remarks about the instrument, Kirby said that although it had originally been devised by Revisio Bruti, its present form was due to 'the ingenuity of that excellent workman Mr.

GEORGE ADAMS, Mathematical Instrument Maker to His Royal Highness the PRINCE of WALES, in Fleet Street, London; who makes these Sectors in silver, ivory, or wood; and who can with great accuracy, by the same principle, lay down on the vacant spaces any other measures, which the instrument is capable of receiving'. The full-size illustrations of the two faces of the sector were also inscribed 'Made by George Adams in Fleet Street London, Mathematical Instrument-Maker to His Royal Highness the Prince of Wales'. Confirmation that Adams did make such instruments while holding this appointment is provided by the existence of one so signed at the Royal Institute of British Architects. The one shown in figure 3.8 is slightly later, as Adams calls himself mathematical instrument maker to his majesty in the inscription.

At about this time Adams obtained some indirect publicity through his association with the Devon mathematician Benjamin Donn (1729–98), son of a Bideford schoolmaster. Donn became interested in astronomy and mathematics at an early age, and from 1749 was a frequent contributor to the *Gentleman's Magazine* and the *Mathematical Repository*. In 1759, aged thirty, he decided to undertake a detailed survey of the County of Devon, in response to an advertisement by the Society for the Encouragement of Arts, Manufactures and Commerce (known today as the Royal Society of Arts) offering a premium for such a project. Early in 1760 he issued a printed broadside entitled *Proposals for Surveying and Making a New and Accurate Map of the County of Devon, by Benjamin Donn, Author of the Mathematical Essays*, in which he outlined his intentions and said that subscriptions (1½ guineas for the map in sheets, or 2 guineas pasted on canvas with rollers) were taken in by no fewer than sixteen named booksellers in various places, plus 'Mr. George Adams, Instrument-Maker to his Royal Highness the Prince of Wales, in Fleet-Street, London'.[22] In subsequent correspondence with the society, Donn said that he would begin the survey as soon as the necessary instruments had arrived from London.[23] As Adams is the only instrument maker named in the proposals, it is reasonable to assume that he was the supplier. The society's conditions for awarding a premium specified that a theodolite or equivalent, not a circumferentor (a surveying instrument in which angular information is derived solely from a magnetic compass), was to be used, as the latter device was not sufficiently accurate. Distances along roads were to be measured with a perambulator (see figure 3.9). The project also involved measuring the latitudes and longitudes of the principal

3.9 A perambulator signed 'G.Adams Fleet Street London' with a table of basic linear measures on its dial giving the relationship between miles, furlongs, poles, yards, links, and chains, suitable for a surveyor's use. See figure 7.29 for a later version with a simpler dial. (*Christie's, London, 30 June 1988, lot 148*)

places and coastal features depicted, so the instruments probably included at least one Hadley's quadrant as well as a theodolite, plain table, perambulator, and measuring chain. As Donn subsequently stated that he employed several qualified assistants in the work (which took about five years), it is quite likely that Adams supplied him with more than one of each type of instrument.

Adams maintained a loose connection with Donn for many years. His name appears in the imprint to Donn's *Description and Use of the Variation and Tide Instrument* (1766, sold also by B. Martin and Heath & Wing), and his *Description and Use of the Navigation Scale* (1772, sold also by Martin, Heath & Wing, Nairne, and Watkins).

In October 1760 King George II died and was succeeded by his grandson, the Prince of Wales, then aged twenty-two, as King George III. Amongst the numerous appointments to the new royal household made in the next few months was that of George Adams as mathematical instrument maker to the king, dated 15 December 1760.[24] Although this was not a salaried post, it was one of a number of skilled craft appointments which came under the lord chamberlain, who was also responsible for 'above stairs' officials such as the royal physicians and surgeons; mere 'purveyors' of victuals and household supplies, on the other hand, came under the lord steward.

In the middle of 1761, a few months after receiving this appointment (the most prestigious available to a mathematical instrument maker), Adams was engaged, presumably by the Earl of Bute, to compile a catalogue and valuation of some scientific instruments purchased by the earl from the estate of his late uncle, the Duke of Argyll. The duke died on 15 April 1761, but there was a lengthy period of lying-in-state at Holyrood House and his funeral did not take place until the end of May. Adams's ten-page catalogue is dated on the last page 'Aug. 19. 1761', and three supplementary sheets are dated 24, 25, and 29 August respectively.[25] The last of these sheets lists the 'Instruments in the late Duke of Argyle's Library' (five large items including a grand orrery and an armillary sphere by Wright), and 'Instruments at Adams's house in Fleet Street' (thirteen items, mostly barometers and thermometers but including a pantographer by Adams). These items were not valued individually, but the total value of the main collection was assessed by Adams at £353 7s 0d, plus £24 15s 6d for the items on the other two supplementary sheets. Thirty-two years later, in 1793, the earl's own collection of instruments, of which the above formed the nucleus, was sold by auction after his death for a total of £1,337 9s 0d.[26] The auctioneer's catalogue at that time identified at least eighteen items as 'by Adams', but some of these – such as lucernal microscopes – were probably supplied by George Adams junior rather than his father. The 1793 sale included a grand orrery and an armillary sphere by Wright, which presumably were those mentioned in Adams's third supplementary list in 1761; they made £68 5s 0d and £15 4s 6d respectively. By then, of course, they were at least half a century old.

The autumn of 1761 witnessed an unusual concentration of 'state occasions', when life in London was enlivened by the pageantry of ceremonies and processions. On 8 September, the seventeen-year-old Princess Sophia Charlotte of Mecklenburg-Strelitz arrived from the Continent, to be married to the king the same evening. Then there was the coronation on 22 September, the first for over thirty years (and there would not be another for more than sixty), followed at the beginning of November by the swearing-in of the new Lord Mayor of London at Westminster Hall. This year, the newly-installed lord mayor subsequently entertained the king and queen and their retinue at Guildhall, an event that traditionally took place once in each reign. Adams's shop in Fleet Street, though remote from the court in St James's and the government in Whitehall, was on the main land route from Westminster to the City of London and only a stone's-throw from Temple Bar, the gateway to the latter (see figure 3.10), so it was well placed for observing ceremonies linking Westminster and the City. Also, Temple Bar was one of the places where proclamations were read by the heralds.

The royal visit to the City in November 1761 received extensive coverage in contemporary newspapers and journals. The *St James's Chronicle* for 12/14 November printed a long piece signed 'James Hemming' giving an eyewitness account of the processions, the decorations in the streets, and the festivities at Guildhall, which was subsequently reprinted (as 'a letter from a gentleman in the country') in the *Annual Register* for 1761. The whole of London, it seems, had turned out to watch – and indeed participate in – this event; the shops were shut, and every house from Temple Bar to Guildhall was crowded with sightseers. Many householders had erected scaffolding in front to accommodate more. The crowds in the streets were so dense that it took the king and queen, with thirty or forty coaches in attendance, four hours to reach Cheapside from St James's. Adams, though a member of the livery, held no official post in the City at this time, and he does not appear to have been personally involved in the celebra-

3.10 A map of London published in the *London Magazine* in 1761 and copied in the same year for the *Scots Magazine*, from which this illustration is taken. It shows the extent of the built-up area north and south of the Thames at the start of King George III's reign. Adams's shop in Fleet Street was almost exactly in the centre of this map, equidistant from the Court in Westminster and the Office of

(3.10 continued)   Ordnance in the Tower. Although Blackfriars Bridge, providing a link between the eastern end of Fleet Street and the south bank of the Thames, is shown, construction had only just begun at this time: it was completed in 1769. (Author's collection)

tions; however, his shop is actually mentioned by name later in this eyewitness account. Because of delays caused by the large turnout, the proceedings at Guildhall not only started late but went on much longer than intended, and it was past midnight by the time the royal family left. The houses in the streets along their route were all illuminated, and 'some of them were adorned with curious transparent devices of the initial letters of their majesties' names, ... particularly Mr. Adams's, his majesty's optician'.[27] Although Adams's royal appointment was actually mathematical instrument maker, not optician, the fact that his name is mentioned suggests that it was most probably his shop, rather than Dollond's, that the writer saw.

The principal products connecting George Adams senior with the king which are still in existence today are, of course, those made for demonstrating the principles of experimental philosophy, forming part of the assembly now known as the King George III Collection and housed, for the most part, at the Science Museum in South Kensington. The origins of the collection go back to the previous reign, and it was still being augmented in the early nineteenth century, so the dates of individual items, by various makers, cover a wide range. Some of them were originally the property of Dr Stephen Demainbray, and were used by him for the instruction of the Prince of Wales and his younger brother prior to the death of George II; but Demainbray was still giving public courses of lectures in the winter of 1761–62, presumably with his own apparatus,[28] and as there is good evidence that Adams made several items for the king in the year 1761, there must have been some overlap between the king starting to acquire his own collection and Demainbray still using his. By the spring of 1763, or possibly the previous year, Demainbray had given up teaching and found employment in the Office of Excise.[29] About five years later he was recalled to the king's service to take charge of the private royal observatory built at Kew specifically for observations of the forthcoming transit of Venus (June 1769), and thereafter his own instruments (if he still had any) became merged with the king's. Kew Observatory continued to house the royal instruments (and other scientific collections, for example of minerals and biological specimens) until 1841, when the government of the day decided it could no longer be maintained at public expense. The mechanical and physical apparatus was presented to King's College, London (established in 1829), and was subsequently (in 1926) transferred to the Science Museum on permanent loan.

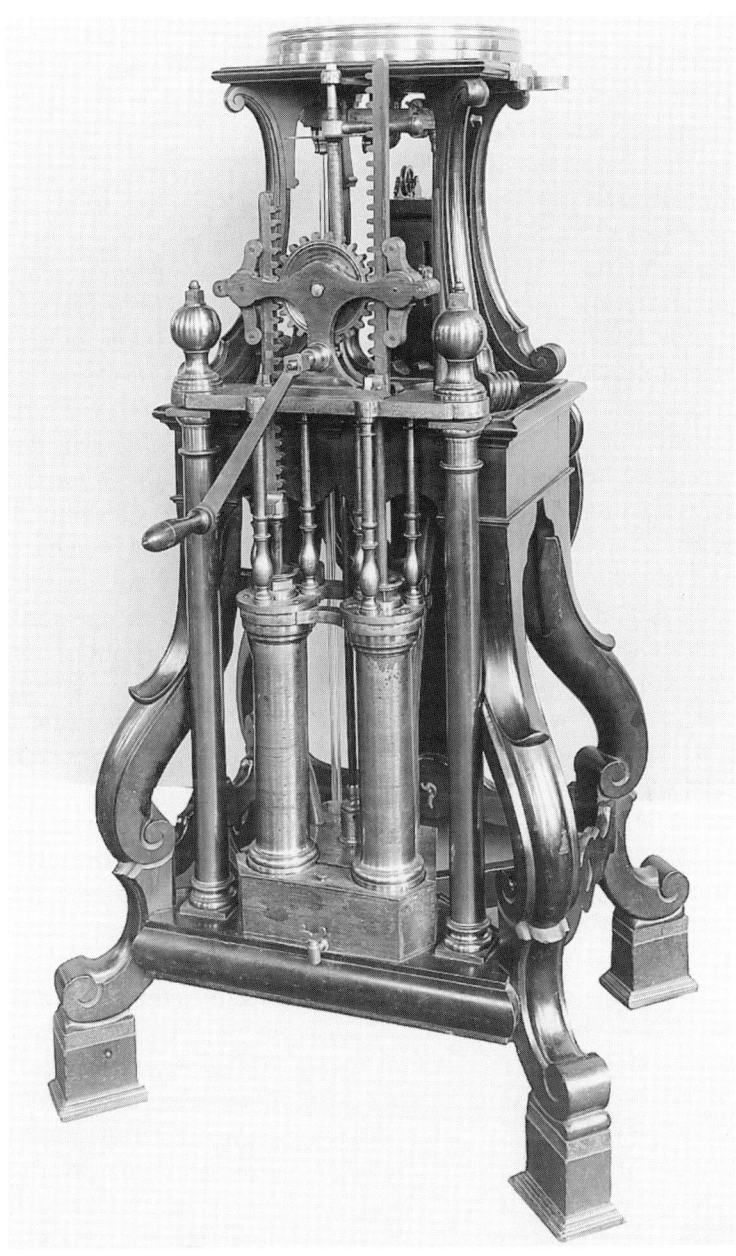

3.11 The large double-barrelled exhausting and compressing air-pump made by George Adams senior in 1761 for King George III, from a design by Smeaton. The bell-glass receiver is not shown in this view. (*Science Museum, London, Inv.1927–1624; Neg.5758*)

Associated with the apparatus now in the museum are several manuscripts by Adams, one of which, the 'Pneumatics Manuscript', describes a large standing air-pump and its accessories. The text makes it clear that this pump was made in 1761, though the manuscript itself may be slightly later.[30] Commenting that the design was based on a pump by Smeaton, except that it had two barrels instead of Smeaton's one, Adams wrote: 'This instrument is the First double-Barrel Air Pump on this Construction, ever attempted, and was made for his present Majesty, King George the Third, in the Year 1761.' The accompanying drawing by Adams depicts the air-pump with a name-plate on the front which includes the date 1761, though the instrument itself is not dated (figure 3.11).

Smeaton's improvements to the air-pump had been described to the Royal Society ten years earlier and published in the *Philosophical Transactions* in 1753.[31] Their effect was to increase substantially the degree of vacuum obtainable, primarily by use of a form of valve which occupies very little space at the bottom of the barrel. Adams said that his pump would rarify by a factor of 1,000 when first operated, and although moisture reduced this to about 500 after working, a common air-pump could seldom achieve better than 150. Another advantage of Smeaton's design was an arrangement of valves which enabled the pump to be switched instantly from exhausting to compressing by turning a single cock.

Adams's 'Pneumatics Manuscript' combines a description of the construction of the pump itself with directions for its use with the numerous accessories that were supplied with it, with frequent references to 'sGravesande's *Mathematical Elements of Natural Philosophy*. Although the manuscript is effectively an instruction manual, it is by no means clear why, or for whom, it was written. Certain passages in the text suggest that it is the draft for a book on pneumatics that Adams intended to publish, rather than a manuscript written specifically for the king himself. For example, after devoting thirteen pages to describing the construction of the pump, Adams comments that he thinks that should be 'sufficient for any intelligent workman' – hardly the language one would use in addressing the king. When he came to the experiment demonstrating the absence of sound in a vacuum, which uses a clockwork-driven bell, he described the one 'made for his Majesty King George 3d in the Year 1761'. This suggests that Adams was using the king's apparatus by way of example to illustrate a more general text. The book in which this manuscript is written also includes several pages of miscellaneous notes, a six-page price list of Adams's air-pump accessories (about forty-five items), and at the end, details of two accounts with customers named Rogers and Cuthbert, in which the entries are priced in code.

This particular manuscript book is not mentioned in Chaldecott's handbook of the collection published in 1951, because it was not discovered in the museum's files until later, though the accompanying folder of rough sketches by Adams was present.[32] Both the text and illustrations did then exist in fair copy form, however,[33] together with a similar fair copy of another text by Adams generally known as the 'Mechanics Manuscript'.[34] The latter describes the apparatus that Adams made about 1762, based on 'sGravesande's *Mathematical Elements*, for demonstrating the principles of mechanics. The heart of the apparatus is the 'Philosophical Table', to which numerous attachments can be fitted for carrying out various experiments in mechanics; they were itemized and described by Chaldecott in 1951, and more recently in the fully illustrated volume by Morton and Wess describing the collection and its background.[35]

These two fair copy manuscripts are written in copperplate script and were evidently transcribed by a professional calligrapher from Adams's rough originals. For whom this was done is not clear. Adams's wording is followed exactly, so the text does not read as though it was written specifically for the king. If Adams was intending to publish a book on pneumatics and mechanics himself, as seems likely, he would have had to take into account what other published works were then available. The book by 'sGravesande (first translated into English by Desaguliers in 1720) that Adams used as the basis of his own writings was last printed in a sixth English edition in 1747, and by the 1760s was almost certainly out of print: this may have been what prompted Adams to think of replacing it himself. However, in the spring of 1760 James Ferguson's *Lectures on Select Subjects in Mechanics* appeared, with a dedication to Prince Edward, and proved to be a popular work: it was still being reprinted in the early nineteenth century. This by itself might not have discouraged Adams from entering this field, as it was directed primarily at readers who had no mathematical knowledge; but a far more serious contender for the market left open by the absence of 'sGravesande's book was a republication of Desaguliers's own *A Course of Experimental Philosophy*, which appeared in a third two-volume quarto edition in 1763. This would effectively have prevented Adams filling the gap himself, at least until the edition was exhausted, and may be the reason why no physics textbook by Adams materialized.

How the 'Pneumatics Manuscript' and the 'Mechanics Manuscript' came to be associated with the King George III Collection remains obscure. A probable explanation for the existence of the fair copies is provided by an oblique reference in a letter written by Dr Birch on 13 November 1762, where he comments that (he had been informed by a friend) a famous writing-master, Mr Champion, had recently been engaged to transcribe them.[36] While this could account for the fair copies becoming part of the king's collection, it does not explain why Adams's rough originals were not returned to him after the writing-master had completed his task. In 1810 Adams's youngest son, Dudley, claimed to have all his father's manuscripts, 'which are numerous', in his possession.[37] Attempts to trace the movements of the manuscripts, both original and 'fine', have been frustrated by lack of detail in nineteenth-century catalogues of the king's collection. The earliest list that is still extant, dating from the last quarter of the eighteenth century, does not mention manuscripts at all;[38] two other eighteenth-century lists that were in King's College Library in 1926 could not be found when Chaldecott compiled his handbook in 1951, and are still lost. Apparently no inventory at all was made in 1841, when the college received the bulk of the physical apparatus. The Adams manuscripts now in the Science Museum first appear in an inventory dated 1880, so it does not follow that they came to the college with the relevant apparatus forty years earlier.

Apart from the pneumatics and mechanics apparatus, probably the best-known item by Adams that was formerly part of the king's collection is his elaborate and impractical silver microscope (figure 3.12). As mentioned earlier, the microscopes became separated from the rest of the apparatus and passed into private ownership. The silver microscope, like the 'Prince of Wales' instrument, was one of the effects of the late Sir Francis Crisp sold by auction in 1925. It was purchased for £360 and presented by Sir John Finlay to the recently-established Old Ashmolean Museum at Oxford (now the Museum of the History of Science). In 1932 it was fully described and illustrated by Clay and Court,[39] who dated it 1761 on the basis of a statement by Adams in the fourth edition (1771) of his *Micrographia Illustrata*.[40] Adams, however, was specifically referring to the eyepiece micrometer when he wrote that 'it was made for His Majesty in the year 1761', so although the inference is that the microscope itself was made in that year, this is

Royal Connections 103

3.12 This elaborate silver microscope is signed 'Made by GEORGE ADAMS in Fleet Street LONDON'. It incorporates a 'wheel' of eight objectives as in Adams's '1746' design, and an eyepiece micrometer which Adams later said was first made for King George III in 1761. An almost identical instrument, known to have been formerly the property of the Prince Regent, is in the Science Museum, London. The date of manufacture of these two silver microscopes is uncertain, but at least one of them was certainly in existence in 1763, when it was examined by Lalande: see the text on p. 104 for comments. (*Museum of the History of Science, Oxford, inventory no. 35086*)

not necessarily so. Furthermore, Clay and Court were unaware that two such microscopes are in existence. The other, which is almost identical except for a different body tube length, was transferred from Windsor Castle to the Science Museum, London, in 1949.[41] It appeared in King George IV's 'Pictorial Catalogue' in the 1830s, as his personal property,[42] but whether it was made at the same time as the one now located at Oxford, and if so, for whom, has not been determined.

It is curious that no contemporary description of the silver microscopes has so far been discovered. Adams, in his brief reference to the micrometer in 1771, added that 'with it was then [1761] presented a manuscript of its use, of which the following is an extract, with some alterations'. Unfortunately, no trace of this manuscript (or of the copy which Adams must have kept himself) has since been found. The only near-contemporary reference to the microscope that has been noticed is a brief mention in the diary of Jerome Lalande, of his visit to England in the middle of 1763. Some excerpts from this were published (in English) in 1926, but the full original French text only became available as recently as 1980, when it was published, with editorial annotations, by the Voltaire Foundation.[43] As well as visiting the usual tourist sights of London, Lalande called on several scientific instrument makers including Short, Sisson, and Watkins; he also obtained introductions to places such as Greenwich Observatory and Buckingham House (then only recently purchased by the king for Queen Charlotte). On 24 May 1763, Lalande wrote, after lunching with Dr Pringle and others they had gone to see the king's microscope, which had cost 'trois mille pièces, de M. Adam'. He noted that it had all the (objective) lenses on a rotating circle, and the setting (*monture*) was of silver. An eyepiece micrometer enabled a movement of 1/10,000th of an inch (*pouce*) to be distinguished.[44]

From these details, there can be little doubt that the instrument seen by Lalande in 1763 was that shown in figure 3.12. But how did he know of its existence, and what it was said to have cost? One might have expected a work of art of this nature and value to have been worthy of at least a brief mention in contemporary journals, such as the *Gentleman's Magazine*, but none such has been found.

As well as the silver microscope, Lalande also managed to examine Adams's air-pump and apparatus. This involved a special dispensation from the king (who, according to Dr Pringle, was familiar with Lalande's publications). On Saturday 27 May the king left for Richmond with the queen at 9.30 a.m., having given instructions that the air-pump was to be produced at 10 a.m. for Lalande's benefit. He duly examined it with Dr Pringle and others; Adams himself was present on this occasion, though (Lalande noted) 'il ait beaucoup de peine à marcher'. Lalande noted briefly that the air-pump was to Smeaton's design, had two pistons, could be changed from a vacuum-pump to a compressor at will, and rarified the air by a factor of 2,000 where a common pump could achieve only 300. (Both of these figures were twice as large as those claimed by Adams in his 'Pneumatics Manuscript'.) He also noticed in passing that there were many cupboards full of instruments, machines, and books, in the room where he saw the air-pump.[45] This observation suggests that the king's collection by this time (mid-1763) incorporated Demainbray's apparatus as well as the items supplied by Adams.

In addition to the silver and 'Prince of Wales' microscopes, air-pump and apparatus, and Philosophical Table, at least twenty other items in the King George III Collection in its present form are either signed by Adams or can be firmly attributed to him or his son.[46] They include various individual mechanical demonstration models (compound motion, pulleys, screw, compound lever, central forces, and so on); compound steelyard; windlass and capstan; absorption hygrometer; standard thermometer; small vacuum pump; hydraulic forcing pump; air-gun; Gregorian reflecting telescope; microtome; universal sundial; drawing instruments; magnetic toys; cylindrical electrical machine; and the pair of 18-inch globes described in chapter 4. Some of these items were supplied after 1772 by George Adams junior. According to Lalande, in a survey of the principal astronomical observatories of the world written in 1792, Kew Observatory itself (where the collection was housed from *c*. 1769 to 1841) contained a transit telescope of 8-feet focal length by Adams.[47] As this instrument apparently has not survived, its attribution to Adams cannot now be checked; possibly he supplied the mounting only, as one would expect Dollond, as Optician to his Majesty, to have supplied the optics.[48]

Apart from the 'Pneumatics' and 'Mechanics' manuscripts, another relatively short extant manuscript by Adams relates to an item in the collection, namely a description of a pantograph. Still at King's College, London, it matches an unsigned instrument in the Science Museum.[49] Possibly also relevant is a three-page manuscript by George Adams junior in the British Library. Dated August 1773 and entitled 'The Universal Compasses, invented by G.Adams Mathematical Instrument maker to his Majesty', it describes, with a sketch, a beam compass which can be used for drawing ellipses as well as circles. Though no instrument matching the description is currently in the Collection, the presence of this manuscript amongst the king's manuscripts suggests that it was originally written for the king's benefit.[50]

4 Adams's Globes

For many people the name Adams is associated primarily with two products: microscopes and globes. As previous chapters have shown, George Adams senior had many other interests besides these, and although microscopes were certainly amongst his products almost from the commencement of his business in the mid-1730s, globes did not assume much importance until twenty years later.

The 'catalogue' appended to *Micrographia Illustrata* in 1746 included 'Globes, celestial and terrestrial, of all Sizes ... viz. of 3, 9, 12, 17, and 28 Inches Diameter', and also (as a separate entry) celestial globes mounted in such a way that the precession of the equinoxes could be shown. The sizes listed, and the reference to the latter form of mounting, indicate that the globes in question were of Senex's make, which Adams would have supplied as a retailer if his customers asked for them. John Senex had superseded Joseph and James Moxon as the principal English maker of globes in the early eighteenth century. When Adams commenced trading in Fleet Street, Senex was located in the same street in premises facing St Dunstan's church, where he sold books, maps, and prints as well as globes.[1] He was both a noted engraver and a scholar in his own right, who was elected to fellowship of the Royal Society in 1728. (He was also, incidentally, a prominent Freemason.)[2] His method of mounting a celestial globe so that its axis of rotation could be moved round the poles of the ecliptic, to show the precession of the equinoxes, was described to the Royal Society in 1738 and published in the *Philosophical Transactions*.[3] Following his death in 1740 his widow Mary carried on the business at the same address; it would have been globes supplied by her that Adams had in mind when compiling his catalogue in 1746. In 1749, when some improved foreign globes threatened Mary Senex's livelihood, she wrote to the president of the Royal Society pointing out that she could still supply globes designed by her late husband, which 'will be found, upon examination, as truly made, as accurate, and as well adapted for the purpose of Astronomy and Geography, as any now extant'. Her letter was read to the society in January 1748/9 and published in the *Philosophical Transactions*.[4]

In 1754 Adams was one of several instrument makers who subscribed for, and presumably sold, Daniel Fenning's *New and easy Guide to the Use of the Globes*. This had apparently been written largely (though not exclusively) for Mrs Senex's benefit, as the subscribers list shows that she took fifty copies.[5] A popular book on globes in the mid-eighteenth century was Joseph Harris's *Description and Use of the Globes and Orrery*, first published in 1731 and in its sixth edition by 1745 (it reached a twelfth and last edition in 1783); but this was published jointly by the instrument maker Thomas Wright and the Cushee family, so the globes depicted in it were those made by the Cushees, and would not have helped Mrs Senex's business. In 1755 she retired, and the entire range of globe plates and associated jigs and tools used by her late husband was put up for sale in October.[6] All except the smallest 'pocket' size were purchased by James Ferguson (1710–76), who was then trying to establish himself as an astronomical lecturer.[7] One of Ferguson's friends was the marine surveyor Murdoch Mackenzie, through whom he had been introduced to various people who could help with revising coastlines and so on, and it seems likely that he seized the opportunity presented by the Senex sale to embark on a globe-making venture in the hope that it would provide a more secure income than lecturing.

Ferguson drew new gores for his 'pocket' (3-inch) globes, which were engraved for him by J. Mynde, the engraver who did most of the plates in the *Philosophical Transactions* at this time. The original 3-inch plates by Senex apparently passed to George Adams, though precisely when has not been determined: he may have purchased them later through a third party rather than in 1755 at the sale. He does not appear to have used them for another ten years at least, but comparison of some later 3-inch globes bearing Adams's name with those of Senex's make reveals that they were almost certainly printed from the same basic plates.[8]

The Senex sale received extensive advance publicity in the *Daily Advertiser*, beginning on 12 September and repeated almost every day up to the morning of the sale, 15 October. This activity prompted

Nathaniel Hill, globe maker and engraver of Chancery Lane, to advertise his own range of products in the same paper. Although he quoted prices, it is doubtful whether the full range was ever actually made:

	£	s	d
3-inch in a case		7	6
Ditto in frames	1	5	0
6-inch	1	10	0
9-inch	2	0	0
12-inch	3	0	0
14-inch	3	13	6
15-inch	5	5	0

Hill added that a new pair of 18 inches diameter, 'engrav'd by the best Masters, with all the modern Improvements', were in preparation. His advertisement first appeared on 13 September and was repeated eight times up to 25 September.

Adams, seeing all these advertisements by Hill and by Mrs Senex's auctioneer, must have thought the globe market was well worth breaking into. Having failed to purchase Senex's plates (except possibly the 3-inch) at the sale on 15 October, he determined to produce his own. Perhaps he had intended to do this anyway, as it must have been known in the trade that Mrs Senex was thinking of giving up. A fortnight after the sale, on 29 October, he inserted the following announcement in the *Daily Advertiser*:

> GEORGE ADAMS, Mathematical Instrument-Maker, at the Corner of Racquet-Court in Fleet-Street, begs Leave to inform the Publick, that he is now finishing a Set of Copper-Plates for a new Pair of twelve-inch Globes, on which will be inserted many new Improvements, and all the new Discoveries. He is also forwarding several other Sizes. A Book of the Use of the Globes, on a new Plan, is almost ready for the Press, and will soon be publish'd.

Despite his references to 'finishing' and 'forwarding', Adams was clearly not ready to launch his new globes at this time. After an interval of more than a year, in mid-January 1757, the following advertisement appeared several times in the London papers:

> Soon will be publish'd
> NEW GLOBES, Caelestial and terrestrial, of several Sizes. Carefully laid down from the Works and Observations of the most eminent Astronomers and Geographers, as well Foreign as Domestic; the latest Discoveries are thereon inserted, with many other essential Improvements.
> By GEORGE ADAMS,
> Mathematical Instrument Maker to his Royal Highness the Prince of Wales, At the Corner of Racquet-Court, Fleet-Street, London. Where Gentlemen

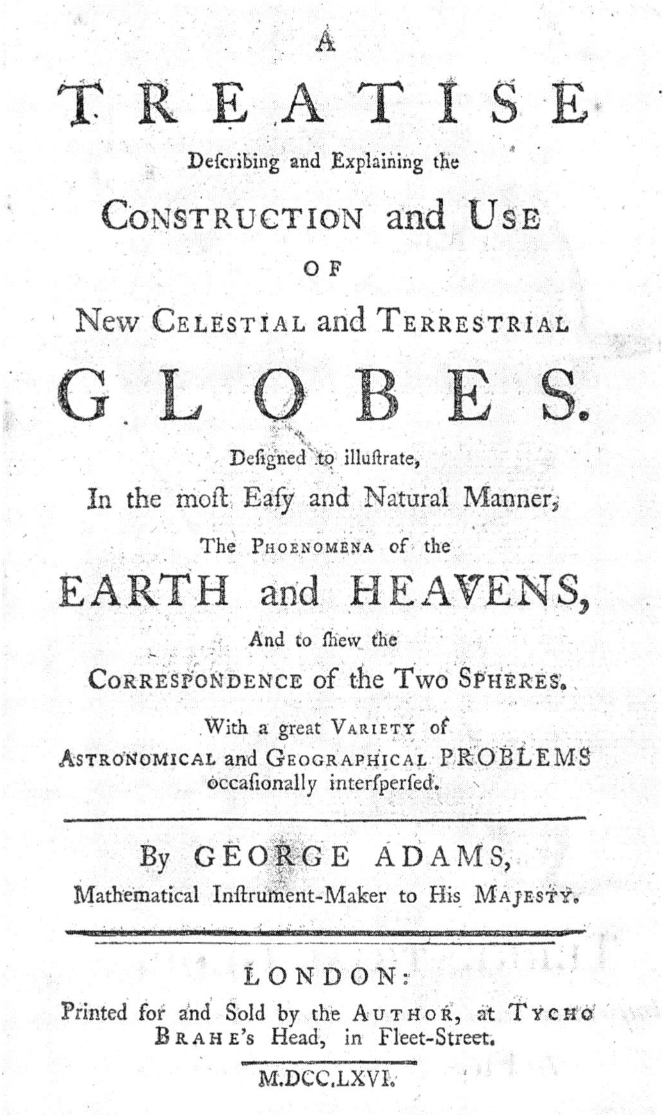

A

TREATISE

Defcribing and Explaining the

CONSTRUCTION and USE

OF

New CELESTIAL and TERRESTRIAL

GLOBES.

Defigned to illuftrate,

In the moft Eafy and Natural Manner,

The PHOENOMENA of the

EARTH and HEAVENS,

And to fhew the

CORRESPONDENCE of the Two SPHERES.

With a great VARIETY of
ASTRONOMICAL and GEOGRAPHICAL PROBLEMS
occafionally interfperfed.

By GEORGE ADAMS,
Mathematical Inftrument-Maker to His MAJESTY.

LONDON:
Printed for and Sold by the AUTHOR, at TYCHO
BRAHE's Head, in Fleet-Street.

M.DCC.LXVI.

4.1 Title-page of the first edition of George Adams senior's book on the globes, 1766. For the second (1769) and later editions the wording was altered slightly to read ... *Describing the Construction, and Explaining the Use* ...; also, all later editions mention (below the edition statement) that the contents include a view of the solar system and the use of the globes to solve spherical triangles. This work was superseded in 1789 by George Adams junior's *Astronomical and Geographical Essays*, but was reissued in 1810 by Dudley Adams as the 'thirtieth edition' (figure 10.19). (*Author's collection*)

4.2 Frontispiece from George Adams senior's book on the globes, showing his improved form of mounting. The absence of an hour circle outside the brass meridian (a time scale was printed round the equator instead) enabled the globe to be completely inverted if required. NHS is a movable meridian, for reading time off the equatorial scale; the wire TWY beneath the broad wooden circle BAC represents the limits of twilight. (*Author's collection*)

may be supplied with all Sorts of Mathematical, Philosophical, and Optical Instruments, of the newest Construction.

Only a few days later the disastrous fire occurred that nearly brought Adams's business to an end. When the reopening of his business at his new address in St Dunstan's was announced later that year, globes were not mentioned, and indeed nearly a decade elapsed before he again advertised their impending publication. This delay may have been partly due to preoccupation with other matters, especially the provision of instruments for George III and the Office of Ordnance, but it seems probable that Adams lost all his new jigs and tools, if not the copperplates themselves, in the fire. He may also have been inhibited by the continuing presence of other globe makers in the market. As well as the full range of Senex's globes, from 3-inch to 28-inch, sold by Benjamin Martin in Fleet Street from mid-1757,[9] there were British-made globes by the second generation of the Cushee family, and also by Nathaniel Hill, to whom Leonard Cushee had been apprenticed,[10] though neither the Cushees nor Hill produced such a wide range of sizes as Senex had done. In particular, nobody but Senex made the largest size (28-inch), the plates and tools for which represented a considerable capital investment.

Whatever the reason, it was not until the middle of 1766 that Adams launched his own globes, together with a book written by himself describing their construction and use. During June and July 1766 advertisements on the following lines appeared several times in at least five London newspapers:[11]

> This Day were Published,
> Inscribed to his MAJESTY,
> NEW GLOBES of 18 Inches, and of 12 Inches Diameter,
> By GEORGE ADAMS,
>
> Mathematical Instrument Maker to the King, at Tycho Brahe's Head in Fleet-Street.
> Designed to illustrate, in the most easy and natural Manner, the Phaenomena of the Earth and Heavens, and to shew the Correspondence of the Two Spheres.
> A Treatise describing and explaining their Construction and Use, Price 5s. bound and lettered, may be had as above, and of all Booksellers in Town and Country.

In a catalogue appended to his *Treatise* (see appendix I), Adams quoted basic prices of 9 and 5 guineas per pair respectively for the two sizes, 18-inch and 12-inch, or 11 and 6½ guineas if mounted in better-quality mahogany stands. A more elaborate brass and carved mahogany version was available for the larger size only, at £26. These figures were considerably greater than those quoted above for Hill's globes, and also Senex's, which Martin sold at 6 guineas for the 17-inch (Senex did not make an 18-inch) and 3 guineas for the 12-inch sizes. Adams sought to justify his higher prices by claiming that his globes were superior to others on the market, both in their accuracy and in the new form of mounting that he had devised. Since the latter differed from what he called the 'common' form, he was obliged to write his own textbook to explain how to use globes fitted with it. It may be presumed, therefore, that his globes and *Treatise* were launched simultaneously.

As Instrument Maker to the King it was natural that Adams should dedicate his book to him; but instead of writing a dedication himself, he enlisted the services of Dr Samuel Johnson.[12] This was a wise move, for it must be admitted that Adams was not particularly literate himself. Though quite capable of writing straightforward descriptions of instruments and apparatus, he did not attempt (as his rival Benjamin Martin did) to enliven the text with poetical digressions or quotes from classical authors. Johnson's dedication ensured that the book would be noticed by people in the literary world who might not have troubled to read a scientific text. In its number for 11/13 September the *London Chronicle* published an anonymous complimentary letter (which could, of course, have been written by Adams himself, or on his behalf) praising the dedication in Adams's *Treatise* for being 'modest, concise, and sensible', and not 'loaded with fulsome flattery', as was generally the case. The dedication itself was appended in full.[13]

The first edition of Adams's *Treatise* (see figure 4.1 for the full title) made an octavo volume of 242 text pages, with only three plates. In later editions both the subject matter and the number of illustrations were considerably extended. Two of the plates depicted Adams's new globes (one terrestrial, one celestial) in elaborately-carved stands consisting of a short pillar on a tripod foot, surmounted by a bowl-shaped support for the brass meridian ring and wooden horizon.[14] The terrestrial globe plate is reproduced here as figure 4.2. In comparison with 'common' globes, Adams's had the following distinctive features:

a) There was no hour circle and pointer outside the brass meridian; instead, a scale of hours and minutes was printed on the globe gores round the equator. The absence of an external scale and pointer enabled the globe axis to be tilted to any angle, while the comparatively long hour scale enabled the globe to be set more accurately to time.

b) A thin brass 'movable meridian' pivoted at the north and south poles enabled the time at any place to be read off the equatorial scale. A semicircular wire adjacent to the equator, passing through the brass meridian and pivoted at the horizon at either end, was provided to enable a marker to be attached at any point on the time scale without defacing the globe surface. On the celestial globe the movable meridian became a movable circle of declination. A small movable disc or annulus could be slid up and down the movable meridian as a latitude or declination marker, or to serve as a 'horizon' indicator on the terrestrial globe. Because of this, Adams called what was normally known as the 'horizon' – the horizontal platform at the top of the stand, in the plane of the centre of the globe – the 'broad paper circle' instead, to avoid confusion when describing the use of the globes.

c) A horizontal circular wire was fixed to the frame, about 18 degrees below the level of the centre of the globe, to indicate the approximate limits of twilight.

Examples of Adams's 18-inch globes are preserved in several countries in continental Europe as well as in Britain, and occasionally appear in the marketplace. The particularly fine pair shown in figures 4.3 and 4.4 are part of the King George III Collection at the Science Museum, London; their stands incorporate brass quadrant supports for the wooden 'horizon', and presumably represent the most expensive form of mounting listed by Adams in his 1766 catalogue (£26).

Because the amount of detail given in museum and sale catalogues varies widely in scope and reliability (sometimes even the dimensions are inaccurate or misleading, referring to the stands rather than the balls), it is hardly practicable to compile a census without personally examining each item; but globes of this size by George Adams (not necessarily George senior, and not necessarily in matching pairs) are certainly located in Arolsen, Kalundborg, Leiden, Lisbon, and Paris.[15] In October 1768 the astronomer Johann Bernoulli the younger saw a pair in the library of the University of Göttingen which he said had been presented by the Queen of England; they were the first globes of this type that Bernoulli had encountered, and he noted with approval how the use of an equatorial scale instead of the conventional hour circle enabled the globes to be set to an accuracy of ¼ of a degree.[16] The celestial globe (only) of this pair is still in the Niedersächsische Staats- und Universitätsbibliothek.

How the pair in Leiden got there is revealed by some extant manuscripts. Many of the dubious marketing techniques employed today are older than is often realized. The episode of the provincial agents quoted in Adams's publicity for his *Micrographia Illustrata*, most probably without their approval, and the launching of his 'mariner's bow' quadrant in direct competition with an almost identical instrument by Cole, indicate that he was prepared to go to any lengths to sell his products. Distribution of free samples to gain penetration of overseas markets was another technique that he employed, in this case to make his new globes known on the Continent. Towards the end of 1766 he sent a pair of 18-inch globes to J.N.S. Allamand (1713–87), who had been tutor to the children of W.J. 'sGravesande and at that time held the post of Professor of Philosophy at the University of Leiden. On 28 December Professor Allamand wrote to Adams to inform him that the globes had arrived safely via Mr Muilman (a London agent). Adams's reply dated 23 January 1767 has survived: it is now in the library of the Wellcome Institute, London,[17] though how it got back to England is not clear. He thanked Professor Allamand for his approbation and for placing the globes in the most conspicuous part of the university, where they would be seen by 'every Gentleman of Science'. Actually, the minutes of the meetings of governors of the university show that the gift was not officially accepted until 2 February, when the governors agreed to place them in the library, and instructed Professor Allamand to write a grateful letter.[18] The globes were subsequently (in the twentieth century) transferred to the Museum Boerhaave, where they are now on display, having undergone restoration in 1990. They were described (before restoration) by Peter van der Krogt in 1984.[19] From the documentation cited above, these globes must have been made no more than six months after Adams launched them in mid-1766.

In a postscript to his letter to Professor Allamand of 23 January 1767, Adams said that he had also written to Mr Brouwer, apparently a Dutch bookseller or instrument dealer who he hoped would sell his globes on the Continent, listing their prices in the 18-inch and 12-inch sizes in various types of stands. He said that he had also told Mr Brouwer what discount he normally allowed to retailers. He repeated the prices for Professor Allamand's benefit, but unfortunately did not say (in this letter) what his rate of discount was, so we have no

George Adams Senior

4.3 An 18-inch terrestrial globe by George Adams, mounted in a carved mahogany frame with brass quadrant supports to the wooden horizon. Adams dedicated his globes, and his book on their use, to the king. As this example is in the King George III Collection it was probably one of the first to be made, before they went on sale to the general public in mid-1766. (*Science Museum, London, Inv.1927–1700; Neg.177/49*)

Adams's Globes 117

4.4 The 18-inch celestial globe matching the terrestrial globe in figure 4.3.
(*Science Museum, London, Inv.1927–1700; Neg.178/49*)

indication of his profit margin on these products. The prices quoted were the same as in his 1766 catalogue, except that the most expensive £26 version of the 18-inch size was omitted and an intermediate version inserted, namely 16 guineas in mahogany 'fluted and carved' stands, or 20 guineas if fitted with a compass in a brass box as well.

A pair of 18-inch globes in the 11-guinea version (that is, in plain mahogany frames) was ordered by the Board of Ordnance on 1 January 1767 for use at the Royal Military Academy, Woolwich. Adams's bill dated 16 February 1767 came to a total of £13 8s, including a pair of green covers at 14s, packing cases at 18s, and a copy of his *Treatise* at 5s.[20] Nearly two years elapsed before this bill was settled, on 29 November 1768. Such delays in payment by government departments were quite usual, and meant that contractors who wanted government business had to have sufficient capital resources to cope with them.

Prior to supplying these new globes, in 1764 Adams had carried out repairs to two globes belonging to the academy.[21] The celestial cost £7 15s, a substantial sum which suggests that the globes concerned were probably the largest size then available, namely Senex's 28-inch. A brand-new pair of 17-inch globes could have been obtained from Benjamin Martin for 6 guineas, so the board must have regarded Adams's new globes at 11 guineas as being a significant improvement over Martin's, which probably had not been updated since he purchased the plates from Ferguson ten years earlier.

Adams dedicated his new globes, as well as his book on them, to the king – in Latin on the 18-inch size and (a different text) in English on the 12-inch. Possibly Dr Johnson was the author of these dedications also (it is difficult to imagine Adams composing them himself), but Boswell is silent on this point. George III was addressed in the Latin version on the terrestrial globe as 'Scientiarum Cultori pariter et Praesidio', which may be loosely translated as 'equally devotee and protector of the sciences'. On the celestial globe he was called 'Astronomorum Patrono Munificentissimo, Celeberrimo', or 'most generous and renowned patron of astronomy'.[22] The cartouche containing the dedication on the celestial globe was judiciously placed close to the new constellation Machina Pneumatica, depicted as the double-barrelled air-pump in the king's collection.

In his newspaper advertisements for his globes and *Treatise*, Adams identified his address as 'at Tycho Brahe's Head, Fleet Street'. The address on his 18-inch globes, and his 12-inch celestial, is simply 'Fleet Street', while that on his 12-inch terrestrial gives his shop sign as well.

Adams's Globes 119

4.5 A 12-inch Adams celestial globe incorporating his movable meridian for reading time off the equatorial scale, but in a much plainer (and cheaper) wooden stand than the larger globes shown in figures 4.3 and 4.4. (*Science Museum, London, Inv.1914–557; Neg.714*)

Inclusion or omission of the latter in inscriptions on his products of this period seems to have been dictated largely by the space available. By the 1760s his name must have been well known to the instrument-buying public, and (as an advertisement quoted in chapter 3 indicates) his shop was identified in the late 1750s by having the prince's arms over the door, as well as the 'Tycho Brahe's Head' sign. When the plates for Adams's globes were engraved, Fleet Street had not been numbered, but that situation was about to change. By the time the globes were on sale in July 1766, Adams's shop had been allocated number 60.[23] However, although some shopkeepers adopted numbers almost immediately, it was a long time before they replaced signs completely; indeed, some traders – including the younger members of the Adams family – continued to use their old signs as well as numbers, just as trade marks are used today.[24]

Adams's 12-inch globes had less decorative stands than the 18-inch, and with their English dedication and considerably cheaper price were probably aimed at the educational market, while the 18-inch were expected to find a place in gentlemen's libraries. An example of a 12-inch celestial globe, in the Science Museum, London, is shown in figure 4.5. Others are located at (amongst other places) the Bibliothèque Nationale, Paris[25] and the Nederlands Scheepvaart Museum, Amsterdam.[26]

A known customer for Adams's 12-inch globes, though not until February 1774, when the business was being run by his widow and eldest son, was Josiah Wedgwood of pottery fame. He purchased no fewer than three pairs of the 12-inch size, apparently at a special discount as an extant bill in the Wedgwood archives shows that the globes themselves were charged at 4½ guineas per pair, though the catalogue price was still 5 guineas, as when they were launched in 1766. Covers at 8s 0d, a packing box at 6s 6d, and a copy of Adams's *Treatise* at 5s 0d, brought the price for each set to £5 14s 0d. One set was purchased as a present for Wedgwood's friend Bentley, another was sent to a school at Manchester where one of Wedgwood's children, 'Sukey', was being educated (an additional 7s 6d carriage was charged for this), while the third was apparently sent to his nephew Tom Bryerley who had gone to New York as a schoolmaster.[27]

Adams's entry into the globe-making field had an immediate effect on established traders, who saw this intrusion as a threat to their livelihood. First to react was Samuel Dunn, who advertised his 'Planispheres' in the London newspapers in June and July 1766 almost as widely as Adams did his globes.[28] Dunn claimed that his planispheres, or 'Globes in Plano', which were engraved on five copperplates (two celestial, two terrestrial, and the slider or index), were 'complete Substitutes for a large Pair of Globes', and were also, of course, much cheaper, at 6s the set plus 4s for the book of their use. Being less durable than globes, his planispheres do not appear to have survived to the present day, and his descriptive book is also rare.[29] It is not known whether they had any adverse effect on the sale of Adams's globes. At the time of their first introduction in 1757–58 they attracted a great deal of support from teachers and lecturers such as Dr Demainbray, much to the dismay of Benjamin Martin, who hurriedly wrote *An Essay on the Nature and Superior Use of Globes* (1758) denigrating Dunn's product.[30] It was Martin, too, who reacted most strongly to the introduction of Adams's new globes in 1766, as they posed a direct threat to his sales of Senex's globes of similar dimensions (17-inch and 12-inch).

Martin's first action was to place an advertisement in the *Daily Advertiser* on 24 June, saying that he was preparing an appendix to his own book on the use of Senex's globes, 'with some Animadversions upon a new Set of Globes, and a Treatise of their Use, lately published'. By this means, he added, 'the judicious Purchaser will be more capable of comparing their Merits, and determining whether these or Mr. Senex's Globes are constructed with superior Accuracy and Elegance'. When Dunn's planispheres first appeared in 1757–58 Martin purchased a set, and kept them available in his shop for customers to compare with Senex's globes (to the advantage of the latter, he naturally assumed), and he adopted the same technique now. Until his appendix was available, he said, 'any Gentleman or Lady may satisfy themselves, by Inspection of a Pair of Globes of each Sort [Adams's and Senex's], at his Shop in Fleet-Street, whenever they please'.

Martin's *Appendix to the Description and Use of the Globes* was an octavo tract of eighty pages with one folding plate. Its contents included a description of a new orrery by Martin, an abstract of Hornsby's account of the forthcoming (1769) transit of Venus, and a description of an instrument for finding the latitude, in addition to various matters pertaining to globes. The section most relevant to the subject of this book, however, was the first thirty-one pages, headed 'Animadversions upon the New Construction of New Globes lately published, and a Treatise of their Use, humbly submitted to the Consideration of the Public'. This consisted of a savage attack on Adams's globes, his *Trea-*

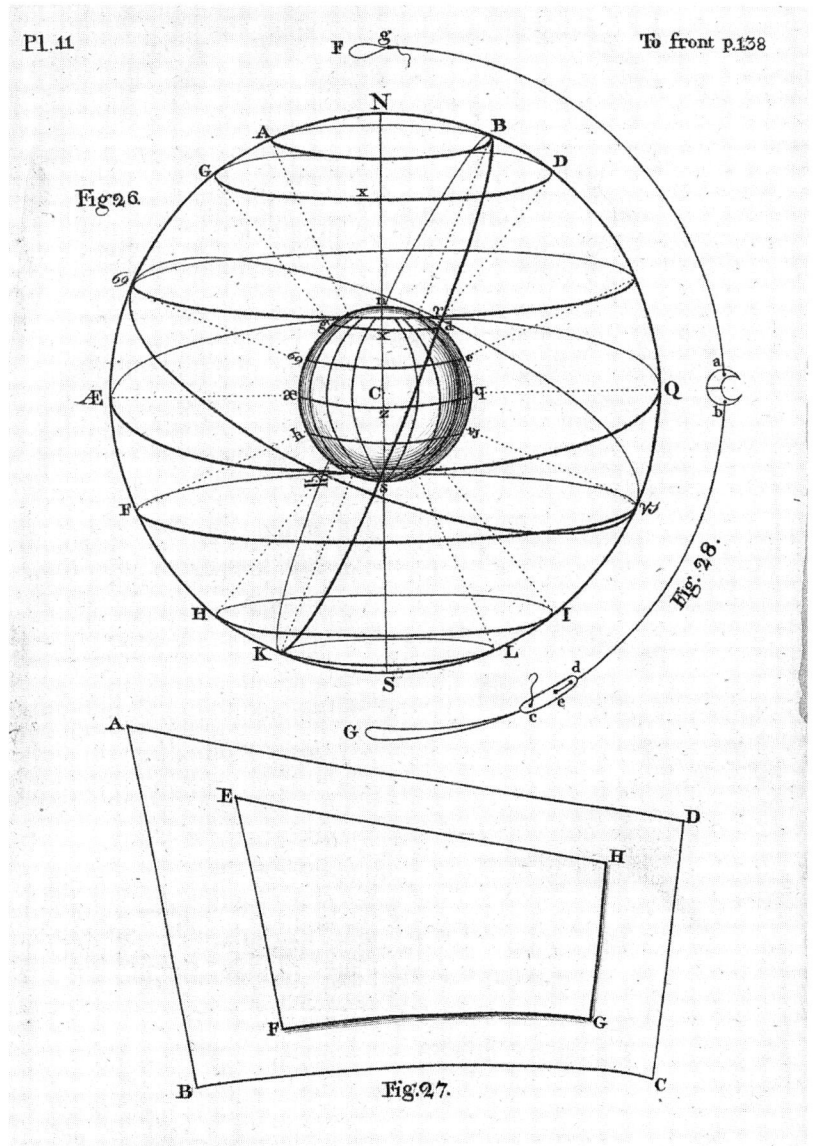

4.6 Because his method of mounting introduced a large gap between the globe surface and the brass meridian or wooden horizon, Adams suggested that purchasers should provide themselves with a piece of card cut to the shapes shown above (for 12-inch and 18-inch globes respectively), to facilitate taking accurate readings off the scales. In the first edition this was plate III; in subsequent editions it became plate 11 (eleven) and the figure numbers were added, as here. (*Author's collection*)

tise, his catalogue therein, and his business methods, prompted (as Martin admitted) by the fear that if these new globes were really so superior to Senex's as Adams claimed, sales of the latter would necessarily suffer greatly.

Adams had described his globes as being of a 'new and peculiar' construction, an unwise choice of terms that enabled Martin to seize upon the word 'peculiar' (by which Adams meant unusual or distinctive) and sarcastically show just how peculiar (in other words, freakish) they really were. His main complaint was that Adams's movable meridian and 1/10th-inch diameter wire adjacent to the equator necessitated an unusually large gap between the surface of the globe and the brass meridian and wooden horizon platform, thereby making it difficult to take angular measurements off the globe.[31] Adams was aware of this drawback: he suggested in his *Treatise* that readers should themselves cut a small piece of cardboard to fit the convex shape of the globe, and use this to take readings off the brass meridian or horizon platform when accurate setting was required (see figure 4.6). Martin expressed astonishment that the purchaser of these globes 'at an excessive Price' should be 'left to put his Ingenuity to the Rack to construct Materials and Implements for using them, himself'. As for the new information that Adams claimed was provided on his globes, Martin dismissed the extra stars on the celestial, and the multiplicity of names of places and geographical features on the terrestrial, as simply adding unusable, confusing, and unnecessary data, thereby making these new globes more difficult, rather than easier, to use. He was particularly scornful of Adams's claim that his celestial globe (to which, Martin said, Adams had added no fewer than fourteen new and unnecessary constellations 'repleat with ... the most vulgar and cumbersome symbols') was more up to date than any existing ones: the precession of the equinoxes meant that corrections had to be applied to all celestial globes anyway with the passage of time.

A paragraph in which Martin comments on the background to the introduction of these globes is interesting, as it confirms that Adams had been preparing them for some considerable time:

> ... if the Public were certified of the Truth of what I have heard from the Workmen, viz. that the Plates were done over and over again, in Whole or in Part, and many Changes and Sets of Hands successively employed for many Years past; and that at last it would have been impossible to have got them compleated, had they not hit upon the Expedient of copying Mr. Senex's Papers of the 17 Inch Celestial Globe: I say if they believe these common

Reports, there will be very few competent Judges found to have any extraordinary Opinion of the superior Accuracy of copied Plates ...

As for Adams's *Treatise*, in addition to criticizing its astronomical contents Martin picked out numerous specific cases of spelling or usage to complain about, such as the fact that Adams, evidently determined to be right 'by Hook or by Crook', had spelt the 'hackney'd word Phaenomena' in three different ways: Phoenomena, Phaenomena, and Phenomena. His catalogue, too, came in for criticism, mainly because Adams had used phrases like 'invented or improved by himself', and terms like 'manual orrery', which Martin claimed had been copied from his publications and advertisements. In effect, Adams was once again being accused of plagiarism, though Martin did not mention the previous occasions when Baker and Cole had made similar complaints. (Possibly he did not know about them, as they occurred before Martin settled in London.)

Adams made no public reply to this onslaught. Secure in his appointments as instrument maker to the king, to the Office of Ordnance, and to the Royal Mathematical School (Christ's Hospital), he could afford to ignore the upstart Martin's remarks. By this time (1766) Adams had been in business in London for over thirty years, while Martin, though older than Adams by four years, had spent most of his life travelling and lecturing and had only become a shopkeeper in the metropolis in his fifties. In the absence of detailed archives for either Adams's or Martin's businesses, it is impossible to say for certain what effect the introduction of Adams's globes had on Martin's sales; but the fact that Adams's globes have survived in somewhat greater numbers than Martin's suggests that after 1766 much of the globe trade was captured by Adams.

In 1769, three years after its first publication, Adams produced a second edition of his *Treatise*. This version was considerably enlarged, with 345 text pages and 14 plates (compared with 242 and 3 respectively in the first edition). The additional material was mainly concerned with 'a comprehensive view of the solar system' and the use of the globes for solving problems in spherical trigonometry. Adams evidently hoped that these additions would transform his book into a general treatise on astronomy, but in practice it has always been known for short as his '*Treatise on the Globes*'. In the catalogue appended to the second edition, 18-inch globes were offered with a choice of four grades of stands: the plainest at 9 guineas, mahogany at 11 guineas, carved mahogany at £24, and mahogany and brass at £36. Prices for

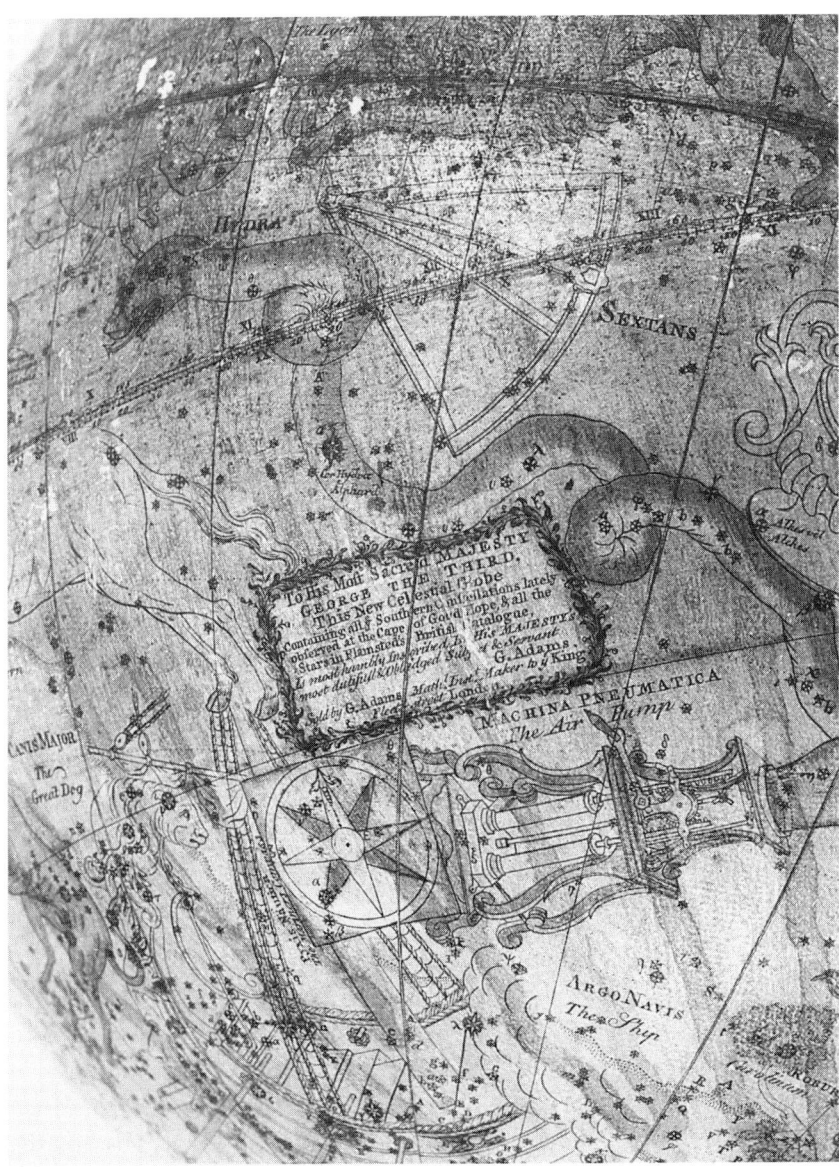

4.7 Detail of the inscription on a 12-inch celestial globe by George Adams. On the 12-inch size, unlike the 18-inch, the dedications and imprint are in English. For later variations of this inscription, by Dudley Adams, see figures 10.7 and 10.21. (*Christie's, London, 29 November 1990, lot 10*)

the 12-inch size were the same as in 1766. The book was reissued, with little alteration, in the year of his death (third edition, 1772) and then twice more by George Adams junior (fourth, 1777, and fifth, 1782) before it was superseded by the latter's *Astronomical and Geographical Essays* in 1789. In 1810 George senior's youngest son, Dudley, who by then had inherited his father's manuscripts and copperplates through his mother, published a further edition which he absurdly called the 'thirtieth', thereby giving rise to the quite erroneous statement in some reference books and catalogues that Adams's *Treatise* 'went through thirty editions'.

Adams's globes themselves – especially the terrestrial ones – were updated from time to time as new discoveries were made, but precisely how many different issues were produced is difficult to say. Whereas a letterpress text has to be printed in distinct editions, generally of at least several hundred copies, after each of which the type is broken up for use elsewhere, the engraved copperplates for globe gores are permanent items which can relatively easily be modified and a few copies printed off as required. Details of how Adams's globe-making business was organized in the mid-eighteenth century are not available, but one might expect that only a small number (if any) of completed globes would be held in stock, the gore prints being run off in small batches to meet specific orders: by this means the globes could be kept continuously up to date. But this is just conjecture, and it does not necessarily follow that Adams adopted this procedure.

The most important change in the cartographic information on Adams's terrestrial globes in the decade after their introduction was the insertion of Captain Cook's discoveries on his voyages of 1768–71 and 1772–75. Any globe which shows the track of the first voyage but not the second can be dated with confidence to the period 1771–75. An 18-inch terrestrial globe by G. Adams in the Bibliothèque Nationale, Paris, which is assigned the date 'c.1771' in a published list is presumably of this type; the matching celestial globe is said to be dated 1775.[32] An Adams 18-inch terrestrial globe in Denmark is said to be dated 1782, and presumably incorporates the results of Cook's last voyage in 1776–79.[33] Major additions or alterations such as this would probably have been announced in the newspapers when they were introduced. Unfortunately, the large number of titles published in the second half of the eighteenth century precludes a comprehensive search, but one Adams globe advertisement of this nature that has been noted appeared in the *Daily Advertiser* on 29 June 1774, stating that 'the LATE DISCOVER-

IES made by Commodore Byron and Captains Wallis and Cook are accurately laid down' on both the 18-inch and 12-inch sizes. On 27 March 1775 George Adams junior placed an advertisment in the *Daily Advertiser* which mentioned not only the usual 18-inch and 12-inch sizes, but also 6-inch, at 3 guineas per pair. All three sizes are listed in the considerably extended catalogues appended to the third (1772) and fourth (1777) editions of Adams's *Treatise*, but without a price for the 6-inch size. These catalogues also include 3-inch globes, 'in frames' at 1½ guineas per pair or 'in black cases for the pocket' at half a guinea (10s 6d).

Bernoulli, who visited several London instrument makers in January 1769, said of Adams: 'On trouve chés lui des globes de cette espece, de 18 pouces, de 12 pouces & de 6 pouces, de diametre.'[34] However, no examples of the 6-inch size by Adams have been located. If it was made at all, its introduction – despite Bernoulli's statement – seems to have been after the spring of 1772, for at that time Matthew Boulton was unable to find a 6-inch globe for an elaborate astronomical clock that John Whitehurst was making for him, and eventually had to settle for an appreciably smaller 5.4-inch globe by Nathaniel Hill, despite the mechanical modifications that this entailed.[35] Perhaps Bernoulli saw a prototype which was not put into production until some time later.

The 3-inch size, on the other hand, was certainly made and numerous examples are extant, though its history is obscure. Two versions are known: firstly an unsigned terrestrial globe inscribed 'A Correct GLOBE with the new Discoveries', attributed to Adams because it is often found in orreries or tellurians made by him (that is, George Adams junior); this globe has the track of Cook's first voyage only, and can probably be dated *c.* 1772.[36] Secondly, a terrestrial globe with a cartouche reading 'A Terrestrial GLOBE G.Adams No.60 Fleet Street LONDON'. (See figure 4.8). The basic gores of this version appear to be printed from Senex's plates; they do not show the eastern coast of Australia, so are before 1771, while the shop number indicates a date after mid-1766. If George Adams senior purchased the plates for this size at Senex's sale in October 1755, apparently he made no use of them until at least ten years later, so perhaps they were in somebody else's hands during the intervening period. Both versions of the 3-inch terrestrial globes are sometimes found in association with a celestial pair, which also appears to be printed from plates which were originally Senex's.[37]

Globes in the 18-inch and 12-inch sizes, updated from time to time, continued to be sold by Adams's elder son, George junior, until his

4.8 A nominal 3-inch pocket terrestrial globe by George Adams in a case lined with the celestial gores. The globe has a cartouche in the northern hemisphere inscribed 'A Terrestrial GLOBE G.Adams No.60 Fleet Street LONDON'. Inclusion of a shop number indicates that this inscription was engraved no earlier than 1766, but the basic gores appear to have been printed from Senex's plates of c. 1730, with some minor updating (and substitution of Adams's name for Senex's). The celestial gores lining the halves of the hemispherical case appear to be unchanged from those used by Senex & Price c. 1710. (*Christie's, London, June 30 1988, lot 17*)

death in 1795. The 3-inch size also, in frames or pocket cases, were listed in George junior's catalogues, and the 6-inch size may have been sold by him, as indicated above. When he wrote his *Astronomical and Geographical Essays* (discussed in chapter 7) in 1789, to replace his father's *Treatise*, he said the 'great and increasing' sale of his father's globes

> may be looked upon at least as a proof of approbation from numbers; to this I might also add, the encouragement they have received from the principal tutors of both our universities, the public sanction of the university of Leyden, the many editions of my father's treatise of their use, and its translation into Dutch, &c. The recommendation of Mess. Arden, Walker, Burton, &c. public lecturers in natural philosophy, might also be adduced ...[38]

At least two pairs of 18-inch globes by George Adams, and one pair of the 12-inch size, were still in the University of Oxford when R.T. Gunther carried out a survey in the early 1920s. The larger pairs were at New College and University College, while the Bodleian Library had the smaller.[39] The globes formerly at the University of Leiden have been mentioned earlier in this chapter. Of the public lecturers cited, John Arden's philosophical apparatus sold after his death in 1791 included a large orrery by Adams, and a double-barrelled air-pump, but only an unnamed pair of 9-inch globes; perhaps by that time he had replaced a larger pair by the more portable 9-inch size then in vogue.[40]

In 1788 George senior's youngest son, Dudley, born in 1762, who had been apprenticed to his brother George junior, applied for his freedom. By the middle of that year he had established his own business at Charing Cross (see chapter 10). Dudley seems to have taken over the globe-making side of the Adams business almost immediately, as several globes positively dated 1789 with Dudley's imprint are extant, while no globes signed 'G.Adams' that can be positively assigned a later date than that are known.[41] Both George Adams junior and Dudley included globes from 18-inch down to 3-inch in their separate catalogues up to the former's death in 1795, so presumably Dudley supplied them to his brother for retail sale (as well as selling them himself) during the few years that their respective businesses overlapped.

It must not be assumed, of course, that either George Adams senior or junior actually made globes themselves, though they may have taken some part in drawing and engraving the copperplates for the gores.

When Benjamin Martin purchased the Senex globe business from James Ferguson in 1757 he advertised that he had engaged the same hands who had made the globes for Senex himself. The actual manufacture of Adams's globes (that is, making the globular wood-and-plaster cores and fixing the paper gores) would have been carried out by specialist workmen (or women) highly skilled in this particular art. It seems that for a while at least the manufacture of Adams's globes was put out to a subcontractor, Thomas Pattrick, who had his own premises, first in Orange Street, Leicester Square, and then (c. 1802–1803) in King Street, Covent Garden. On his trade card citing the former address he described himself as 'Globe Maker and Optician', and claimed to be 'Manufacturer of Adams & Senex's Globes from 3 to 28 Inch Diameter', as well as tubes for telescopes and so on.[42] Precisely what the relationship was between Pattrick (and possibly other similar subcontractors) and the Adamses has not been ascertained: presumably George junior and/or Dudley would have at least inspected and approved globes bearing their name before passing them on to customers.

Dudley's globe-making activities, which were not limited to Adams's 18-inch and 12-inch globes but covered the full range of sizes including Senex's 28-inch, are discussed in chapter 10; they continued certainly until 1815 and possibly to 1817. Long before then, however, the superiority that Adams's globes enjoyed from the 1760s to the 1780s had evaporated. This was largely due to the entry of two specialist firms into the globe-making field: Cary and Bardin.

John Cary (c. 1754–1835), of 181 Strand, was a map engraver who first advertised globes 'from entire new Plates' of 21, 12, 9, and 3½ inches diameter in 1791.[43] His 12-inch size were priced at 3½ guineas, or £5 10s in mahogany frames, appreciably cheaper than Adams's of this size. His 21-inch, at 9 guineas (or 13 guineas in mahogany) were a little more expensive than Adams's 18-inch but were noticeably larger – about 10 inches longer round the equator. Cary's globes continued to be made for at least forty years (latterly by his sons G. & J. Cary of 86 St James's Street); an 18-inch size was eventually added to the range, though not until about 1817, after the demise of the Adams business.

William Bardin (c. 1740–98), originally a leather worker, entered the globe-making field a decade earlier than Cary in 1782, in association with Gabriel Wright, who had previously worked for Benjamin Martin (and probably learnt how to make globes while employed there).[44] Wright/Bardin globes were made in only two sizes, 12-inch and 9-inch, in simple plain stands, and were aimed at the educational market, so they probably did not have much effect on George Adams junior's globe trade at first. In fact, he sold the 9-inch size himself, to fill a gap in his range.[45] Shortly after his death in 1795, however, William Jones of Holborn started to design new globes of 18 and 12 inches diameter, to be made by the Bardin firm (by then W. & T.M. Bardin, father and son). These were eventually launched in 1799/1800 entitled 'New British Globes', dedicated to Sir Joseph Banks (terrestrial), and the Astronomer-Royal (celestial), and soon received favourable comments from users. Advance publicity was provided by Samuel Vince of Cambridge in his *Complete System of Astronomy* (1797), who (ignoring both Adams and Cary) wrote that:

> Since the time of Senex, no new globes have been made of a large size. Messrs. W. & S. Jones, Mathematical and Philosophical Instrument Makers in Holborn, London, are therefore now preparing a very fine set of new globes of 18 inches diameter, containing all the latest astronomical and geographical discoveries. The places upon the terrestrial globe are all laid down by an eminent Geographer, with their names in English, and upon the celestial globe there are about 6000 stars very accurately laid down for the year 1800, from the best observations; to which are added, many clusters of stars and nebulae, with the figures of the constellations. ... These globes will be a very valuable acquisition to the public.[46]

'New British Globes' of 18 and 12 inches diameter were made in a wide variety of stands from 1799 until about 1860, signed after 1832 by S.S. Edkins, successor to the Bardins by marriage to William's granddaughter. By then the Newton firm had also entered the field, but Newton globes were made mainly in the smaller sizes (3-inch to 12-inch) until after the closure of the Adams business, so they may not have had so much effect on the latter as the more prestigious New British Globes.

John Bransby of Ipswich, in his *Use of the Globes* published in 1791, thought it necessary to describe the improvements that George Adams senior had made in his globes launched in 1766, because 'many of Adams's Globes are now used'; but in the second edition of this work, published seventeen years later in 1808, he commented: 'The globes made by Adams and Cary are very good ones; but the New British Globes, manufactured by Bardin, under the direction of W. & S. Jones, possess superior accuracy.' This opinion seems to have been shared by other writers on 'The Use of the Globes', such as Thomas Keith.

Thus, although updated versions of the 18-inch and 12-inch globes introduced by George Adams senior in 1766 were still being made and sold half a century later, their dominance of the market lasted no more than 25–30 years and probably somewhat less.

5 George Adams Senior's Last Few Years, 1767–72

In 1766, when George Adams senior launched his new globes, he was aged fifty-seven and had only six more years to live, though having passed unscathed through the danger periods of infancy and adolescence and reached that age he might reasonably have expected to survive for somewhat longer. His contemporaries and neighbours in Fleet Street, James Ferguson and Benjamin Martin, lived to 66 and 77 respectively. His predecessor as mathematical instrument maker to his majesty (George II), Thomas Wright, lived to 67, and his former master, Thomas Heath, to 75.[1] The only clues to possible poor health are the statement by Lalande in May 1763 (chapter 3) that he had 'beaucoup de peine à marcher', which might indicate the onset of gout or arthritis, and the mystery over his delayed acceptance of a seat in the court of the Grocers' Company in 1767–69, mentioned below.

Adams's eldest son by his second wife, George junior, born in 1750, was formally apprenticed to him in the Grocers' Company at the age of 15 in 1765, when his youngest son and last child, Dudley, was only 2½ years old. In between he still had several daughters from his second marriage, aged (in 1766) from 5 to 13, but his second son, William, born in 1751, had died at the age of 12 in September 1763. His elder surviving daughter from his first marriage, Sarah, had married Robert Blunt and left the family home in 1759; by the mid-1760s she already had five children at Charing Cross, and was eventually to have no fewer than eight sons and four daughters – a family rivalling in numbers Adams's own by his two wives. Some of her direct descendants are alive today.[2] Her younger sister, Ann, was in her mid-twenties, unmarried, and presumably still living at home. Probably she had to assist in

GEORGE ADAMS,

MATHEMATICAL INSTRUMENT-MAKER TO HIS MAJESTY,

At Tycho Brahe's Head, in Fleet-Street, London.

Makes and Sells all Sorts of the most curious Mathematical, Philosophical, and Optical Instruments, in Silver, Brass, Ivory, or Wood, with the utmost Accuracy and Exactness, according to the latest and best Discoveries of the modern Mathematicians.

HADLEY's Quadrants, with the latest Improvements, in the most exact Method, with Glasses whose Planes are truly parallel.

Azimuth and Steering Compasses, invented by Dr. *Gowin Knight*, F. R. S. approved of, and used by his Majesty's Royal Navy. *N. B.* These Compasses are all examined and certified by Dr. *Knight*.

Large Astronomical Quadrants, Transit and Equal Altitude Instruments, for observing the Transits of the Sun and Stars over the Meridian, &c.

Sun Dials Horizontal, for Pedestals in any Latitude; with Variety of Portable ones, either Universal, or for several different Latitudes, with new Improvements.

Choice of curious Cases of Drawing Instruments, in Silver, Brass, &c. containing a Sector, Scales, Proportionable, and other Compasses; Drawing-Pens, a Protractor, Parallel Rules, &c.

A new-invented Portable Microscope for viewing all Kind of Minute Objects, as well Opake as Transparent, in so conspicuous and concise a Manner, as to comprehend all the Uses of all the other Sorts of Microscopes in one Apparatus; and magnifies to so great a Degree, as to discover the Circulation of Blood in Animals, the Peristaltic Motion of Insects, the Farinae of Vegetables, and many other surprising Phænomena, otherwise not perceptible.

The double Constructed Microscope; Mr. *Ellis*'s Æquatic Microscope; Solar Microscopes; Magellescopes, &c.

The New Acromatic Telescope, with a compound Object Glass, approved by all the Curious in Optics; with all other Sorts of Refracting Telescopes; Night Telescopes, &c.

Reflecting Telescopes, of the latest Improvement.
Micrometers of the newest Construction, elegantly fitted to Refracting or Reflecting Telescopes.
Orreries and Planetariums, greatly improved.
Instruments proper for Gunnery, Fortification, &c.
Pantographers, for reducing Drawings and Pictures of any Size, in the most complete Manner.
Instruments for taking the true Perspective of any Landscape, Building, Gardens, &c. and others for copying of Drawings.

New Globes, mounted in a peculiar Manner, whereby the Phænomena of the Sun, Earth, Moon, &c. are exhibited according to Nature.

Air-Pumps, or Engines, either for exhausting or condensing the Air, and this by turning one Cock only, with all their Appurtenances; whereby the Properties of that most useful Fluid are discovered and demonstrated by undeniable Experiments; Hydrostatical Balances, nicely adjusted for determining the specific Gravity of Fluids and Solids, &c.

Curious Barometers, Diagonical, Wheel, Standard, or Portable, with or without Thermometers. Also the so much famed Quicksilver Thermometers, made after any of the Forms.

Theodolites, of the latest Construction; Water Levels, which may be adjusted at one Station; Measuring Wheels; Pocket and Coach Way Wizers, for measuring the Way, &c.

Spectacles ground on Brass Tools, in the Manner approved of by the Royal Society, set in Variety of convenient Frames: Also Reading Glasses of all Sorts, set in Silver or other Metal, to turn into Cases of various Kinds.

Prisms, for demonstrating the Theory of Light and Colours.

The Camera Obscura for drawing in Perspective, in which all external Objects are represented in their proper Colours and exact Proportions.

Concave, Convex, and Cylindrical Mirrors, Opera Glasses, Multiplying Glasses, Spectacles of the true *Venetian* Green Glass; Magic Lanthorns, &c.

Zoograscopes, for viewing Perspective Prints.

N. B. Gentlemen may have any Model or Instrument made in Metal or Wood, with Expedition and Accuracy, and carefully packed up to be sent to any Part of the World.

5.1 Trade card used by George Adams senior post-1760, probably before 1766 as no shop number is quoted. Dr Gowin Knight (died 1772) is mentioned as still alive and certifying compasses for the Navy. Adams's new globes (launched 1766) are listed but sizes are not given, so they were probably still under development. (*Science Museum, London, Inv.1951–685; Neg.1112/52*) bringing up her younger half-brothers and half-sisters, but unfortunately no personal letters or diaries have survived that might have thrown some light on this point.

There were no apprentices in the household in the late 1760s apart from George junior. Nathaniel Kettle, whom Adams had taken apprentice in December 1754,[3] should have completed his term by the end of 1761, but as he did not take his freedom it is not known whether he dropped out before then. Compared with some other members of the Grocers' Company, Adams took very few raw apprentices: only four in twenty-nine years, including his own son. His former master, Thomas Heath, took ten in twenty-seven years.[4] Adams did, however, take at least one partially-trained man from another guild: Robert Tangate was turned over to him in April 1758 after serving five years of an apprenticeship in the Joiners' Company, and there could have been more instances of this nature that have not yet been discovered.[5] As it is inconceivable that Adams could have run his business single-handed, he must have employed several journeymen who had been trained elsewhere, or else made extensive use of subcontractors working on their own premises. Documentary evidence for either of these possibilities is not easily found, though one unusual source – licences for non-freemen – has been cited in chapter 2. In our present state of knowledge of the instrument-making trade in the mid-eighteenth century we can only make tentative deductions about the size of a workforce from circumstantial evidence, such as signatures on shop bills or references to workmen in correspondence or invoices.

It seems likely that in practice Adams would have made use of both internal and external sources of labour. Certain types of instrument, for example globes, are known to have been made by specialists whose names are mostly unrecorded. The basic skills of a mathematical instrument maker were metalworking and engraving, so one might expect a wide-ranging business like Adams's to employ other specialists, such as joiners for making the wooden parts of instruments, and turners for making the parts which had to be accurately turned to size. The optical parts of microscopes and telescopes would almost certainly have been bought in from specialist grinders who did nothing else, rather than made on the premises. In the early nineteenth century a 'working optician', Charles West, claimed in advertisements that he supplied optical components to all the leading firms including (Dudley) Adams;[6] however, that was fifty years later than the period with which we are dealing here. Similar evidence for the mid-eighteenth century is elusive.

Micrographia Illustrata:

OR THE

MICROSCOPE

EXPLAINED,

IN SEVERAL NEW INVENTIONS,

Particularly of a New VARIABLE MICROSCOPE for Examining all Sorts of Minute Objects;

AND ALSO OF A

New CAMERA OBSCURA MICROSCOPE,

Designed for Drawing all Minute Objects, either by the Light of the Sun, or by a Lamp in Winter Evenings, to great Perfection;

WITH

A DESCRIPTION of all the other Microscopes now in Use.

LIKEWISE

A NATURAL HISTORY OF AERIAL, TERRESTRIAL, AND AQUATIC ANIMALS, &c. Considered as Microscopic Objects.

By GEORGE ADAMS,
Mathematical Instrument-Maker to His MAJESTY.

THE FOURTH EDITION.

Illustrated with Seventy-two Copper Plates, containing 560 Delineations of various Microscopic Objects.

LONDON:

Printed for the AUTHOR, and sold by him at No. 60, in Fleet-Street, and by all Booksellers in Town and Country.

M.DCC.LXXI.

5.2 Title-page of the fourth edition (1771) of George Adams senior's book on the microscope, *Micrographia Illustrata*, in which he introduced the 'variable' pattern (see figure 5.4), the first form of the lucernal, and also a small 'pocket' compound microscope (see figure 5.3). (*British Library, London, 44.c.15*)

There are a few instances of men who had once worked for Adams subsequently becoming well known in their own right. One such was Nikolay Galaktionovich Chizhov (1731–67), who was in London from December 1759 to August 1760, during which period he 'worked with George Adams, the optical and philosophical instrument maker, to improve his technical skills'.[7] Chizhov became Master of the instrument workshop at St Petersburg Academy of Sciences in 1762, and retained this post until his death five years later.

Another man who spent some time in Adams's workshop, probably in the late 1760s, was John Miller, founder of the firm of Miller & Adie which flourished in Edinburgh in the nineteenth century. Precisely when Miller was in London has not been determined, but it must have been before 1769, as he was already running his own business in Edinburgh then. Until recently, the statement that Miller had once worked for Adams was based solely on tradition in the Adie family, but documentary evidence confirming this was found in the Bute archives in the 1980s, in a letter from Dr James Lind to Lord Loudon dated 23 June 1769.[8]

On 17 July 1767 Adams achieved the distinction of being elected to the Court of Assistants, the governing body of the Grocers' Company. Curiously, he did not attend and take the oath until 21 April 1769, over a year and a half later.[9] No reason for this delay is given in the minutes. Ill health is a possible explanation, but there could have been other reasons: he could have been working away from London during this period, or even abroad. From certain clues in their writings, it is known that both George junior and Dudley Adams spent some time on the Continent, though precisely when has not been discovered. In the absence of firm evidence, this point remains unresolved.

After taking the oath in April, on 14 July 1769 Adams was chosen as (third) warden for the year 1769–70. The minutes show that he was present in that capacity at meetings on 27 October and 15 December 1769, and on 11 April, 29 May, 13 July, and 19 September 1770. At the end of the September meeting the next set of wardens, elected for the year 1770–71, took their seats. Adams attended four meetings in his capacity as an assistant during the year 1770–71, the last on 17 October 1771. He does not appear to have attended any meetings of the court during the remaining twelve months of his life.

To prevent confusion, it is worth noting that the procedure for choosing wardens and assistants varied from company to company. In some of the smaller guilds the wardens were chosen wholly from members of the court, election to which was the first step in moving up the hierarchy. In the Grocers' Company (in the mid-eighteenth century), this procedure was reversed: a man was expected to have served at least one term as a junior warden, and thereby proved his reliability, before being considered for election (for life) to the court, which was a permanent self-perpetuating body and the real seat of power, with control of the company's considerable assets.

During the last few years of his life Adams produced two major publications, on globes and on microscopes. Though both were updated editions of existing books, both incorporated a substantial amount of new material and hence involved a lot of writing plus drawings for additional plates.[10]

The second, much enlarged, edition of *Treatise on the Globes*, issued in 1769, has already been mentioned in passing in chapter 4. The new material consisted chiefly of 'A comprehensive view of the Solar System' (forty-six pages), and the use of the globes to solve problems in spherical trigonometry (fifty pages). Eleven new plates were provided with this edition, eight of which depict the various bodies and phenomena of the solar system (the planets, moon phases, eclipses and so on), while the other three illustrate the solution of spherical triangles. The two plates depicting a terrestrial and a celestial globe, and the one showing readers how to make a template for taking accurate measurements off the globe (figure 4.6) were retained.[11] Despite Martin's scathing remarks on this, Adams really had no option but to continue using it in this and subsequent editions, as the distance between the globe surface and its mounting remained a major disadvantage of Adams's globes when they were used as an analogue computer for solving problems in spherical trigonometry.

In the following year (1770) an edition of *Treatise on the Globes* translated into Dutch was published in Amsterdam. This evidently found a ready market, as another Amsterdam edition with a different publisher's imprint appeared in 1771, translated and with extended comments by Jacob Ploos van Amstel, MD.[12] For these Dutch editions the title was changed to give prominence to the astronomical content of the text, *Gronden der Sterrenkunde* (which may be roughly translated as *Foundations of Astronomy*), with the construction and use of the globes relegated to a secondary role. This may have been done to avoid upsetting Dutch globe makers, who would not have taken kindly to Adams's rival English globes being promoted in Holland through the medium of his book.

Towards the end of 1771 Adams issued a fourth edition of his *Micrographia Illustrata*,[13] first published in 1746 as a small quarto (chapter 2). Possibly its appearance at this time was prompted by the publication in 1769 of a fifth edition of Henry Baker's *The Microscope made Easy*. If so, Adams certainly stole a march on his adversary this time, for Baker's fifth edition was a virtually unchanged reprint of his second (1743), with now-obsolete references to John Cuff and illustrations of apparatus now thirty years out of date. Adams's book, on the other hand, included descriptions (in about sixty octavo pages) of the latest types of microscopes and accessories as well as older ones. This was reflected in the altered title, *Micrographia Illustrata: or the microscope explained, in several new inventions, particularly of a new variable microscope for examining all sorts of minute objects* (see figure 5.2). The seventy-two plates in this edition, though repeating many of those in the 1746 edition concerned with microscopic objects, included fourteen of apparatus; Baker's book had only five of apparatus, showing the Wilson screw-barrel of c. 1700, the same on a stand, a Culpeper or three-pillar microscope, Cuff's solar microscope of 1743 with a wooden wheel, and a compass microscope for opaque objects.

The bulk of Adams's text (325 pages), as in the first edition, was concerned with the wide variety of objects that could be seen through the microscope. Of more interest to historians and instrument collectors is the preliminary section containing the descriptions of hardware, as this includes two new designs by Adams. Since the introduction of the Cuff pattern in the 1740s, with its single metal pillar and wooden box foot, several all-brass designs had been introduced by other makers, especially Benjamin Martin and Francis Watkins (the elder). These had a folding brass foot, as did some versions of Adams's '1746' model (see figure 2.4), but different arrangements for coarse and fine focusing.[14] Adams, meanwhile, had made several trunnion-mounted microscopes which combined some of the features of his '1746' model and the Cuff pattern, with the added convenience of provision for tilting the optical axis to any angle (see chapter 3). Watkins achieved a similar facility by incorporating a compass joint at the top of a pillar, to allow the body and stage assembly as a whole to be tilted, but his design was flimsy and unstable. Martin preferred to mount the compass joint low down at the foot, but apparently did not do so until some years after Adams's death. The points of resemblance between various manufacturers' designs, coupled with uncertainty over the precise dates of their introduction, make it difficult to say with confidence

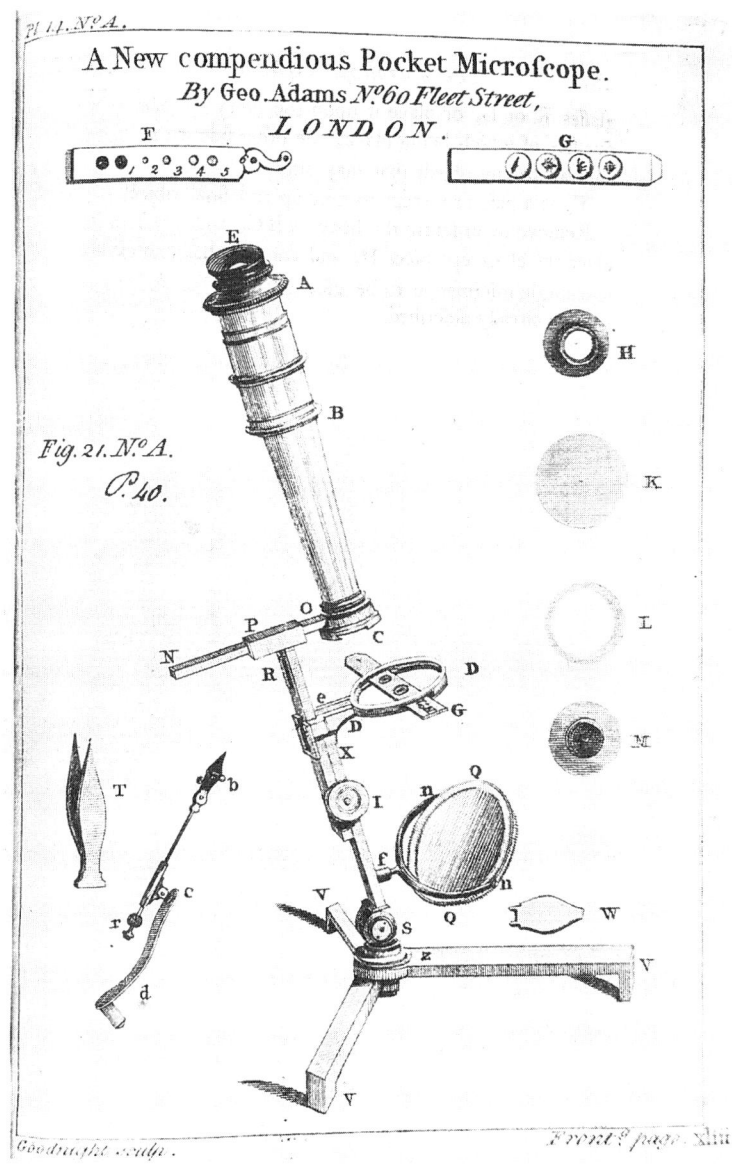

5.3 This 'pocket' microscope, described and illustrated in the fourth edition of Adams's *Micrographia Illustrata* (1771), was intended primarily for use in the field, for selection of specimens to be brought back for examination with a more powerful instrument indoors. (*Museum of the History of Science, Oxford*)

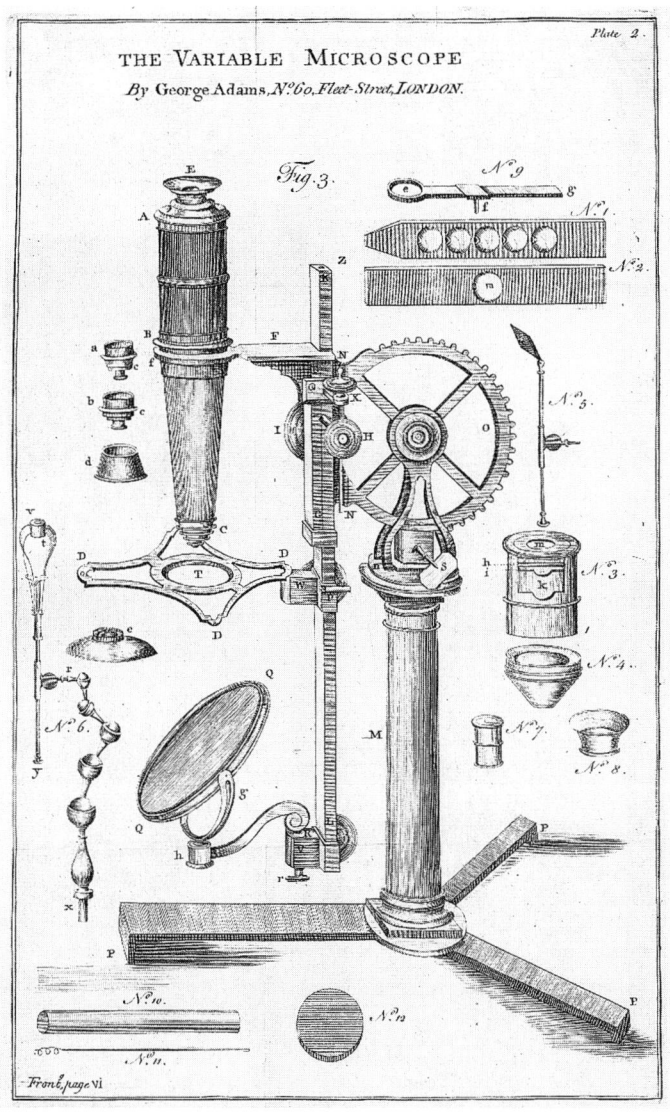

5.4 Adams's 'variable' pattern microscope, as depicted (for the first time) in this plate in John Hill's *The Construction of Timber* (1770), reused in Adams's *Micrographia Illustrata* in the following year. Hill's book was first published in octavo; a folio edition appeared in the same year (1770), for which the legend 'By George Adams, No.60, Fleet-Street, LONDON' was added at the top of the microscope plate, as shown here. Details 'No. 10' to 'No. 12' were added at the same time. (*Author's collection*)

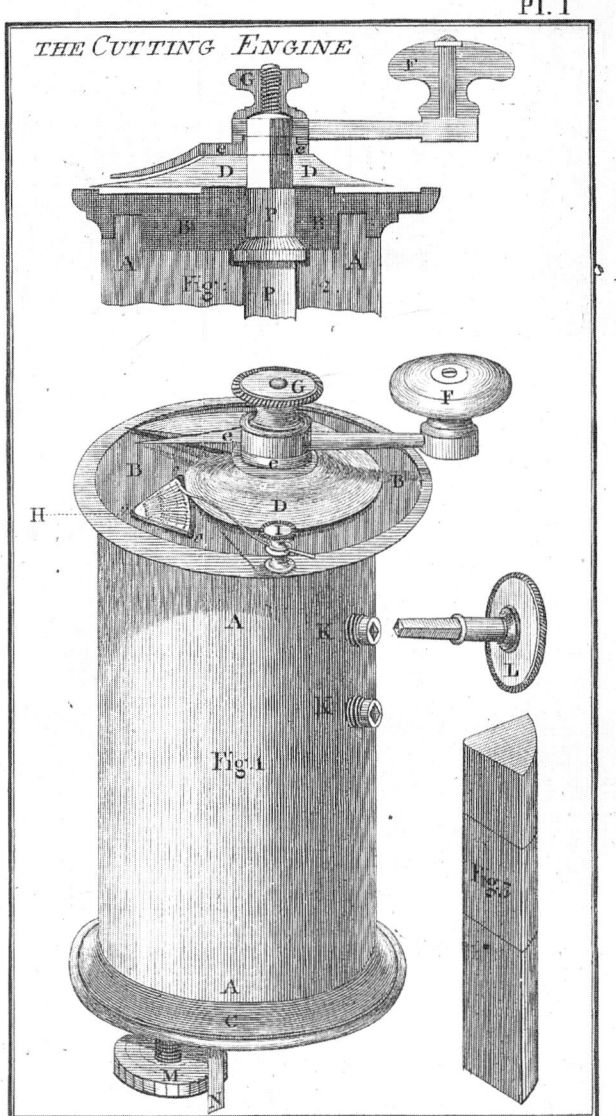

5.5 This form of cutting engine (microtome), devised by 'the ingenious Mr. Cummings' and subsequently made by Ramsden, was used by John Hill to produce thin sections of wood for examination by Adams's 'variable' microscope. The cutting edge is of spiral form, enabling tough materials to be cut, but difficult to sharpen effectively. Plate I in Hill's *The Construction of Timber*. (*Author's collection*)

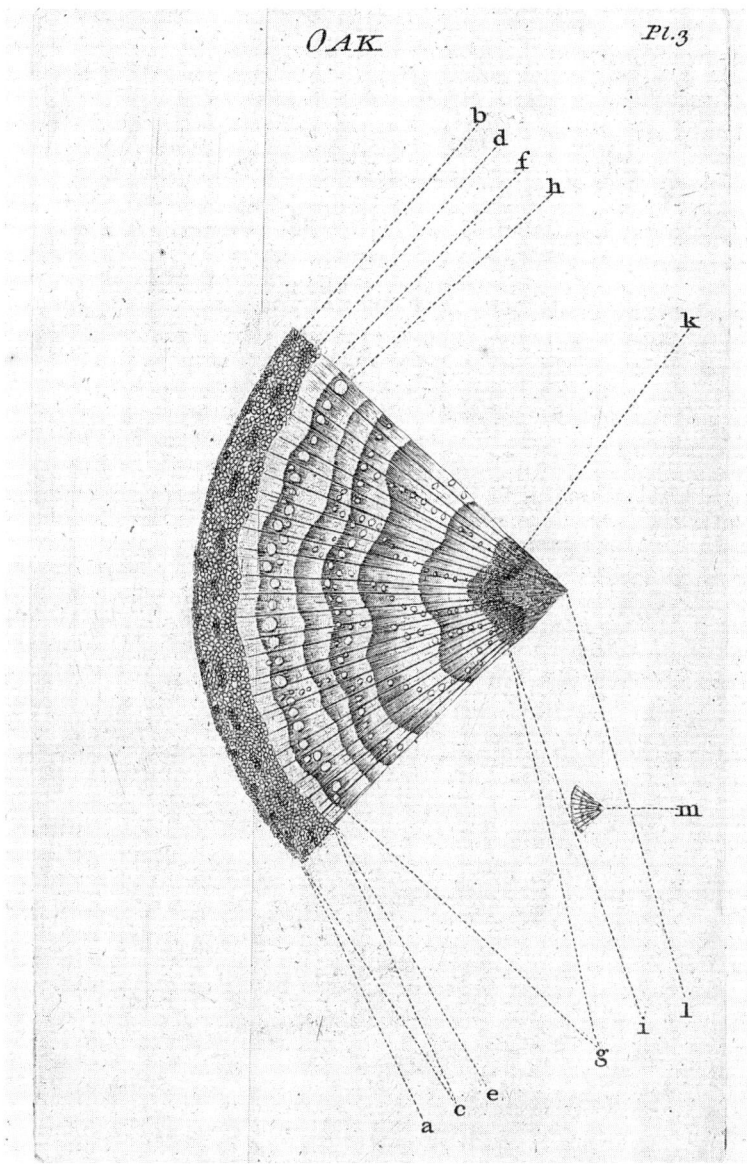

5.6 A typical timber section as observed by John Hill with Adams's 'variable' microscope. The actual-size drawing 'm' indicates that a magnification of only about ×12 was employed. Plate 3 in Hill's *The Construction of Timber*. (*Author's collection*)

who was copying what from whom. In his 1771 book Adams included an illustration of a 'new compendious pocket microscope' with a compass joint at the foot, which has an interpolated number '14A' (see figure 5.3), suggesting that it was an afterthought prompted by seeing another man's design. This 'pocket' instrument, as its name implies, was of small dimensions, suitable for carrying in the pocket, 'that it may be ready when gentlemen and ladies are amusing themselves in their gardens, parks, &c. to inspect and collect objects as may be proper to preserve for a future examination'.

The principal new instrument described by Adams, however, and shown both in a folding frontispiece and in a large folding plate 2, was his 'variable' microscope (figure 5.4). This owes something to Watkins's design, but instead of a simple pivot at the lowest point of the body and stage assembly, the latter is mounted tangentially on the rim of a toothed wheel at the top of a brass pillar, the angle of tilt being adjustable by means of a pinion engaging the teeth of this wheel. This form of mounting bears some similarity to that used for small telescopes, but the geared tilt control is unnecessary in a microscope and the additional height of the wheel makes the instrument top-heavy, as well as inconvenient to use in a near-vertical position. The wheel was omitted in later models, made after George senior's death, which reverted to a simple pivot but retained the idea of mounting this near to the centre of gravity of the body and stage assembly instead of at the lower end. These later models are usually known to collectors as Jones's 'most improved', though their origin can be traced back to Adams's 'variable' in the early 1770s.

Some microscopes of the 'variable' pattern were certainly made and sold, as several have survived.[15] This design was also copied by foreign makers: Clay and Court illustrate one example signed 'J.Brun, 1772'.[16] What is now thought to have been the prototype, made in silver, is also extant (see figure 5.7).[17] Apart from a geared tilt control, other features of the 'variable' included two draw-tubes, which enabled the distance between the objective and the eyepiece to be varied, and a set of objective lenses whose mounts could be screwed into each other, so that they could be used individually or in series combination. These two features, enabling different magnifications to be obtained, were the justification for calling this model the 'variable' microscope. Several of the extant examples, including the silver prototype, are provided with an eyepiece micrometer reading in theory to 1/1,000th of an inch, as fitted to the elaborate silver microscopes made for the king. The

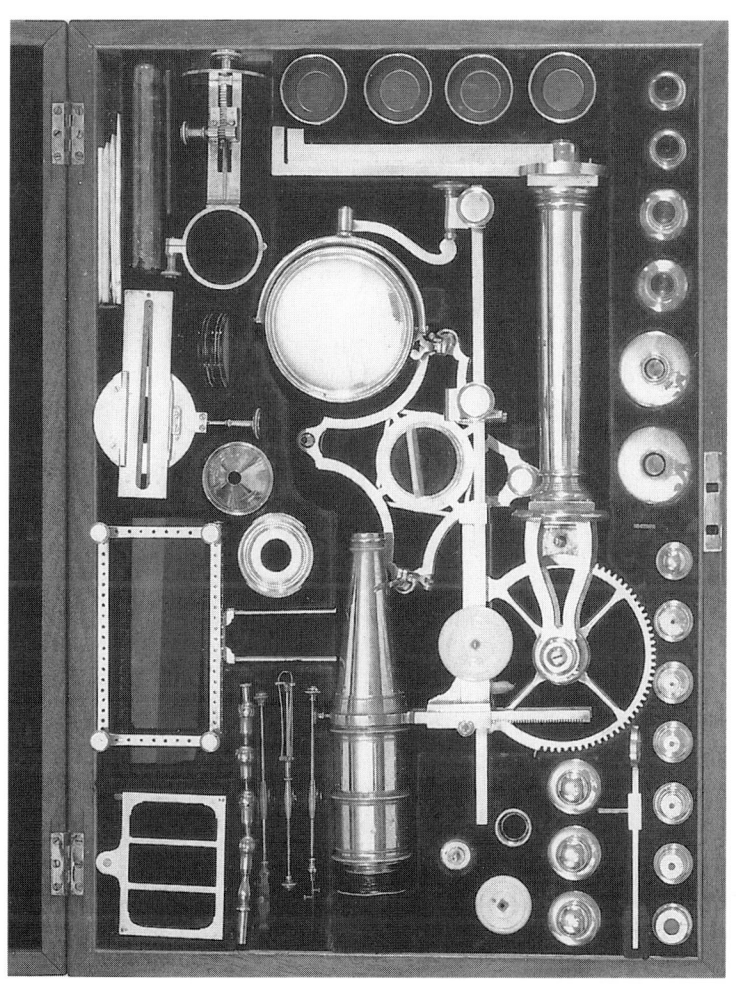

5.7 The silver 'variable' microscope packed into the case in which it is currently housed, probably provided by C.W. Dixey in the mid-nineteenth century. This instrument is now located at the Royal Museum of Scotland, Edinburgh. (*Christie's, London, 19 November 1987, lot 433*)

micrometer, shown in detail in another plate in Adams's book (plate 14 figure 28), was depicted as an attachment for any microscope, not specifically for the 'variable' model.

In his preface, Adams said that he owed the construction of the first 'variable' microscope to 'the ingenuity and generosity of a noble person'. The identity of the latter is revealed indirectly by a slightly earlier publication, *The Construction of Timber*, by John Hill, dated 1770, which includes a description and plate of Adams's instrument. Hill said that the 'composition' of his work was due to his patron, not named but meaning the Earl of Bute, under whose direction the special microscope was made by Adams. The latter's association with the earl went back over twenty years, to his compilation of the inventory of the late Duke of Argyll's instruments in the 1740s, so this statement is quite plausible. Amongst the optical items in the posthumous sale of the earl's instruments in 1793 was Lot 46, 'An elegant compound silver microscope and apparatus, with rackwork motions, the weight of silver computed to be about 100 ounces, packed in a mahogany case'; it was bought by 'Bellamy' for £31 10s.[18] Prior to the recent emergence of the silver 'variable' microscope this lot description was thought to refer to the second of the elaborate silver microscopes made by Adams for the Prince of Wales later George III (see figure 3.4) but it now seems probable that the silver 'variable' was the instrument concerned, and that it was the prototype for copies subsequently made by Adams in brass for commercial sale.

Less than a year after publication of the fourth edition of *Micrographia Illustrata*, on 17 October 1772 George Adams senior died – of an 'apoplectick fit' according to one report.[19] His death was noticed briefly in most of the London newspapers, but there were no long obituaries: these were generally limited to royalty and the nobility. The *Gentleman's Magazine* simply stated, in the deaths for October, '17th. Mr.Adams, mathematical instrument maker to his Majesty, suddenly'. He was buried in the churchyard of St Bride's, his old parish, on the 24th, the entry in the burial register giving his age as sixty-three, consistent with a birth date in early 1709.

He had been running his own instrument-making business in Fleet Street for thirty-eight years, presumably with some financial success despite charges of plagiarism in the early years from Baker, Cole, and later Martin. His appointments to the king, the Royal Mathematical School, and the Office of Ordnance brought him a significant amount of official business, quite apart from his dealings with the general public, but in the absence of detailed records it is impossible to say to what extent it was profitable. One can only assume, from the money and property mentioned in his will, and the fact that, unlike some of his rivals, he managed to avoid bankruptcy, that he was doing rather better than just making ends meet. Glimpses of his financial affairs emerge from time to time indirectly; for example there is the statement that he sold £100-worth of patent telescopes in a few days in the 1750s (chapter 2), but the only complete records of his dealings with specific customers are those relating to Christ's Hospital and the Office of Ordnance. Even these give only the value of his sales, not his costs or profits.

The ledgers of Christ's Hospital show that the total that he received from that source, from 1750 until his death in 1772, when his son took over, was £945. For twenty-three years this is equivalent to an average of about £41 per year, but the value of the payments in any given year fluctuated between two or three pounds and a maximum of £71, and some years were omitted altogether. This may have been due partly to the accounting system used at the Hospital. Much of the institution's income came from charitable bequests by long-dead benefactors, and the accounts were so organized that expenses were charged to specific named charities in order to use up all the available income. Consequently, in some years bills from one supplier were split between several different accounts; also, in the early years separate entries were made for 'Instruments for the Mathematical Boys' and 'Instruments &c. for Boys going into Sea Service'. The latter included books, which are sometimes specifically mentioned, though not by title.

Though £41 per year may not sound very much, it was almost enough to keep one workman fully occupied, and the worldwide dissemination of Adams's instruments by mariners must have generated useful publicity for his business.

The Ordnance trade was larger in value, averaging nearly £100 per year, but irregular, depending on whether Britain was at war or enjoying one of the brief intervals of peace which – for taxpayers – must have seemed all too rare in the eighteenth century. Table 5.1 shows the year-by-year totals of Ordnance bills submitted by Adams.[20] In the early stages of the Seven Years' War, in the four years 1756–59, Adams was submitting bills worth on average about £325 per year, while after the Peace of Paris in 1763 some years would have been completely blank had it not been for exceptional items. In 1765, for example, the only order Adams received was for six sets of drawing instruments for use as prizes at the Royal Military Academy, and in 1767 his principal

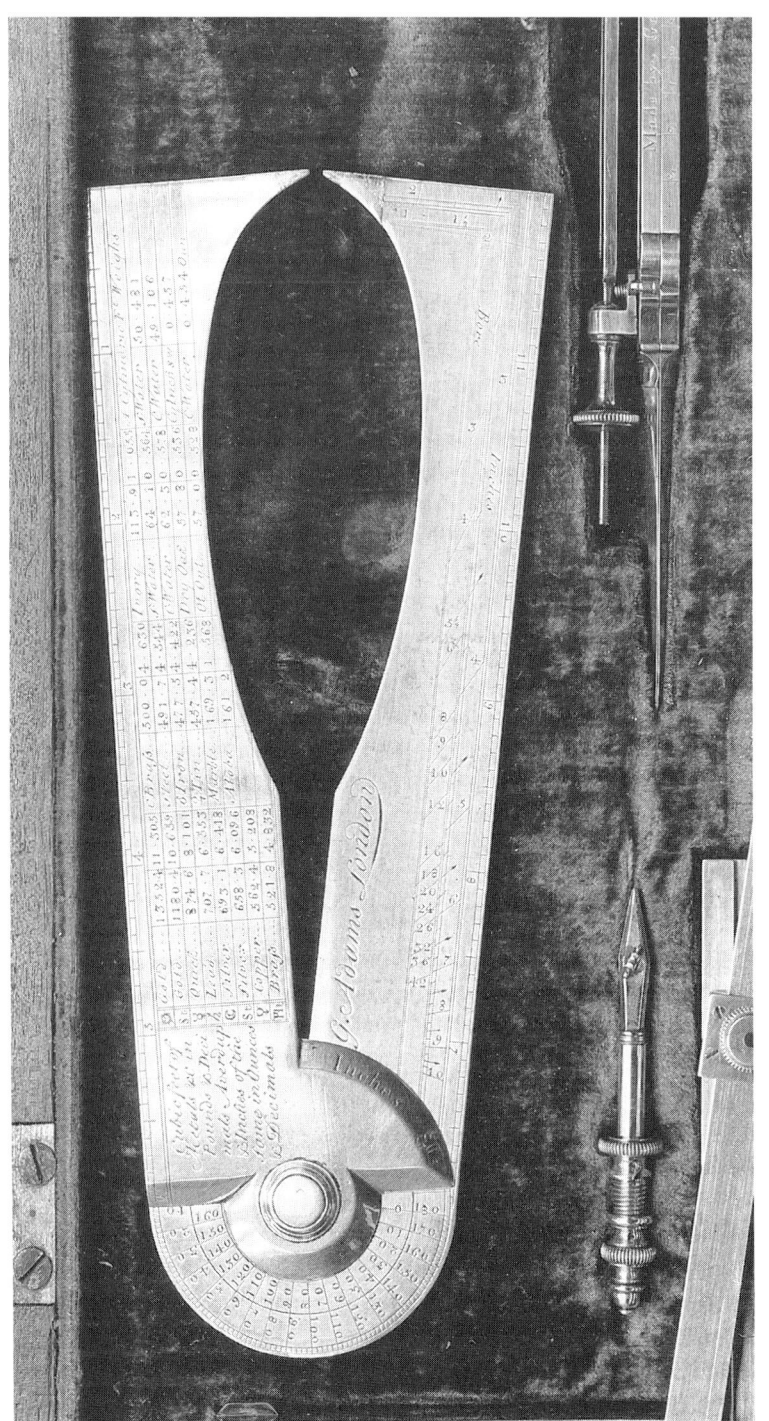

5.8 A pair of brass gunner's calipers signed 'G.Adams London', forming part of a magazine case of drawing instruments. The faces shown have tables of weights of metals and some other materials on the upper limb, and 'crossover' scales for bores of cannon from 1½-pounders to 42-pounders on the lower limb. A linear scale of 12 inches is provided on the outer edges of these faces for use when the limbs are opened to 180 degrees. The degree scale at the hinge enables the two limbs to function as a protractor. (*Science Museum, London, Inv.1918–13; Neg.1099 [part]*)

5.9 A pair of brass gunner's calipers signed 'G.Adams London' but in a different style to that of figure 5.8, with an individual pocket case. The faces shown have tables of weights of powder required for brass and iron guns up to 42-pounders, while the calibrations at the hinge show the weight of iron shot corresponding in diameter to the opening between the points. (*Tesseract, Cat.19 [1987/88], item 52*)

5.10 A gunner's perpendicular (or level), 'new pattern', with a spirit level instead of a plumb line, signed 'G.Adams London'. A steel marking plunger is fitted in a groove in the brass backplate, and the feet have steel tips dovetailed in. (*Whipple Museum, Cambridge, Inv.Wh.1216*) For later examples of perpendiculars by Adams, differing slightly in proportions and minor details, see figures 10.9 and 10.10.

order was for a pair of his new globes for the same place. By contrast, in 1758 he submitted seventeen bills totalling almost £450, mostly for military instruments such as gunner's calipers, quadrants, and shot gauges (see figures 5.8 to 5.10). Unless his civilian trade was largely neglected at this time, he must surely have taken on extra staff or put much of the business out to subcontractors, to cope with this 'bulge'. It was particularly unfortunate that the fire in February 1757, necessitating a move to new premises, occurred when Adams was not only heavily committed to Ordnance work but also making instruments for the Prince of Wales.

Another peak in military orders occurred near the end of the war in 1762, when Adams was writing the 'Pneumatics' and 'Mechanics' manuscripts for the king; on the other hand, the sudden collapse of Ordnance work when peace was declared in 1763 provided a break which probably enabled him to concentrate on making the numerous undated items in the King George III Collection. It also enabled him to resurrect his plans for launching new globes, which came to fruition in the mid-1760s (chapter 4). In the early 1770s military orders were just beginning to build up again, due to unrest in the American colonies, when Adams's death in 1772 put an end to the Ordnance connection for the time being.

Analysis of the 146 bills that Adams submitted to the Office of Ordnance between 1748 and 1772 shows that he supplied a total of 178 gunner's quadrants (mostly at £2 12s 6d each), 152 perpendiculars (mostly at £1 10s 0d each), 67 pairs of gunner's calipers (£1 15s 0d or £2 12s 6d per pair), and 59 sets of shot or shell gauges (mostly at £1 15s 0d per set, though some large shell gauges cost as much as £14 12s 0d per set). These four types of product, which were purely military in application, together accounted for approximately £1,000 of the total sum shown in table 5.1. Many of the remainder, however, had civilian as well as military uses, and would have been included in Adams's range of products even without the Ordnance connection. Major items amongst these were 20 theodolites (at 10 or 20 guineas, according to type), 27 plane tables (mostly at 5½ guineas), 65 measuring chains (about equally divided between 100-foot at 18s 0d and 50-foot at 9s 0d), and 33 magnetic compasses for surveyors (mostly at £1 11s 6d). In addition Adams supplied numerous drawing instruments of all types for the Drawing Room in the Tower, as well as fifty sets in cases for officers in the artillery and engineers, at various prices. The latter in some instances would have included gunner's calipers doubling as sec-

Table 5.1
George Adams senior's Ordnance bills

Year	Bills	£	s	d	Remarks
1748	7	35	8	6	End of War of the Austrian
1749	2	2	7	0	Succession
1750	4	17	19	6	
1751	3	9	5	0	
1752	5	87	8	6	Activity in Newfoundland
1753	3	4	11	0	
1754	9	99	0	6	Expeditions to Virginia
1755	6	57	2	0	
1756	13	299	11	0	)
1757	18	331	14	0	)
1758	17	449	8	0	)
1759	7	221	3	4	) Seven Years' War
1760	4	84	11	6	)
1761	7	81	18	6	)
1762	12	225	6	0	)
1763	2	9	2	0	Peace of Paris
1764	3	31	15	0	
1765	1	15	0	0	Prizes for RMA
1766	1	12	12	0	
1767	2	14	13	0	Globes for RMA
1768	2	22	10	6	
1769	3	61	16	6	
1770	3	34	5	6	Activity in Dominica
1771	9	83	17	6	Build-up of military stores
1772	3	58	2	6	Drawing Room, Portsmouth
Total	146	£2,350	8	10	

tors, otherwise all of these drawing instruments would have been standard products as sold in his Fleet Street shop and listed in his catalogues. They included 125 pairs of compasses, 85 drawing pens, and 183 parallel rulers of various sizes.

The Ordnance trade, significant though it undoubtedly was, may have constituted only a small part of Adams's total output. It left untouched some important areas in which he was a major producer, for example microscopes. The board bought a few telescopes for artillery officers (a total of fifteen were mentioned in his 146 bills), but this government department had no use for microscopes, nor (except to a limited extent for the Royal Military Academy) for philosophical apparatus such as air-pumps and electrical machines, and mechanical demonstration equipment with a purely educational function. It bought only one pair of globes (for the academy), another of the products with which the Adams name is particularly associated today. Astronomical models (spheres, orreries, and planetariums) were entirely outside its scope, as were navigational instruments such as octants, though some of the latter would probably have been sold to Christ's Hospital. Despite the detail available in the financial records of the Office of Ordnance, therefore, the full extent of Adams's output still remains virtually unknown.

154 *George Adams Senior*

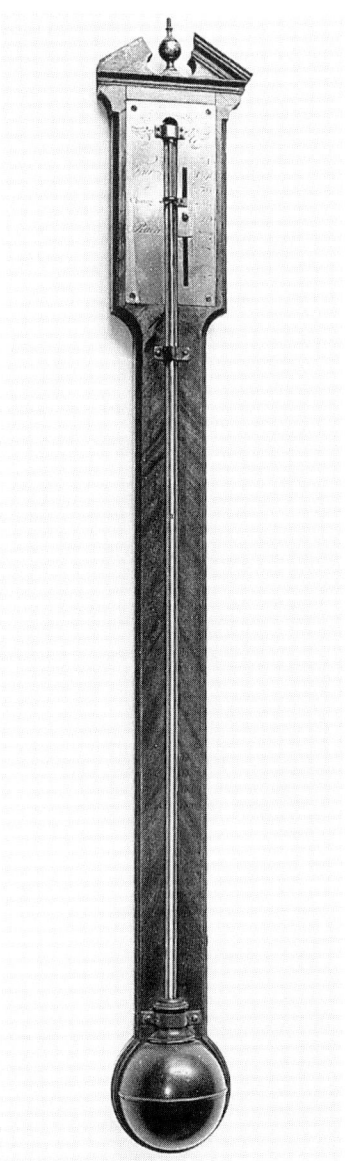

5.11 A stick barometer signed 'G.Adams N° 60 Fleet-Street London Ins^t Maker to his Majesty', in a flat mahogany frame, overall height about 38 inches. It has a brass register plate, and is provided with a vernier reading to 1/100th of an inch. From the signature this instrument could have been made anytime between 1766 and 1787, when George junior added Optician to the Prince of Wales to his appointments. (*Museum of the History of Science, Oxford, inventory no. 25732*)

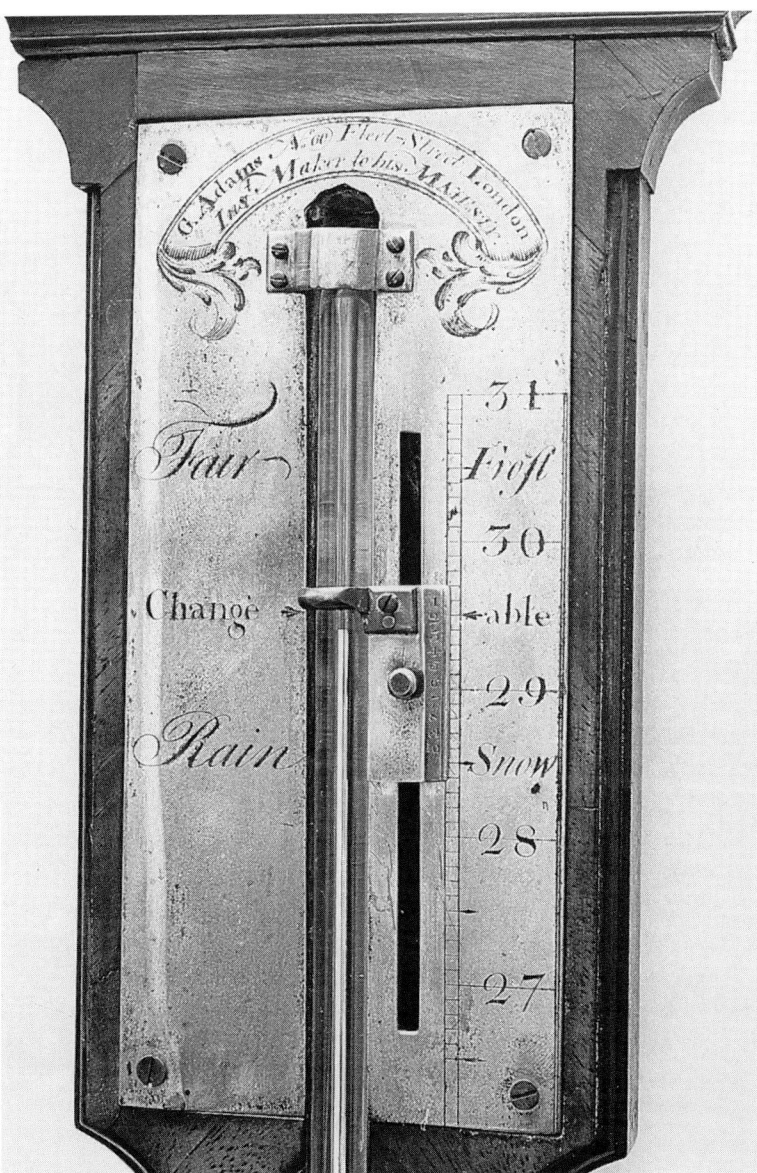

5.12 Detail of the head of the barometer shown in figure 5.11. The scale is calibrated from 26½ to 31 inches. Only three weather states, instead of the seven sometimes found, are provided, but this probably reflects the price of this instrument (which has an exposed tube) and is not a reliable guide to dating. (*Museum of the History of Science, Oxford, inventory no. 25732*)

156 *George Adams Senior*

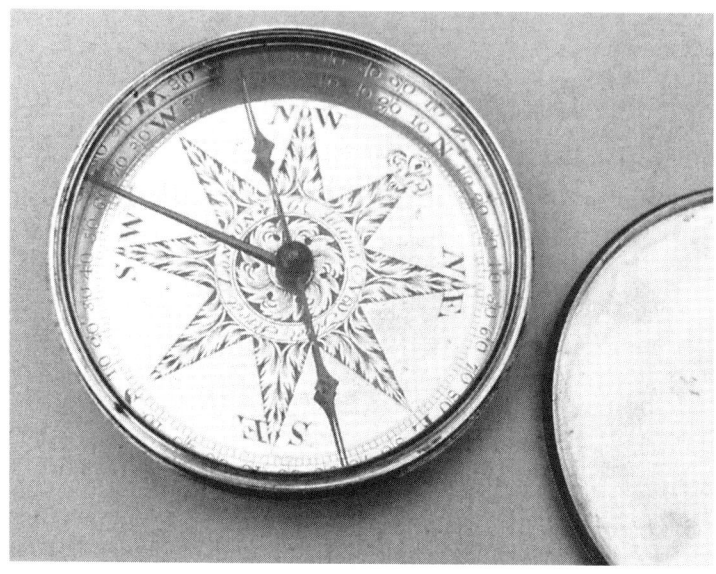

5.13   A magnetic compass in a silver box 3¼ inches in diameter, signed 'G.Adams No.60 Fleet Street LONDON'. (*Tesseract, Cat.27 [1989–90] item 40*)

Part II:

George Adams Junior

6   Continuation of the Business

At the time of his father's death George Adams junior was aged twenty-two and had served a little more than the seven years required to complete his apprenticeship, commenced on 5 February 1765, but had not applied for his freedom. As it was vital, for the maintenance of his unmarried sisters and younger brother Dudley, that the business should continue without a break, his mother Ann lost no time in applying for probate of his father's will, which was granted on the same day as the funeral, 24 October 1772.

The will (PRO PROB/11/981 Q.348) reveals that George Adams senior owned two freehold premises: the shop with residential accommodation at 60 Fleet Street, where he lived, and a country house called Nutting Grove in Buckinghamshire. The latter comes as something of a surprise, for Adams was a Londoner born and bred, with no Buckinghamshire connections so far as is known. Perhaps he acquired the property through a marriage settlement, or in payment of a debt, though the possibility that he bought it with the intention of eventually retiring to the country cannot be ruled out.[1] As neither George senior nor George junior ever lived there, a detailed account of the property (which consisted of a homestead with outbuildings, and a few fields and meadows, in the parish of Langley Marish) will be deferred to chapter 10, when an Inclosure Act provided Dudley with an opportunity to increase the size of the estate.

Of more immediate importance was the provision made in the will for continuation of the Fleet Street business. At the time the will was signed, in March 1772, George Adams's eldest son was still an apprentice, and all his other surviving children of his second marriage were minors. After leaving some money to his two grown-up daughters of his first marriage (£500 and £200 respectively), he therefore took care

6.1 Detail of the head of a mahogany stick barometer signed within a banner with floral decorations 'Geo.Adams Fleet Street London'. This instrument has a glazed door at the top, with lock and key. The register plate, inscribed with seven states, is silvered. These refinements indicate that this was a more expensive instrument than that shown in figures 5.11 and 5.12. (*Christie's, London, 18 July 1985, lot 11*)

Continuation of the Business 161

to ensure that his widow Ann, the life tenant of his estate, would have full power to manage it as she chose, without being hampered by a need to conserve its value for eventual transmission to their children after her death.

On Wednesday 28 October 1772, four days after probate was granted, the following announcement appeared in several London newspapers:[2]

> ANN and GEORGE ADAMS, Mathematical Instrument Makers to his Majesty, &c., at Tycho Brahe's Head, No.60, in Fleet Street, beg leave to acquaint the Nobility, Gentry, Merchants, and others, that they continue to carry on the business of the late Mr. Adams, in the Mathematical, Philosophical, and Optical branches, as in his lifetime, and humbly solicit the continuance of their former favours. Those gentlemen and ladies who honour us with their custom, may depend on their orders being executed with the utmost fidelity, accuracy, and dispatch. Of whom may be had all sorts of instruments in the various sciences of the Mathematics, Philosophy, Astronomy, Geography, Geometry, Drawing, Surveying, Navigation, &c. each of the most perfect construction, and at the most reasonable rates.

This date falls in a gap in the sequence of surviving records of appointments in the lord chamberlain's office: after a lapse of twelve years, the records resume in 1773.[3] Consequently it is not known whether Ann asked for the appointment of mathematical instrument maker to his majesty to be transferred from her late husband to herself, or whether – as seems highly probable – she simply assumed that it applied to the business rather than the person. The '&c.' after 'his Majesty' in the above advertisement could be taken to mean the Royal Mathematical School at Christ's Hospital, as the latter's ledgers show that the Adams business continued to supply instruments to that institution without a break. It did not include, however, the Office of Ordnance. Though Ann moved quickly to ensure the continuation of the Fleet Street business, she was not quite fast enough to secure the Ordnance appointment. At a meeting of the board on 22 October, a few days after Adams's death but before his will had been proved, an order to Jeremiah Sisson for some instruments for the Royal Observatory was discussed, and a letter from Sisson asking for an advance payment was considered. Probably at the same time (though this is not actually stated in the minutes) Sisson seized the opportunity to petition the board for Adams's appointment to be passed to him: he was perpetually in financial trouble and the extra business would no doubt have been very welcome. At any rate, later at the same somewhat lengthy board meeting it was 'Ordered that Jeremiah Sisson be entertained as

Instrument Maker to this Office in the room of George Adams deceased and that a Signification be made out to him accordingly'.[4]

It was fortunate for Sisson that the Board of Ordnance met frequently (several times per week was usual), and that he was already known to the board through his work for the Royal Observatory. (Instruments and their makers for the latter were chosen by the astronomer-royal; though paid for by the Office of Ordnance from 1765 onwards, they did not come within Adams's mandate as the office's mathematical instrument maker.) Business meetings of the Grocers' Company were relatively infrequent: the next one after George Adams senior's death was on 3 November. In the presence of the wardens, one of whom was Thomas Heath, George senior's former master (then aged seventy-four),[5] George Adams junior was admitted a freeman on the oath of his mother Ann, widow and executrix, having served the full term of his apprenticeship before his father died. The entry in the 'Grocers' Freemen and Apprentices Book' concludes: 'Mr. George Adams, Fleet Street, Mathematical Instrument Maker',[6] indicating that he was now a fully-qualified member of the company. If the normal practice was followed, he and his mother would then have gone along to the Guildhall to obtain and register his freedom of the City of London, after which he would be designated in official documents 'citizen and grocer'.

Six weeks later, on 17 December 1772, 'A Petition of George Adams a Freeman of this Company praying to be admitted to the Livery' was read.[7] His father had waited sixteen years for admission to the livery (chapter 2), but George junior was in more of a hurry, needing perhaps to establish himself in the eyes of his competitors and customers as a senior member of the instrument-making community despite his relative youth. Consideration of his petition was deferred until a later meeting, but this was not intended as a snub: it was the correct procedure, laid down by a resolution passed on 7 November 1771. At the meeting on 25 February 1773 his petition was considered and granted, 'upon which he appeared in Court and was cloathed with a Livery Gown and paid the usual Fine and Fees'.[8]

On 18 June 1773, George junior returned to Grocers' Hall to bind his first apprentice, Fowler Bean, son of a Scarborough surgeon (then deceased).[9] Though he subsequently bound three more apprentices, two of whom were his own relatives, George junior does not appear to have had any ambition to participate in the government of the company. No mention has been found of him serving in any of the offices. Nor did he take part in running his parish (St Dunstan's); when called upon to serve as constable in 1781 he paid the fine of £12 instead, and in 1785 he paid £10 to be excused serving as Overseer of the Poor.[10] The only reference to him carrying out parish duties that has been found occurs in 1779, when he was one of the three people chosen each year to collect the new Paving Rate.[11] It seems that George junior, unlike his father, preferred to concentrate his efforts on managing his own business, though as he died in his forties one cannot tell whether his attitude might have changed as he grew older.

An alphabetical inhabitants list for the parish, compiled initially in 1771 with later additions and amendments, has 'Adams, Widow and Son, Fleet Street', and then 'Adams, George, commenced Christmas 1773'.[12] These entries suggest that George junior officially became the householder and ratepayer at 60 Fleet Street a little over a year after his father's death, taking over from his mother. Evidence from newspaper advertisements and shop bills is inconsistent on this point, 'Ann & George' appearing in advertisements in April 1773,[13] and in a bill of February 1774,[14] but George alone in the advertisements of August 1773 discussed below.

At the time of his father's death a number of bills for goods supplied to the Office of Ordnance were awaiting approval. Also, several completed instruments, destined for the new powder magazine at Purfleet, were awaiting shipment; they had been ordered from Adams by a warrant dated 30 April 1771, eighteen months earlier. On 1 December 1772 George junior wrote to the board explaining that the instruments had been completed by his father some time ago, but could not be mounted at Purfleet as the walls of the new building were not sufficiently dry. His letter was considered at a meeting on the same day, when it was minuted that he would be informed as soon as the magazine was ready.[15] This happened about three months later: at a board meeting on 26 February 1773 it was 'Ordered that Mr. Adams send to Purfleet the Barometer and Hygrometer which were before ordered and that he be desired to see them properly fixed'.[16] On 9 March 1773 he submitted his bill for £42 19s, comprising £14 12s for a wheel barometer, 14 guineas for two upright barometers with thermometers, 10 guineas for a Dr Pullen's hygrometer, plus 3 guineas for 'Expense of Mr. Adams and two men going to Purfleet'.[17] At the same time he submitted a bill for 2½ guineas for repairing and altering an old barometer belonging to the Ordnance Department. According to the bill book entries, these two bills were eventually passed for payment on 20 October 1773, just a year after George senior's death.[18]

Wheel barometers are commonly regarded as domestic, rather than scientific, instruments, but the price charged for the one supplied to Purfleet (14 guineas) was considerably higher than for the usual domestic cistern-tube barometers, and twice as much as for the 'upright' barometers with thermometers supplied on the same order. Possibly relevant is an Adams wheel barometer described and illustrated in Goodison's *English Barometers*.[19] In this particular instrument attempts have been made to improve its accuracy and reliability beyond domestic requirements, for example by using a larger diameter tube than normal, and by paying attention to smooth running of the cord and counterweight. The mahogany and brass case, on the other hand, is noticeably – as Goodison puts it – 'austere'; the corners of the square dial are left quite plain, where in an expensive instrument one might expect to find decorative spandrels (as in a clock). Instead of ornamentation, the upper part of the trunk is provided with an engraved panel describing the weather corresponding to dial readings, divided into thirty-three states from 'Settled, fine clear sky' to 'Violent Hurricane'. The panel is signed at the foot 'GEO. ADAMS No 60 Fleet Street, LONDON'. This could well be the instrument supplied to Purfleet in 1773, or at least one made to the same design.

On 18 November 1772 a 'warrant of justification' (that is, retrospective) was made out to Adams's executors for two wooden Gunter's scales, delivered to the Royal Laboratory at Woolwich some time previously.[20] The corresponding bill entry is dated 4 January 1771 for 6s (two scales at 3s each).[21] Apparently it had been overlooked at the time. Presumably the executors were going through the order books carefully and chasing up any unpaid bills, however small. This bill was passed for payment on 11 February 1773, along with three other bills submitted by George Adams senior in 1771 and 1772 totalling £25 2s, entered consecutively in this particular bill book. Examination of the other bill books covering the last few years of George senior's life reveals that another bill, entered in a different book, was allowed on the same day (11 February 1773); it was dated 8 October 1772, and was probably the last that he submitted before his death a week later. It covered a variety of instruments for the Drawing Room at Portsmouth, apparently a new Ordnance (as distinct from naval) establishment as it is not mentioned in earlier bills.[22] The principal items were a theodolite at £21, and a 'Protractor with Ruler and Nonius Divisions moveable about its Center' at £4. The rest of the fourteen items were drawing instruments of various types (see figure 6.2) including compasses, pens,

Continuation of the Business 165

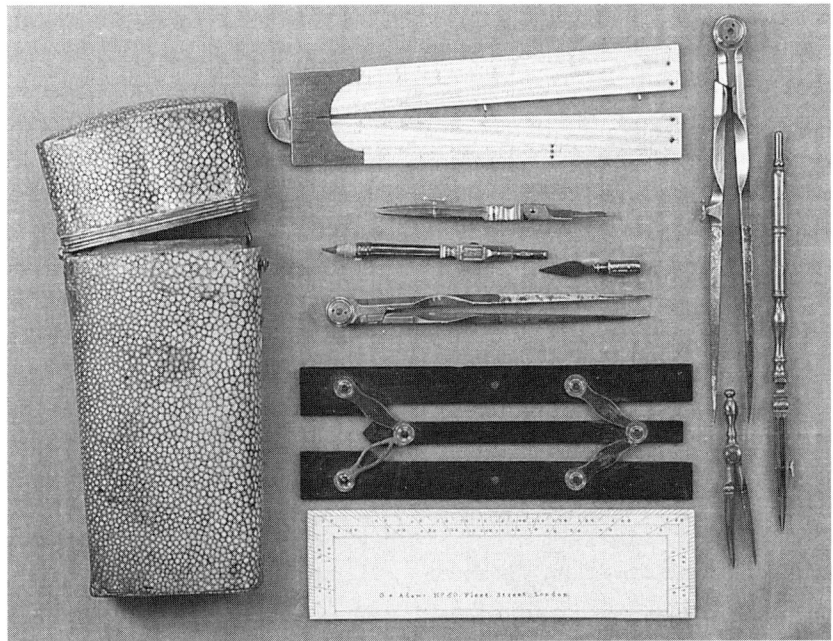

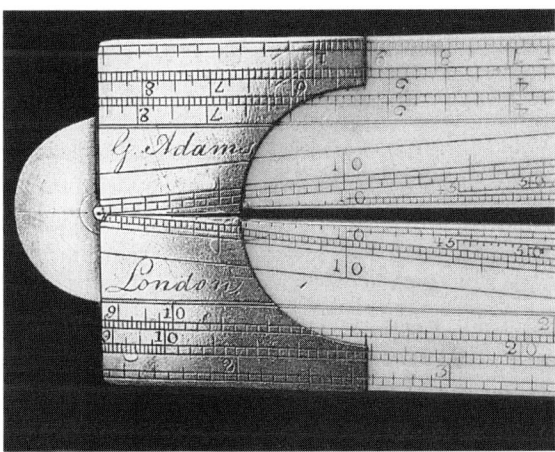

6.2 A pocket set of drawing instruments signed 'G.Adams London' on the brass hinge of the sector, in a silver-mounted shagreen covered case 7 inches high. The principal instruments are of the 6-inch size most commonly used by Ordnance draughtsmen; this particular set includes also a small pair of ink compasses and a medium pair of dividers. The ivory rectangular protractor is stamped 'G*Adams. N° 60 Fleet Street. London'. (*Andrew Alpern Collection, Inv.109; previously sold at Sotheby's, London, 19 May 1983, lot 12*)

rulers, parallel rulers, and a sector (£1). The latter is the only one mentioned in all of George Adams senior's Ordnance bills, perhaps because the functions of a sector would normally be performed by gunner's calipers of the flat calibrated 'English' pattern, and none had been ordered specifically for Portsmouth at this time.

Another small bill which had been submitted on 4 January 1771, for two pairs of compasses for the Modeller at Woolwich, priced at 4s each (a little cheaper than compasses for the Drawing Room), was eventually tracked down and allowed on 4 May 1773.[23] This appeared at the time to be the last of the outstanding Ordnance bills. Counting those for the Purfleet instruments the total amount received from the Office of Ordnance following George Adams senior's death was just over £105. The early 1770s were a fairly slack period for military orders, and had it not been for the exceptional items ordered for Purfleet and Portsmouth the amount outstanding would have been only about £30. If Adams had died at the height of the Seven Years' War, on the other hand, the bills awaiting payment could easily have amounted to a significant part of his estate.

Probably unknown to George junior and his mother, there were actually a number of other quite expensive items outstanding for which bills had apparently not been submitted. This only came to light almost a decade later. During the last four years of his life, George senior supplied several pieces of demonstration apparatus to the Royal Military Academy which had been ordered by the Professor of Fortification and Artillery, Dr Allan Pollock, who had succeeded John Muller on the latter's retirement in 1766. This proved to be a somewhat unsatisfactory appointment: Dr Pollock's attitude and behaviour gave rise to many complaints, with the result that in 1777 the board dismissed him with a small pension of £50 per year. (Muller's pension at this time was £200 per annum, later increased to £250.) Dr Pollock felt that he had been unfairly treated, and during the next few years frequently wrote to the board seeking redress for his grievances. It would appear, from comments in the minutes, that he regarded the demonstration apparatus provided for him to use at the academy as being his own property, and had taken it with him when he left. In 1780 he was ordered to return to Woolwich any instruments, and so on, remaining in his possession. Only then, apparently, was it realized that several of those supplied by George Adams senior had not been paid for. If Dr Pollock had neglected to certify to the storekeeper at Woolwich that the goods had been received, as seems to have been the case, Adams would not have been able to submit a bill for them to the office at the Tower, so his executors in 1772 would not have known about these items.

In compliance with an order of the board dated 4 April 1780, a bill for instruments supplied to the academy by George Adams senior between May 1769 and September 1772 was submitted by George junior. The bill book entry gives the principal items and dates as:

		£	s	d
30 May 1769	A 12-inch astronomical quadrant	31	10	0
ditto	An instrument to illustrate the ranges of cannon balls by the various parabolas	12	12	0
27 October	A Sutton's quadrant and staff	1	18	0
ditto	An instrument to illustrate the ranges etc. of a larger size	20	0	0
28 August 1771	An instrument for finding the resistance of air	18	7	6
5 Septem. 1772	A pendulum for seconds	8	15	0

When the charges for smaller items and packing-cases were added, the total came to almost £100. This bill was passed for payment on 18 October 1781, when George junior must have been pleased to find himself unexpectedly richer by that amount, nine years after his father's death.[24]

Though Ordnance orders for mathematical as well as observatory instruments were given to Jeremiah Sisson from November 1772, this situation lasted for less than three years. It will be seen later that the Adams family renewed their association with the Office of Ordnance in the late 1770s, and an extensive trade with that department was subsequently handled by first George junior and then Dudley until the early nineteenth century. At the time of the events related above, George junior was already a major (though not the sole) supplier of mathematical instruments to the Office of Ordnance in his own right.

Reverting to the period following George senior's death, in April 1773 Ann and George Adams placed several advertisements in London newspapers under their joint names for the fourth edition of *Micrographia Illustrata*. Though this edition had been published in 1771 in George senior's lifetime, the advertisements were still headed 'This Day is Published', a common and misleading practice amongst booksellers. This and the third edition of *Treatise on the Globes* (1772) were the only works by George senior then in print, and it is probable that the primary purpose of these advertisements was to draw attention to the continued existence of the Adams business at 60 Fleet Street, rather than the books themselves.

Shortly after this, another product suddenly assumed great importance: a coin balance, or rather, a number of coin balances of different types. With numerous foreign coins, particularly of gold, constantly in circulation, prudent shopkeepers had always made use of measuring devices specially calibrated to show the value by weight of any given coin. Scale-making was a distinct branch of the instrument-making trade, embracing small balances for use by apothecaries, and large beams for heavy weights, as well as special devices for weighing coins. In the middle of 1773 the government suddenly decided to take action to eliminate the widespread practice of clipping coins to obtain small pieces of gold, thereby reducing their gold content but not their face value. An Act of Parliament prohibiting the 'counterfeiting, clipping, and diminishing' of gold coins received the royal assent on 1 July 1773. Amongst other provisions it gave authority to anyone who was offered payment in gold coins to reject them if they had been diminished other than by reasonable wear. Apparently the Act was instigated by the Treasury and passed through both houses of parliament with little discussion, and little thought for the consequences.

The immediate effect on trade was disastrous, as the Act failed to define 'reasonable wear'. Many gold coins in circulation were up to a century old, and were well below their proper weight although they had not been clipped; these now became virtually unusable, as nobody wanted to be found in possession of coins that were too light to be legal tender. Everyone handling gold coins suddenly needed to weigh them, but since nobody knew what the limits were, arguments about legality were frequent. Newspapers were full of comments such as 'The Legislature by their great Penetration and Sagacity have produced a most marvellous and extraordinary Effect: People of all Professions, Occupations, and Denominations throughout London, are for some hours in the day, busy in the weighing of Gold.' But not all tradesmen were unhappy. As one commentator wrote, 'The makers of Gold Weights and Scales have been so busy, ever since the passing of the Act, that they have been obliged to work night and day, and cannot procure half hands enough; from which it is supposed that there is not one of this profession in London but what will make a fortune.' One reporter calculated that at least 600 money scales were in use in Smithfield Market, and another wrote that 'many thousands' had been sold. 'Thanks to the Premier, our great Master of the Equilibrium, we shall have Balance-Masters in every shop and every house in the Kingdom.'[25]

Although scales specifically for money were not listed in Adams's catalogues, the opportunity to pick up some of this extra trade was too good to miss. His rival Benjamin Martin of 171 Fleet Street was one of the agents for the patent 'Index Balance' made since 1772 by Anscheutz & Schlaff of Denmark Street,[26] and shortly after the passing of the Act this device was extensively advertised in the London papers although it was a general-purpose instrument not specially designed for weighing coins. What was really needed was a pocket balance that traders and the general public could always carry with them, just as today one would wear a watch.

Martin seems to have been the first to market such a device of his own design – a 'Portable Money Steelyard'. Advertised on 14 July,[27] only a fortnight after the passing of the Act, it consisted of a pencil-sized ivory beam, calibrated in terms of money values rather than weight, suspended by a cord near one end (which was made of, or encased in, metal), and having forceps hanging from that end to grip the coin that was to be weighed. A metal balance weight sliding on the ivory beam gave a direct reading of value when the beam was level, indicated by a small pointer at right angles to the beam at the point of suspension. The price for this device was from 4s 6d to 10s 6d, depending on whether the metal parts were made of brass or silver. It was calibrated for coins up to 36s 0d in value.[28]

Similar devices by other makers quickly made their appearance and were extensively advertised. Martin warned customers to beware of counterfeits, saying that every one made by him was sold with printed directions for its use, 'subscribed with my Name, in my own Handwriting'. In several advertisements he said that he had taken on extra staff to cope with the demand. Henry Pyefinch – not normally associated with weighing devices – was one of those who advertised a coin balance sounding suspiciously like Martin's.[29]

Early in August, advertisements for similar devices by Adams started to appear. Curiously, the first (on 4 August), for a 'commodious and portable Instrument for weighing of Money', omitted his name, though it gave his address and shop sign; perhaps there was a difference of opinion over whether it should be headed 'Ann & George' or just 'George' Adams.[30] From 20 August, the text always included the name 'Mr.Adams', for example:

6.3   An equal-arm hydrostatic balance signed on the beam 'Made by Geo. Adams at Tycho Brahe's Head in Fleet Street LONDON'. Though probably made twenty years before George junior became involved with coin balances in the 1770s, this instrument indicates that the Adams firm had some expertise in the balance field. For a full description see the Whipple Museum's *Catalogue 2: Balances and Weights*, item 17. (*Whipple Museum, Cambridge, Inv.Wh.1871*)

GOLD COIN. Mr.Adams, at Tycho Brahe's Head, No.60, Fleet-Street, informs the Publick, that he can now accommodate them with his Weighing Rules, Price 4s. the Case included. The Weighing Rule discovers with Accuracy and Ease the proper Weight or Deficiency in Weight of the current Gold Coin, is commodious and portable, and turns with a smaller Weight than a common Balance.

On the following day (21 August) 'Tin balances for weighing with exactness Gold Coin, on Mr.Martin's Principle' were advertised by a tinsmith, Arnold Finchett, at only 2s or 3s.[31] The main contenders for the coin balance market, however, were Martin and Adams. After the latter's advertisement on 20 August, Martin advertised his portable steelyards on 21 August; 22 August was a Sunday; Adams advertised again on the 23rd, Martin on the 24th, and Adams on the 25th, so for five consecutive weekdays rival advertisements by Martin and Adams alternated in this paper (the *Daily Advertiser*). On 26 August Martin advertised jointly with Mr Dawson, of 49 St Martin's Lane, a newly-invented hydrostatic coin balance which was sensitive to 1/30th of a grain; this was rather more expensive than portable steelyards, at 7s 6d with a case.[32] By this time Martin had published a tract, *The Monied Man's Vade-Mecum*, price 6d, which must have given him a temporary advantage over Adams, who does not appear to have published anything on this subject himself.

A week later Adams must have been startled to see an advertisement in the newspapers for a 'Tycho Brahe Coin Balance', blatantly using his shop sign in what appears to have been a deliberate attempt to create confusion in the minds of the public.[33] Priced at 3s 6d, this device undercut both Martin and Adams. It was sold at Courtauld & Cowle's toyshop near the Royal Exchange, Young's toyshop in Fleet Street, and by Pinchbeck in Cockspur Street, Haymarket. (Toyman in this context meant a dealer in small trinkets, not playthings.) Later advertisements for this product stated that genuine ones had 'CP' and a serial number on the back,[34] so it presumably emanated from Christopher Pinchbeck, who had not only a shop but also an exhibition room at his premises in Cockspur Street. His first advertisement stated that 400 'private orders' had been completed before the device was offered to the public, and later ones mentioned 'large Demands from Publick Offices, Turnpikes, &c.'

By November, Martin was claiming sales of between 5,000 and 6,000 for his portable steelyards,[35] a figure that is confirmed by serial numbers on extant examples. No information has been discovered on

Adams's sales. On 2 September, the day after the first appearance of the 'Tycho Brahe' coin balance advertisements, Adams placed a longer advertisement for three different types of weighing devices:[36] the weighing rule, price 4s with a case; the hydrostatic sliding rule, accurate to 1/30th of a grain; and a dial steelyard of 'elegant and convenient' form. No price was quoted for the last two instruments.

By the beginning of October, Adams had joined forces with Thomas Hatton, a watchmaker and author,[37] perhaps to counter Martin's *Monied Man's Vade-Mecum*. A joint advertisement by Martin and a Mr Dawson of 49 St Martin's Lane has been mentioned above; from 9 October, Thomas Hatton advertised from that address 'Common Scales improved, and made compact for the Pocket', plus several 'new-invented' ones which were made and sold by Hatton and by G. Adams at 60 Fleet Street. The wording is a little ambiguous: it is not clear whether both men made and sold the devices concerned, at their separate addresses, or whether they were made and sold by Adams only, to Hatton's design.[38] A treatise on gold currency was promised for the following week. This duly appeared, after a slight delay, and was advertised on 22 October. It was entitled *An Essay on Gold Coin*, by Thomas Hatton, sold by him at 49 St Martin's Lane, G. Adams at 60 Fleet Street, Nairne & Co. opposite the Royal Exchange, and J. Gilbert on Ludgate Hill.[39]

A copy of the first edition of Hatton's treatise has not been located in Great Britain, but the second edition (1774) links Hatton more positively with Adams by including an advertisement leaf listing five types of weighing instruments for gold coins, 'invented, made and sold by T. Hatton, and by Mr. George Adams'.[40] They were (abbreviated): hydrostatic sliding rule; pencil steelyard improved; dial steelyard; joint rule for weighing gold and other purposes; and various other constructions of the steelyard.

By this time Martin had begun his usual thrice-weekly winter lectures on experimental philosophy, advertised alternately in the *Daily Advertiser* and the *Gazetteer*. Several of his lecture advertisements included a reference to his money steelyards, and as Adams was apparently content to rely on Hatton's book for publicity the direct advertising battle between Martin and Adams for the coin balance trade petered out. 'Tycho Brahe' balances, however, continued to be advertised extensively by Pinchbeck, who stated in an advertisement shortly after Christmas 1773 that his exhibition room had examples of 'most of the new-invented Machines for weighing Gold Coin, from the Index Balance to the Tycho Brahe'. These could be seen, tried, and minutely examined for sixpence, which would be refunded to anyone making a purchase.

The coinage legislation was eventually clarified by a proclamation in June 1774, which laid down that after 15 July 1774 no guineas coined before 1772 would be valid if they wanted more than sixpence in value, and other gold coins in proportion. Anyone possessing light gold had to dispose of it before 1 September, at an official exchange rate, to authorized collectors who were appointed in all the major towns. In theory, this should have removed all questionable coins from circulation, and the need for everyone to carry a portable coin balance lessened, though the necessity to weigh foreign coins to check their value remained. Coin balances continued to be made and used for many years, especially after the introduction of sovereigns and half-sovereigns in the nineteenth century, but after the frantic upheaval of the mid-1770s had subsided the demand was met by specialist scalemakers; major instrument makers such as Martin, Adams, and Nairne & Blunt ceased to be involved in this branch of the trade.

Adams had, in fact, apparently decided that it was time for him to revert to his normal business around Christmas 1773. On 7 January 1774 (repeated 11 January) an advertisement for Adams's New Globes, of 18, 12, and 6 inches diameter, appeared;[41] like the advertisement for *Micrographia Illustrata* in April 1773, this also referred to the third edition of *Treatise on the Globes* as 'This Day is Published', though this edition had first appeared in 1772. Also like the previous advertisement, this one referred to 'Ann and George Adams, Mathematical Instrument Makers to his Majesty'. As mentioned in chapter 4, an invoice for globes in the Wedgwood archives, dated February 1774, is also headed 'Ann & George'. Since Ann was the owner for life of the globe plates and the copyright of the accompanying *Treatise*, it was logical that her name should appear in connection with them. The various coin balances mentioned above, on the other hand, were presumably devised by George junior after the death of George senior, so 'Mr.Adams' or 'George Adams' alone was quoted in the relevant advertisements and in Hatton's book. No extant examples of Adams's coin-balance products have been located, so it is not known how they were signed. An equal-arm balance by George Adams at the Whipple Museum, Cambridge,[42] is almost certainly much earlier than the 1770s, but it indicates that the Adams firm had some expertise in the balance field (see figure 6.3).

Adams's globes and the third edition of *Treatise* were advertised again on 29 June 1774, still citing 'Ann & George' as makers and sellers. A few days later, on 3 July 1774, George junior married Hannah Marsham, by licence, at St Dunstan's, Fleet Street. She was described in the register as 'of St.Saviour's, Southwark'.[43] Her age is not given (being over twenty-one), but from the records of a manorial estate in Essex mentioned in George junior's will it appears that she was about the same age as her husband, that is, twenty-four at the time of their marriage.[44] She had an elder sister Mary, and a brother Thomas, who was most probably the Thomas Marsham who subsequently became secretary and treasurer of the Linnean Society,[45] but otherwise little is known of her background.

This marriage must have given rise to some complications in both the business and the household, but one can only speculate on the domestic and financial arrangements. George junior's mother was now aged about fifty-two, and her youngest child, George's brother Dudley, was eleven and a half. There were also at least three, possibly four, girls in the family, all of whom were dependent on the Adams instrument business for their support – and now there was the prospect of another family (George and Hannah's children) competing for their share of the available resources. Perhaps it was fortunate that George and Hannah did not, in fact, have any offspring, though whether this was by design or through some biological inadequacy cannot now be determined. Hannah was just too old, at forty-five, to have children by a second husband, had she so wished, when George junior died in 1795. References to several 'God-children' in his will (one named George Adams Rogers) prompts the query whether these were his illegitimate offspring. On the other hand, the presence of 'Warner on the Testicles' in the Adams library sale in 1796 suggests that the problem – if problem there were – lay not with Hannah but with George. In any case, George junior's marriage seems to have brought to an end the 'Ann & George' era in the Adams story: henceforth it was George (or perhaps Hannah) who was running the Fleet Street business.

A few years later an opportunity arose for George junior to recover some of the Ordnance trade lost when his father died in October 1772. Jeremiah Sisson, who had taken over the appointment of mathematical instrument maker to his majesty's Office of Ordnance, held it for less than three years. Perpetually in financial difficulties, by February 1775 he was bankrupt. A hint of trouble to come occurred as early as November 1773, when Maskelyne (the astronomer-royal) asked the

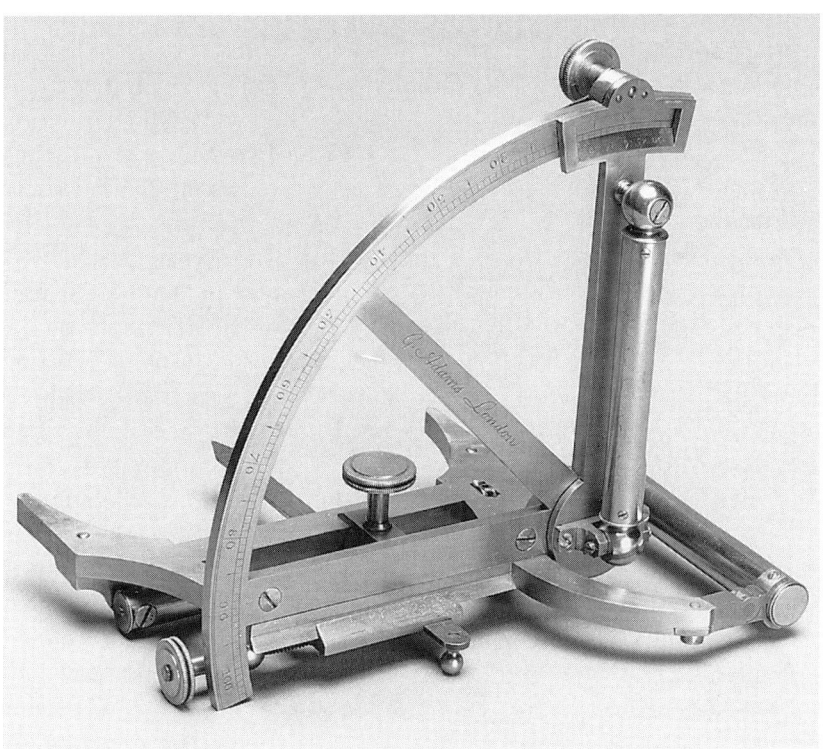

6.4 Both George Adams senior and junior were sometimes asked by the Board of Ordnance to make experimental instruments for trial at Woolwich. The one shown here is a combination of a gunner's perpendicular and a quadrant. It has three spirit levels mutually at right angles, and the marking plunger of the perpendicular can be moved sideways on a dovetail slide by a micrometer screw. The purpose of the latter movement is obscure. (*Science Museum, London, Inv.1975-179; Neg.630/91*)

board to advance £75 to Sisson in part payment for work carried out at the Royal Observatory, 'as he stands in great need thereof on account of his private affairs'.[46] In October 1775 Maskelyne wrote to the board saying that he had personally advanced £29 6s 6d to Sisson to settle one of the latter's Ordnance bills,[47] and had obtained a letter of attorney authorizing him, instead of Sisson, to receive this sum from the office when the bill was eventually passed for payment. The board agreed to make the bill payable to Maskelyne as requested.[48] However, long delays were commonplace in the Office of Ordnance at this time, and in February 1776, before the treasurer's department had dealt with the bill, the board received a letter from a Mr Bassett of Clement's Inn claiming that Sisson had been bankrupt for over a year, Maskelyne's letter of attorney was void, and the payment must be made to Sisson's assignees. The board had to agree,[49] and presumably Maskelyne lost his money. Meanwhile, the board had begun to place instrument orders elsewhere. In December 1775 Benjamin Martin was asked to make some mechanical demonstration apparatus for the Royal Military Academy, consisting of 'Pullies, Levers single and compound & other Machines ... as explained in Mr. Ferguson's Treatise on Mechanics'.[50] In November 1775 an order was issued for instruments for engineers going on foreign service;[51] the supplier was not named in this minute, but on 6 February 1776 a bill from Jesse Ramsden for £29 19s for instruments for Captain D'Aubant, engineer, was approved, which was probably the one concerned.[52] A few weeks later Captain Archibald Robertson, engineer extra, submitted a list of instruments that he considered necessary for the foreign service that he was ordered on; the board agreed and placed an order for them with George Adams, and directed that they were to be sent on the first available ship to Boston.[53]

This seems to be the first mention of George Adams junior as a supplier of instruments to the Office of Ordnance (other than those ordered from his father for Purfleet and delivered in 1773). As these instruments for Captain Robertson were evidently required urgently, Adams probably got this particular order because he was able to supply them quickly, perhaps from stock. It would appear that the board initially contemplated appointing Ramsden to succeed Sisson, for early in 1777 five bills submitted by him were approved, the first of which was dated 25 April 1776.[54] However, Ramsden had a reputation for failing to meet deadlines, and in July 1777 he was officially warned that the board would employ somebody else 'if he don't Comply with the Orders he receives'.[55] In particular, he had failed to supply some drawing instruments ordered six months previously; probably for that reason, an order for a magazine case of instruments for the inspector of the Royal Military Academy was given to Adams.[56] In May 1778 Adams supplied 'a large pullout refracting telescope' to an officer at Portsmouth,[57] and in June that year he was asked to quote for making a brass measure for Purfleet.[58]

Adams's Ordnance bills in 1779 include one dated 28 January worth nearly £100 for a variety of gunnery and surveying instruments, including six box quadrants of General Williamson's pattern at £3 13s 6d each.[59] Presumably this was the article listed (unpriced) in Adams's general catalogues as 'Major-General Williamson's new instrument for elevating pieces of Ordnance'.[60] This bill also included fourteen perpendiculars at £1 10s each, a plane table at £5 15s, and two theodolites, at £14 14s and £29 respectively. The latter instrument was described as 'late improvement'. In Adams's 1777 catalogue,[61] 'Theodolites of the latest improvement' were described at some length: they had parallel plates and screws, a double sextant over the compass box and a telescope with a spirit level, 'so contrived that when the bubble rests in the middle of the spirit tube, the intersection of the hairs in the telescope will cut an exact level'. Though more expensive than the standard £21 instrument supplied by George Adams senior, this improved theodolite evidently found favour with engineer users, as the price of £29 was specifically approved by the board at a meeting in December 1780 when Adams's prices for several other instruments were considered,[62] including £3 13s 6d for 'General Williamson's Instrument for Howitzers &c.'

Other bills submitted by Adams in 1779 included one for £58 1s 6d on 28 February, for a 14-guinea theodolite, a plane table, sets of shot gauges, calipers, rulers, protractors and so on, and four brass gunner's quadrants with spirit levels. Two more bills, totalling £39 4s 6d, for gunnery and surveying instruments, were both dated 17 May 1779. Amongst the items listed was a 'brass scale for fuzees' (7s 6d), and several pairs of compasses with a bow; these were used for measuring the length (time of burning) of fuses inserted in shells.

Meanwhile Ramsden, despite the warning given in July 1777, was paying little attention to Ordnance orders. In April 1780 the chief draughtsman (George Haines) wrote to the board saying that twenty-four sets of drawing instruments had been ordered from Ramsden as long ago as January 1777, but he had delivered only twelve, all of

6.5 A 2½-inch refracting telescope with four brass draw-tubes, signed 'G.Adams London', with a dust cap and a figured mahogany outer tube. The fish-skin covered case is about 17 inches long. Fully extended this telescope has a length of 5 feet 3½ inches (*Christie's, London, 17 October 1985, lot 156*)

which were now in need of repair and unusable. Three dozen sets were now required, as well as sundry other articles for the Drawing Room. Haines's letter was considered at a meeting on 14 June 1780, and the board's immediate reaction was to order the instruments from Adams the same day.[63] Adams's bill, dated two months later,[64] came to a total of £100 16s – business that Ramsden lost through his inability to meet deadlines, possibly because he was preoccupied with work for the Board of Longitude at that time. Each of the thirty-six sets of instruments on this order included a pair of 6-inch brass drawing compasses with ink and pencil points, a brass drawing pen with jointed knibs, a pair of bow compasses, and two parallel rulers; the order also involved a variety of wooden rulers of lengths up to 5 feet, with and without brass edges.

Clearly, Adams was now a favoured supplier of mathematical instruments to the Office of Ordnance, but the board did not abandon Ramsden and appoint Adams in his place. Spot checks on sample bill books over the next fifteen years show that both men continued to supply gunnery, surveying, and drawing instruments until Adams's death in 1795, when his place was taken by his younger brother Dudley. Unfortunately, the number of bill books for the late eighteenth century is considerably larger than for the middle years when George Adams senior was supplying instruments to this office. Furthermore, from about 1775 onwards the volumes of the WO51 series, and the WO52 series which takes over around 1782, are not indexed at all, so they have to be searched page by page. Between 1778, when George Adams junior started to receive Ordnance orders, and Dudley's bankruptcy in 1817, approximately 600 bill books were filled, any of which could contain Adams bills. As each volume contains around 500 bills, or about 300,000 in total for this period, a comprehensive search comparable with that carried out for 1748–73 was deemed to be impracticable. Nevertheless a run of thirty-six consecutively-numbered books covering bills allowed in 1795–96 was searched to determine the change-over point from George junior to Dudley, as will be seen later (chapter 9). This limited search turned up eighty-two Adams bills, confirming that in the early stages of the Napoleonic wars George junior was a major supplier of military and drawing instruments. Only four of these thirty-six books contained no Adams bills at all. The Ordnance business at that time was worth about £800 per annum to Adams, but its value would, of course, have fluctuated with the state of activity of Britain's armed forces. George junior happened to enter the field in the late 1770s at a time when the American War of Independence was in full swing, creating a large demand for military stores, but he must have known that this situation could not last for ever, so it is not clear how he coped with sudden demands for (say) three dozen gunner's quadrants, each of which (at £2 12s 6d) represented about ten man-days of work. For example, one such order, worth nearly £150, received on 13 June 1781, was delivered nine weeks later on 17 August, implying a workforce of around ten men engaged on this order alone.[65]

Though most of the Ordnance orders were for standard instruments at previously-agreed prices, occasionally experimental equipment was ordered which would have necessitated some design and development, possibly by trial and error. The Ordnance minutes contain frequent references to experimental work being carried out at Woolwich, though usually few details are given. (The principal reason for an event being noted in the minutes was to enable the participants to claim travelling and other expenses.) For example, in the 1790s Ramsden made several experimental gun mountings incorporating means for measuring the angle of vibration.[66] A 'stand for a pendulum to take time of a shell's flight' occurs in one of Adams's bills in 1781,[67] and 'alterations to an instrument of Jackson's' in another.[68] An item, 'an instrument with a rack motion', costing £12 in the same bill gives nothing away; this may have been deliberate, for security considerations occasionally intervened in the supply of instruments.[69] An instance of this nature occurred in mid-1781, when another of the instruments for checking the bores of cannon invented by Thomas Desaguliers (son of the lecturer), chief firemaster at Woolwich from 1748 until his death in 1780, was required. An order was placed with Adams, but on 20 June 1781 the board was informed by the master-general that he 'thinks it proper that different men should be employed in making different parts' of the instrument, to prevent it becoming too generally known, and that its construction should be entrusted entirely to the inspector of artillery at Woolwich. Adams was directed to return the relevant warrant for cancellation.[70]

This indicates that Desaguliers's instrument was considered to be highly important at the time, though from the example preserved in the Rotunda Museum at Woolwich it seems to have been a most unwieldy device.[71] The entry for Thomas Desaguliers in the *Dictionary of National Biography*, written towards the end of the nineteenth century, goes so far as to say that it was 'still in use at the royal gun factories for examining and verifying the bores of cannon', but as smooth-bore cannon had then been obsolete for half a century this statement may have been based on out-of-date information. No further references to Desaguliers's instrument, or parts thereof, were found in the bill books searched, but several 'mirrors with round brass frames' appear in later bills by both George and Dudley Adams, described in one bill as 'a round looking glass for inspecting guns'.[72]

A rare instance of Adams supplying a 'philosophical' instrument to the Office of Ordnance occurred in August 1781, when he submitted a bill for £22 14s for an 'Air Pump with Apparatus'.[73] Though probably destined for the Royal Military Academy at Woolwich, this was delivered to the Tower, so its ultimate destination is uncertain. There may have been some irregularity in the ordering of this unusual item: Adams had to wait three years for payment. His bill dated 10 August 1781 does not cite a prior order; it was authorized retrospectively by a warrant of justification dated 23 June 1783 and eventually allowed on 13 July 1784.

The end of the American War of Independence in 1782–83 did not result in a complete cessation of Ordnance orders: there was a continuing if small demand for drawing and surveying instruments even in peacetime. Orders by the dozen for gunnery instruments, however, often worth upwards of £100 per order at the height of the war, ceased abruptly. For the next ten or so years, until the build-up to the Napoleonic wars began in the early 1790s, Adams must have relied almost entirely on civilian sales for his turnover. Information on these is relatively sparse, apart from certain well-documented cases such as the provision of instruments for Teyler's Foundation in Haarlem, covered in detail in chapter 8. It may have been the drop in military orders following the end of hostilities in America that prompted Adams to diversify into writing and selling textbooks from 1783 onwards, an activity covered in chapter 7.

In the year 1777, just before recapturing the Ordnance business lost on his father's death, George junior had taken two apprentices, despite already having one (Fowler Bean) who was then in his fourth or fifth year.[74] The first, apprenticed on 6 February 1777, was his younger brother Dudley, then aged fourteen and a half. Surprisingly, George charged his mother £49 premium for this, which suggests that there was some friction in the household by then, as it was unusual for a master to demand a premium from a close relative.[75] The timing meant – fortuitously – that Dudley was introduced to military work at an

6.6 Trade card of Christopher Stedman senior, prior to 1762 (when the shops and houses on London Bridge were removed). Stedman senior was one of the London instrument makers who took numerous 'foreigners' (journeymen trained outside London) as employees, under licence from the City. His son, Christopher Stedman junior, was apprenticed to George Adams junior in 1777. (*Author's collection*)

early age. The second was Christopher Stedman, apprenticed on 2 October 1777.[76] Christopher was the son of the instrument maker of that name who formerly had a shop on London Bridge,[77] which he was obliged to quit when all the houses on the bridge were removed around 1762 (see figure 6.6). He subsequently occupied premises in Leadenhall Street, but died before his son was old enough to be apprenticed. If the £49 that George Adams charged his own mother was surprising, the premium that he obtained from Christopher Stedman's mother was astonishing – no less than £280. This was by far the highest premium charged by any mathematical instrument maker in the Grocers' Company in the eighteenth century. Its nearest rival was £100, and even that was unusual.[78]

No reason has been found for this extremely high figure. Admittedly the Adams business held the prestigious appointment of mathematical instrument maker to his majesty, but George junior was aged only twenty-seven at the time and had been a master instrument maker for only five years. Possibly there had been some business association between George Adams senior and Christopher Stedman senior which has not yet emerged from the records. In the 1750s Stedman was one of the instrument makers who took advantage of the legislation which enabled them to employ non-free journeymen when insufficient London-trained men were available. In 1759 he engaged no fewer than fifteen such men.[79] As the number of surviving instruments bearing his name is not particularly large, this suggests that his staff were employed mainly on subcontract work for another maker, and as the year 1759 was the peak of the Seven Years' War, the latter could well have been George Adams senior, overwhelmed by Ordnance orders. This is mere conjecture at present, but it illustrates the problems that remain to be solved in unravelling the structure of the instrument-making trade in the eighteenth century. The exceptionally high premium paid for Stedman junior's apprenticeship to Adams junior could have been a hangover from business dealings between their respective fathers in the distant past.

Fowler Bean, apprenticed in 1773, obtained his freedom in 1781 and disappeared from the instrument-making scene.[80] In the following year Adams took another apprentice, bringing the number back to three, namely Robert Blunt junior.[81] Robert was a son of the linen-draper of that name and his wife Sarah, George Adams senior's elder daughter by his first wife, so he was George junior's half-nephew. Robert Blunt junior obtained his freedom in 1789 on completion of his apprenticeship,[82] and was admitted to the livery of the Grocers' Company in 1795,[83] but it is doubtful whether he ever practised as an instrument maker. His father had died in April 1788, leaving the linen-drapery business to his widow (Sarah) on condition that she admitted Robert junior as an equal partner when he came of age. There is just a faint possibility that Robert junior may have worked for Dudley Adams at 53 Charing Cross for a few years while his mother ran the linen-drapery business at number 64, but this is just speculation and no documentary evidence has been found to support it. Robert's membership of the Grocers' Company, obtained through his apprenticeship to George Adams junior, enabled his descendants to claim the freedom by patrimony, and the company's books show that so far six generations in the Blunt family down to the present day have done so. The Master of the Grocers' Company in 1962, Arthur Graham Blunt, was a direct descendant of George Adams senior.[84]

The catalogues of instruments appended to works by George Adams junior in the 1780s and 1790s, the last of which, dated 1795, is reproduced in facsimile in appendix II, show the extensive range of the firm's products in the last quarter of the eighteenth century. Many examples can be found today, both in museums and in the marketplace, though in the case of products which altered little in design over several decades (such as basic drawing instruments) it is often difficult to decide whether they should be attributed to George senior's or junior's period. Some clues to dating provided by details of inscriptions are given in appendix III, but many instruments are signed simply 'G.Adams, London' and can be dated only on stylistic grounds or by reference to contemporary documentation.

Products that were usually positively dated in the inscription itself included terrestrial and celestial globes, the former due to updating to incorporate new geographical discoveries, and the latter due to the need to specify the epoch for which the star positions were calculated. In the mid-1780s Dudley Adams, apprenticed to his brother in 1777, completed his training and shortly afterwards left to start his own business at Charing Cross (see chapter 10). Under the terms of George senior's will, the Adams globe plates and associated tools belonged to his widow Ann for her lifetime: perhaps it was her suggestion that her youngest son (Dudley) should take over this self-contained part of the business, to avoid potential conflict between the two brothers and their wives (Dudley married in 1787) at 60 Fleet Street. At any rate, from 1789 onwards Adams globes were signed by Dudley at his own address, though in at least one instance (the special globes for van Marum

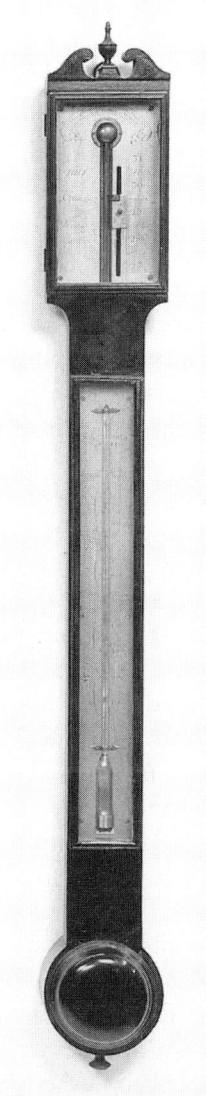

6.7 A stick barometer signed on the rectangular register plate 'G.Adams London Instr. Maker to his MAJESTY and Optician to the PRINCE of WALES'. The appointments cited indicate that this instrument must have been made between 1787 and George junior's death in 1795, so it is contemporary with his tract on the barometer (1790). Although the register plate is inscribed with only three states (as in figure 5.12), this instrument has an enclosed tube and a large Fahrenheit thermometer mounted on the outside of the case, reading from 'Freezing' (32) to 'Fever Heat' (110). (*Christie's, London, 11 September 1986, lot 14*)

described in chapter 8) globes by Dudley Adams were mounted in brass stands signed by George Adams, indicating that the two brothers were associates rather than rivals in this branch of the business.

Globes mounted in a brass meridian and wooden horizon (as virtually all globes other than pocket ones, and those for van Marum, were in the eighteenth century) were usually provided with a 'quadrant of altitude' for measurement of the coordinates of any point on the surface with respect to the horizon or the zenith. This consisted of a flexible brass strip, a little over a quarter of the circumference in length, attached at its upper end to a boss which could be clamped to the uppermost point of the brass meridian once the globe had been set to the desired inclination. It was never a very satisfactory accessory, as its flexible nature was liable to result in damage to the surface of the globe.

In 1783 a Mr George Maling of Scarborough had the idea of using a rigid vertical pillar adjacent to the globe, instead of the flexible strip, the pillar being calibrated with a scale of sines corresponding to degrees of latitude; a horizontal pointer could be slid up and down the pillar to relate any point on the surface of the globe to the vertical scale. The Society of Arts awarded Mr Maling a silver medal for this.[85] Maling's device, which was made of wood, is listed at no. LXXXIV in the 'Catalogue of Models and Machines' published in the Society's *Transactions* in 1784.[86] The next item in the catalogue is 'Two ditto made in Brass and presented to the Society by Mr. George Adams'. The latter are not described, but the implication is that they were copies of Maling's wooden model, made in brass. Another instrument maker, Gabriel Wright, presented his own version of an altitude-measuring device to the society at about the same time; Wright's device is listed in the catalogue immediately after Adams's. The problem also caught the attention of John Smeaton, FRS, who in 1788 devised yet another substitute for the quadrant of altitude; it was described in a paper read to the Royal Society and published in the *Philosophical Transactions*,[87] so it must have been regarded as important at the time. However, in practice none of these ideas was adopted commercially, probably because they would have significantly increased the cost of globes incorporating them.

Adams's two brass devices presented to the Society of Arts in 1783 provide the only mention of him in the indexes to their *Transactions*.[88] From 1785 he was an active member of another, more local, scientific club, the Chapter House Philosophical Society, known today only

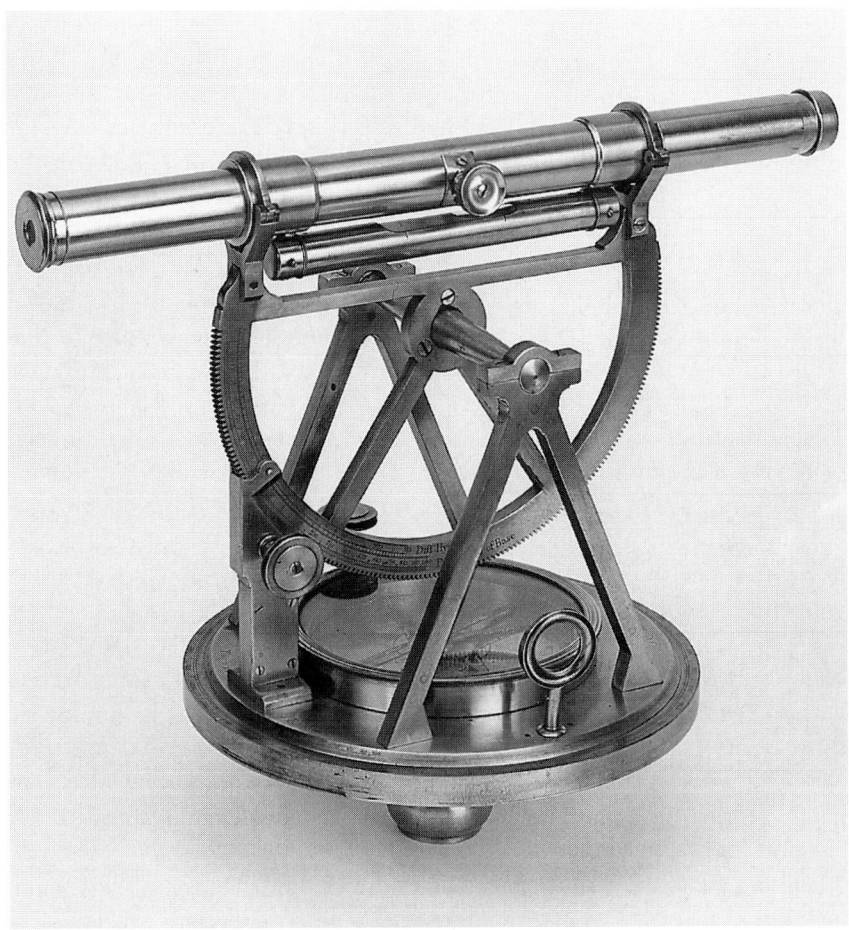

6.8 A brass theodolite with an axle-mounted elevation quadrant with rackwork and a long spirit level beneath the telescope, which is reversible. It is signed on the upper face of the moving platform 'G.Adams London'. The telescope is 13½ inches long. This instrument is broadly similar to figure 1 in plate XVI of Adams's *Geometrical and Graphical Essays* (see figure 7.25). (*Christie's, London, 29 November 1990, lot 182*)

through the survival of a notebook containing copies of their minutes from 1780 to 1787.[89] Compiled by William Nicholson, the club's secretary from 1784, this volume was probably intended for his own personal use as it covers only a small selection of the club's activities: it is not an official minute book. The club was formed in 1780 with the intention of meeting fortnightly at the Chapter Coffee House to discuss 'Natural Philosophy in its most extensive signification'. Amongst the early members were John Whitehurst, Edward Nairne, Richard Kirwan, and J.H. de Magellan.[90] Tiberius Cavallo was elected at the second meeting, and other elections followed from time to time. Details (in this book) of the meetings held in the first few years are sparse, but fuller accounts begin with Nicholson's election as secretary. From March 1783, meetings were usually held at the Baptist's Head Coffee House, Chancery Lane, quite close to Adams's shop in Fleet Street.

Adams was proposed for membership by Magellan at the meeting on 7 January 1785, seconded by Mr Cooper. He was duly elected at the following meeting (the usual practice) on 21 January. Twelve members were present on that occasion: Kirwan (in the Chair), Magellan, Dr Crawford, Dr Wells, Simms, Major Gardner, Walker, Dr Lister, Nicholson, Babington, Dr Cooke, and Dr Lorimer. At thirty-four, Adams was one of the younger members (though Nicholson was his junior by three years): Whitehurst, the oldest member, was seventy-one, and Magellan was sixty-two. Adams himself being in the adjacent coffee room at the time, he was admitted to the meeting as soon as his election had been confirmed.

Almost at once a procedural point arose which gave rise to some heated argument. As mentioned above, at one of the early meetings it had been resolved to discuss 'Natural Philosophy in its most extensive signification'. Magellan now said that he wished to communicate an astronomical paper, whereupon Kirwan (in the Chair) said the constitution of the society did not permit consideration of matters that might lead to 'mathematical disquisition'. As Nicholson succinctly observes, 'This produced a conversation.' The record of the original resolution of 1 December 1780 was called for and read; but Kirwan claimed that the society ought to be governed by usage rather than law. Nicholson agreed, but Wells said usage up to then had always excluded mathematical questions. Kirwan tried to settle the matter by formally moving that astronomical communications be excluded in future. This was seconded, but whether it passed or not is not clear, for someone then raised the question of the society's name. Kirwan said it had originally been decided that the society should have no name, being known only from the place of meeting. Nicholson pointed out that although the first minute book was headed 'Chapter Coffee House Society', in the second book it was called the 'Philosophical Society', and he had once been summoned by the previous secretary (Dr Watkinson) to a meeting of the 'Chemical Society'. By this time Adams must have been wondering what sort of an organization he had joined.

However, after these digressions the main purpose of the meeting was reached. It was the society's custom for each member in turn to propose a subject for discussion at the next meeting, and that chosen for this occasion (by Dr Wells) was 'What Proofs can be brought to Establish the Existence of the Chemical Principle called phlogiston by Stahl and his followers?' Most of the members present seem to have been firm believers in phlogiston (especially Kirwan), though some of the evidence cited in its favour was disputed. Adams did not contribute to the discussion; perhaps he had not thought about the matter at this stage of his life, being primarily concerned with making mathematical instruments. Nine years later, when his five-volume *Lectures* was published in 1794 (chapter 7), he was a strong supporter of the phlogiston principle.

At the next meeting on 4 February 1785 discussion of the phlogiston question was resumed, but no conclusions were reached. Kirwan then spoke about a method of ascertaining the weight of air. This gave Adams the opportunity to mention a practical point concerning the construction of air-pumps,[91] which he said had been devised by 'one of his workmen'. The piston rod was made hollow, and a wire within it was fixed to the lower valve and so contrived that it opened the valve as the rod neared the top of its travel. Nicholson commented that he thought this contrivance would 'limit the power of Exhaustion' of the pump, by leaving a greater cavity between the lower valve and the piston when down.

Later entries show that Adams attended a further nine meetings in 1785, five in 1786, and four in the first three months of 1787, which is as far as this volume goes. This means that in addition to the meeting on 21 January 1785, during which he was admitted to the society, he was present at nineteen of the fifty-two meetings covered by the remainder of this volume, but apart from the air-pump episode only one other contribution by him is recorded. On 28 October 1785 he mentioned that a person at Liverpool had procured a patent for taking impressions on glass from engraved copperplates, but the process had only been applied with success to small pieces so far.[92]

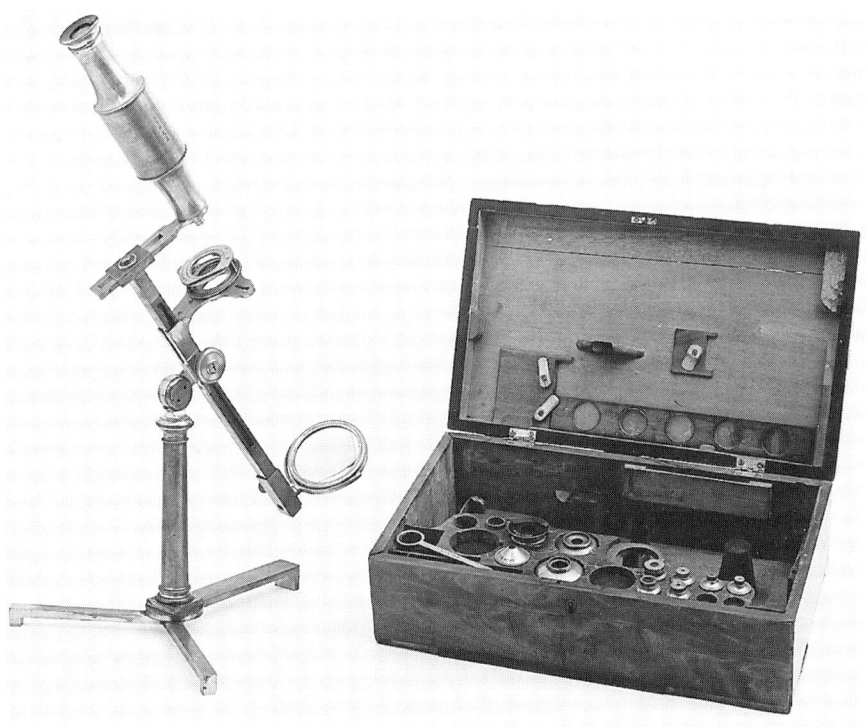

6.9 A compound microscope signed 'G.Adams No 60 Fleet Street London', probably by George Adams junior *c.* 1790. This instrument has a compass joint at the top of a brass pillar on a folding flat tripod base, as depicted in his *Essays on the Microscope*, but the limb is a rectangular bar rather than the tubular assembly shown in his plate. In this respect it is more like the instrument shown in Jones's 1798 replacement plate (see figure 7.8), though the body tube has the form shown in Adams's 1787 version. (*Museum of the History of Science, Oxford, inventory no. 82338; G.L'E. Turner cat. no. 30. Photograph by G.L'E. Turner*)

In the absence of the original minute books there is no way of telling how representative Nicholson's 'minutes' are. They appear to show that the proceedings of the society were predominantly chemical in character, but it is possible that he noted only points that were of particular interest to him personally.

The fact that Adams was introduced to the society by Magellan is significant, for recent unpublished research on the latter's activities (currently in progress) has shown that in addition to procuring instruments from Adams for van Marum (chapter 8), he probably acted in a similar capacity for other overseas buyers. Magellan's bank account contains about thirty entries in the name of Adams between 1776 and 1787, totalling around £3,000, as well as numerous references to other instrument makers such as Martin, Dollond, and Troughton.[93] Some of the Adams entries were probably payments from Coimbra University, Portugal, where many Adams instruments are located.[94]

In parallel with his membership of the coffee house club, in the 1780s Adams became involved with the Swedenborgian movement. Emanuel Swedenborg, the Swedish 'scientist and mystic' as one modern biographer has called him,[95] had died in 1772 (coincidentally, the same year as George Adams senior), in his mid-eighties, having spent the last thirty years of his life claiming to be in constant communication with, and indeed to have visited, the spirit world. From his personal experiences there he wrote a number of theological works which put a new interpretation on certain parts of the Bible, in particular the nature of heaven and hell and the whole concept of life after death. Although he did not found a new sect or religion himself, he was convinced that the Last Judgement had already taken place (in 1757), and that the old Christian church would soon be replaced by a 'New Church' based on the principles expounded in his last work, *The True Christian Religion*.[96]

Swedenborg's works were initially published in England and Holland, in Latin; they were too controversial to be published in his own country. Many people regarded his claims to have visited the spirit world (including both heaven and hell) as visionary dreams, or the products of an unsound mind, not to be taken seriously. Nevertheless he did find a few believers, especially in England, where he was residing at the time of his death in March 1772. Their numbers were limited to a few individual enthusiasts until 1782, when the young printer, Robert Hindmarsh (1759–1835) happened to come across some of Swedenborg's works. He found them so enthralling that he tried to contact other people of similar persuasion to study the writings together. In an account of his experiences written in the 1820s, but not published until 1861 (some years after his death, and nearly eighty years after the events described), he tells how surprised he was to find that so few people were aware of Swedenborg's ideas.[97] An advertisement in the London papers in December 1783, he said, drew a response from only five persons, who met first in a coffee house and then in rooms in the Temple. By the following year (1784) their numbers had grown to a few dozen, who called themselves 'The Theosophical Society'.

One of these new members was George Adams junior. He evidently took an active part in the proceedings, for Hindmarsh relates that at their regular Thursday evening meetings 'the untranslated writings of Swedenborg were read from the Latin either by Mr. George Adams or myself'. In 1785 the society procured from Sweden the manuscript of an unpublished work, 'Apocalypsis Explicata', and four members, one of them Adams, agreed jointly to publish this at their own expense. As the manuscript was not quite complete, the four completed it themselves by reference to an earlier work of Swedenborg's on the same theme, *Apocalypsis Revelata*. A note explaining what they had done, dated 17 June 1790, is given in the fourth volume of the resultant set,[98] printed by Hindmarsh. In 1786 George Adams himself corrected and financed a second edition in English of *Doctrine of the New Jerusalem concerning the Lord*, first published in Latin in 1763 and already translated into English by another follower of Swedenborg.[99]

The Theosophical Society continued to meet in the Temple until 1787, when the question of whether or not the society should separate itself from the established church, and ordain its own ministers, began to be discussed. A document outlining the 'Principles of the New Church' was drawn up at a meeting on 29 July 1787, held at the premises in the Poultry of Thomas Wright, watchmaker to the king, and the New Church effectively came into existence two days later. According to Hindmarsh's account of these proceedings, Adams was not present on either occasion. The name 'Theosophical Society' was dropped in May 1788. Separation from the established church, and acquisition of their own places of worship, gave rise to some disagreement amongst members. Those in favour signed a document entitled 'Reasons for Separating from the Old Church ... by Members of the New Jerusalem Church'; Adams was not amongst the seventy-seven signatories, nor was he involved in the arrangements made in June

1788 for ordination of ministers of the New Church. As Hindmarsh does not mention him again, it seems likely that Adams did not approve (or thought it prudent not to approve, in view of his royal appointments) of separation from the established church.

In March 1790 Hindmarsh began a monthly periodical devoted to the ideals of the New Church, entitled *The New Magazine of Knowledge concerning Heaven and Hell*. Adams's *Essay on Vision*, printed in 1789 by Hindmarsh, received a complimentary review in the first volume.[100] When the American branch of the New Church published its historical *Annals* at the beginning of the twentieth century, under the date 1789 Adams's *Essay on Vision* was mentioned with the comment: 'New Church principles are introduced into this work by the author ... who was a prominent member of the New Church in London'.[101] However, from the foregoing evidence it seems fair to draw the conclusion that Adams was not a practising member of the New Church after its formal separation from the established church in 1788. Nevertheless, the catalogue of his library, sold by auction by his widow in 1796 (chapter 9), shows that he owned numerous volumes of Swedenborg's works, mostly in editions published in the 1780s though he had some earlier ones as well.

References to Adams and/or his business at 60 Fleet Street are occasionally found in travel diaries and correspondence. For example, the Danish astronomer Thomas Bugge, who visited London in 1777, recorded that he bought a set of drawing instruments from Adams for 2 guineas, though the bulk of the £88 that he spent on instruments during this visit went to Nairne & Blunt.[102] In a letter from Philadelphia to Sir Joseph Banks in 1787, Adams is cited as an authority on the magnetic variation and means for measuring it;[103] but the writer, John Churchman, may have been confusing George junior with his father, as George senior is known to have made several large variation compasses in the 1750s. Elsewhere in the Banks correspondence files, Sir Charles Blagden in 1786 refers to Adams's improvements in barometers,[104] but again there is the possibility of confusion between father and son. A more definite reference to Adams's shop occurs in a letter from the same correspondent to Banks in 1791, when the writer saw there one of Blair's fluid achromatic lenses for telescopes.[105] Adams would not let him take the lens out of the tube, but 'I looked through it at some gilt letters in Fleet Street, which it shewed very distinctly, without any colour at the edges'. Blair's fluid lens aroused considerable interest amongst astronomers: Adams mentioned it twice in letters to van Marum

6.10 A magazine case of drawing instruments and paints, by George Adams junior. The lowest compartment contains blocks of colours and a china mixing tray; plus a brass parallel ruler, signed 'G.Adams London', of the 'sliding L-square' type devised by Haywood and illustrated in Adams's *Geometrical and Graphical Essays* plate II (see Maya Hambly, *Drawing Instruments*, p. 112). The removable tray contains the usual assortment of ink and pencil compasses and accessories, dividers, pens, proportional compasses, a small square, and a trammel block. In a compartment in the lid are a sector, rectangular protractor, and scissors-type parallel ruler.

Continuation of the Business   195

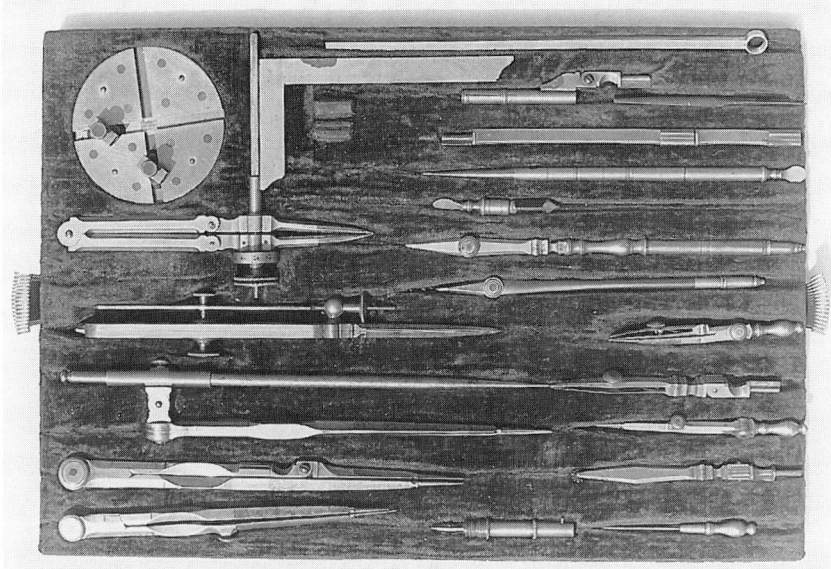

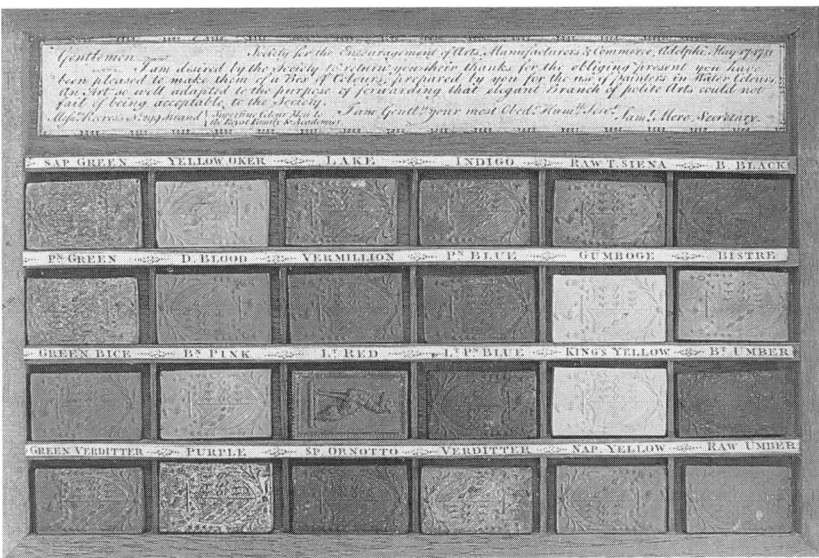

6.10 continued   The outside dimensions of the case are 12½ by 9 inches by 3¼ inches high. The tray of watercolour paints was provided by Reeves, and contains a printed transcript of a letter from the Society of Arts dated 17 May 1781, thanking them for donating a box of colours to the society. Most of the colour blocks are embossed with the coat of arms used by Reeves; one has the Reeves crest only (a sitting greyhound). (*Private collection*)

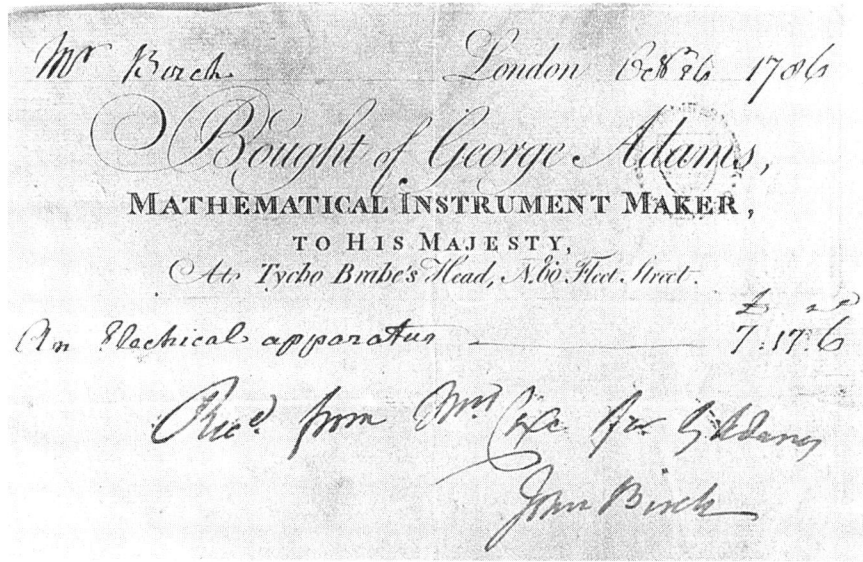

6.11 The engraved heading of this bill dated 1786 cites both a shop sign, 'Tycho Brahe's Head', and a house number, 60. By the following year (1787) George junior was Optician to the Prince of Wales as well as Mathematical Instrument Maker to his Majesty. The purchaser of this electrical equipment was probably the John Birch who contributed a 'Letter on Medical Electricity' to the fourth edition of Adams's *Essay on Electricity* in 1792. Shop bills, when paid, were normally receipted by the shopkeeper or by one of his employees. The inscription on this one apparently indicates that John Birch withdrew 7½ guineas from Messrs Cox for the specific purpose of paying this bill, which presumably was retained by Messrs Cox as evidence that the money had been handed over.

in 1790 and 1791, but it seems nevertheless to have found little practical application.

Apart from manuscript bills in van Marum's files at Haarlem, cited in chapter 8, and of course the numerous clerks' copies in the Ordnance bill books, only a handful of extant invoices by George Adams junior out of the many hundreds or thousands that the firm must have produced have been located. One has been cited earlier (the bill to Wedgwood for globes in 1774); another is illustrated here (see figure 6.11). A few are known to be in the Royal archives at Windsor Castle, in particular two dated 1786, mainly for electrical apparatus,[106] but otherwise little documentary evidence has been found of any work carried out by George junior (as distinct from his father) to justify his title of mathematical instrument maker to his majesty. From 1787 he used in addition the title of Optician to his Royal Highness the Prince of Wales',[107] (see figure 6.7) but what led to this appointment is not known.

Though George junior probably supplied some of the Adams items in the King George III Collection during the 1780s and early 1790s, his last years were dominated by the outbreak of war with the French Republic, which gave rise to a vast expansion in Ordnance work. The avowed intention of the republic to give every assistance to other peoples wishing to overthrow their kings caused alarm in other governments besides the British. Examination of the Ordnance bill books covering the change-over period from George junior to Dudley in 1794–96 showed that in the last twelve months of George junior's tenure, from October 1794 to September 1795 inclusive, twenty-seven bills were submitted in his name, totalling £811, and this may have been less than the annual average (due to his terminal illness), as Dudley submitted several large bills totalling over £500 in the three months following his brother's death. Assuming that the Ordnance trade was profitable (and surely neither George junior nor Dudley would have undertaken it at a loss), the fortunes of the Adams family must have been significantly affected by world events in the last decade of the eighteenth century. It does not necessarily follow, however, that their financial position was improved by the war, as their living expenses, which had remained virtually static since the Adams instrument business commenced in the 1730s, now began to increase alarmingly. To pay for the war, new taxes were introduced and old ones extended. Kearsley's *Annual Tax Tables* for the late 1790s include, in addition to the land tax (which could be redeemed if desired by paying eighteen years' tax in a lump sum), details of taxes or duties on houses, windows, dogs, male servants, bachelors, carriages, horses for pleasure or draught, mules, armorial bearings, salt and tea, not to mention the new income tax (albeit at what seems to us a ludicrously low rate) and death duties. Coupled with increases in the price of basic foodstuffs (Londoners, unlike country dwellers, could not grow their own), these taxes must have made it appreciably more difficult for Dudley than his brother to make ends meet.

7 Essays and Lectures

A few years after Adams became one of the official suppliers to the Office of Ordnance, the American War of Independence came to an end. One result was an immediate reduction in Ordnance spending. The Duke of Richmond, master-general at the time, endeavoured to force reductions in the prices of a wide range of artificers' goods, from muskets to wheelbarrows, to help reduce the enormous debt that had built up during the war.[1] There was much discussion in parliament of the current cost of the Ordnance Department, in comparison with the situation in 1763, when the previous major conflict (the Seven Years' War) ended in Europe. The accumulated debt on the Ordnance account then was £595,000: it was now no less than £1,724,000. The excess of actual expenditure over the estimate submitted to parliament for the year 1782 alone amounted to £819,000. The estimate for the coming year, at £662,000, though much reduced from the war years, was still £112,000 more than in 1763. During the Seven Years' War the Ordnance Department spent a total of about £3 million: the corresponding figure for the American war was £10 million. A drastic reduction in orders for new equipment from suppliers such as Adams was inevitable.

Perhaps it was this reduction in military spending that turned Adams's thoughts to a different line of business: writing and selling textbooks on scientific subjects. So far, he had been content to produce new editions of his father's *Treatise on the Globes* whenever they were needed. A third edition had been issued in 1772, the year of his father's death; this sufficed for five years. A fourth edition appeared in 1777, and a fifth in 1782, with little alteration. Towards the end of 1783 George junior produced the first new work written by himself, on electricity and magnetism, a subject that his father did not tackle at all.

This was his first contribution to what was intended to be (he said later) a comprehensive survey of the whole of the mathematical and philosophical sciences, and their relevant instruments, an ambitious project that he did not live to complete, though he struggled valiantly against failing health for the next twelve years. During this period he wrote on the microscope (1787), astronomy and geography (1789), vision (1789), Hadley's quadrant (1789), barometers (1790), geometrical drawing and surveying (1791), and finally his largest work, *Lectures on Natural and Experimental Philosophy* (five volumes, 1794). Most of these went through several revised editions, so between 1784 and 1795 hardly a year passed without some new or updated work being offered to the public. (See table 7.1.) The publishing history and coverage of each of these titles is outlined below. Like his father, George junior published his books himself, without the assistance of a bookseller, though the imprint on the title-page usually included the printer as well. From 1787 this was Robert Hindmarsh, fellow Swedenborgian and printer to the Prince of Wales. In parallel with his literary activities, of course, George junior had to continue to attend to the daily needs of his instrument-making business, and ensure that he did not lose trade to his numerous competitors in this field.

An Essay on Electricity

Though dated 1784 on the title, this was first advertised towards the end of December 1783:[2] an octavo with six plates, price 5s in boards (that is, ready for the customer to have permanently bound to his own taste). When George Adams senior published his *Treatise on the Globes* in 1766 he said in his preface that he hoped readers would excuse any errors and correct them themselves, a remark that drew scathing comments from Benjamin Martin. George junior used almost identical wording in this work on electricity: 'The various interruptions and avocations, from which, as a tradesman, I cannot be exempt, will, I hope, induce the reader to make some favourable allowances for any errors which he may discover, and kindly correct them for himself.' Martin had died two years earlier, so George junior did not have to endure the line-by-line demolition of his work that *Treatise on the Globes* attracted; indeed, *An Essay on Electricity* must have been favourably received by the public, as a second edition 'corrected and enlarged' was published in the following year (1785), and a third two years later in 1787. A fourth edition with some alterations in coverage

Table 7.1
George Adams junior's publications in chronological order (excluding overseas editions and translations)

1784	*An Essay on Electricity*
1785	*An Essay on Electricity*, second edition
1787	*An Essay on Electricity*, third edition
1787	*Essays on the Microscope* (quarto)
1789	*Astronomical and Geographical Essays*
1789	*An Essay on Vision*
1789	*Description, use ... of ... Hadley's Quadrant*
1790	*Astronomical and Geographical Essays*, second edition
1790	*A short dissertation on the Barometer*
1791	*Geometrical and Graphical Essays*
1792	*An Essay on Electricity*, fourth edition
1792	*An Essay on Vision*, second edition
1794	*Lectures on Natural and Experimental Philosophy* (5 vols)
1795	*Astronomical and Geographical Essays*, third edition

Posthumous editions by W. & S. Jones:

1797	*Geometrical and Graphical Essays*, second edition
1798	*Essays on the Microscope*, second edition (quarto)
1799	*Astronomical and Geographical Essays*, fourth edition
1799	*An Essay on Electricity*, fifth edition
1799	*Lectures on Natural and Experimental Philosophy*, second edition (5 vols)
1803	*Astronomical and Geographical Essays*, fifth edition
1803	*Geometrical and Graphical Essays*, third edition
1812	*Astronomical and Geographical Essays*, sixth edition
1813	*Geometrical and Graphical Essays*, fourth edition

(magnetism was omitted, and medical applications expanded) appeared in 1792, and after his death William & Samuel Jones of Holborn published a slightly updated fifth edition in 1799. Some alterations were made to both the text and plates between editions, as noted below.

Adams made it clear that what he was offering to the public was not an exhaustive treatise on electricity, but rather a practical manual for those who wished to try the experiments themselves – especially, of course, those who purchased electrical machines and apparatus at his shop. He made no claim to be a theorist himself, though in later editions he did not hesitate to comment on the conflicting theories about the nature of electricity then current, and their relevance to specific experiments. From the references cited in the text and footnotes, it is evident that Adams's reading on the subject had been extensive,[3] and the numerous practical hints scattered throughout the book show that he was familiar with the experimental side as well. At one point he mentioned that he had tried some modifications to an electrical machine in 1772, so he must have become interested in electricity shortly after his father died, or possibly even before.

The type of electrical machine that he recommended (and presumably sold) is shown in figure 7.1 (his plate I); the two versions differ only in the means for turning the cylinder, the 'multiplying wheel' arrangement being preferred. In his instructions for the use of machines of this type, he stressed the importance of keeping the glass cylinder and glass insulating pillars clean and free from moisture. Also, it was essential for maximum output that the line of contact between the cushion and the cylinder should be as perfect as possible, and that no amalgam should be placed anywhere but on this line, otherwise it would simply serve to conduct the electricity away. As to the constitution of the amalgam, Adams said that after many trials he had come to the conclusion that there was little to choose between the 'aurum musivum' of the shops and one made of quicksilver (mercury) five parts and zinc one part, melted together with a small quantity of beeswax. (Many electrical experimenters had their own favourite compositions for amalgam: in 1790 van Marum sent a sample of his to both Nairne and Adams, but they were reluctant to try it.)[4]

The electrical apparatus and experiments that Adams described were common to most practitioners at this period: demonstrations of attraction and repulsion, sparks, discharges from points, the Leyden jar, protection of buildings from lightning, and so on. On the question of whether lightning conductors should end in balls or points, Adams did not express a definite opinion. 'The experiments which have been made on this subject are very numerous; but the greater part appear to be very inconclusive.' For the latest assessment of the position, he quoted at length the article on lightning in the *Encyclopaedia Britannica*, volume 6, page 4244.

In a chapter of twenty-four pages on medical electricity, Adams referred to the work of Nollet, Cavallo, Brydone, and others, but he

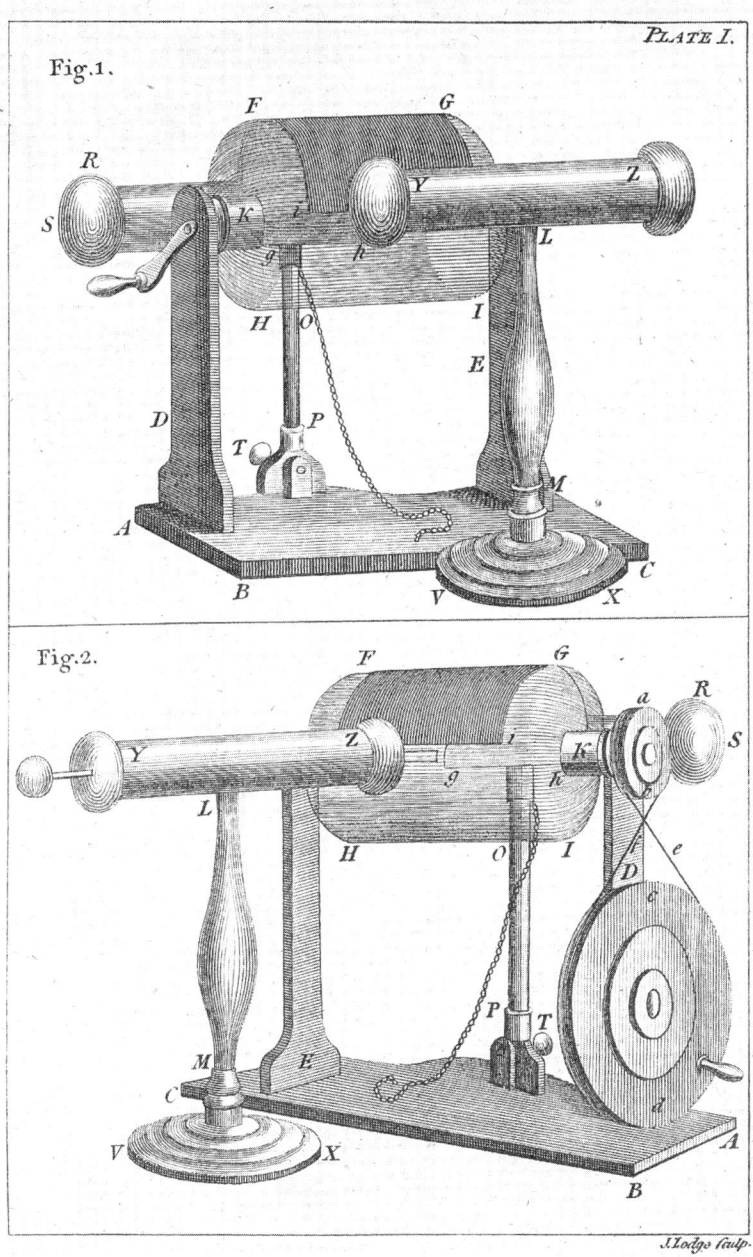

7.1 Plate I of Adams's *Essay on Electricity*, used as a frontispiece for the first edition, showing a cylinder machine with alternative driving means: a hand crank, or a 'multiplying wheel' and cord. (*Author's collection*)

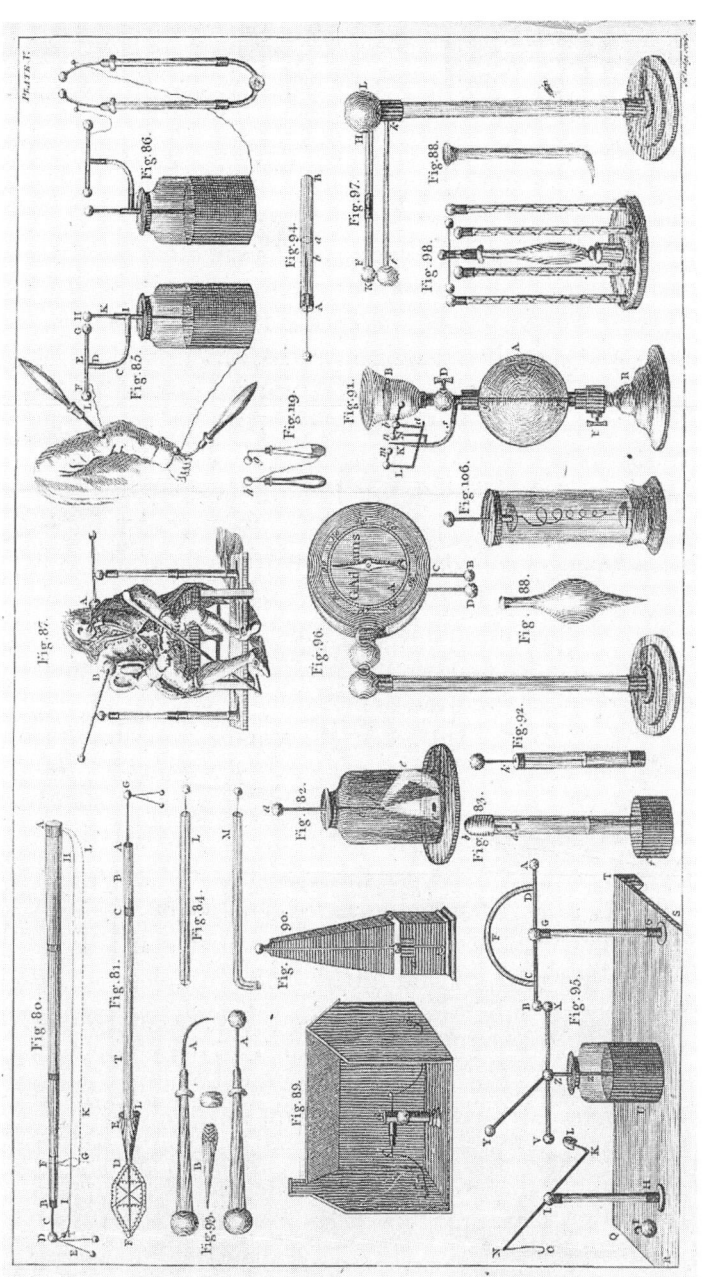

7.2 Two methods of applying the electrical discharge to a patient are shown at the top of this plate (Adams's plate V). At the lower right is an electrometer which measures the repulsion of the upper arm by the lower by means of a sliding weight. The dial electrometer at lower centre incorporates gears multiplying the angle of deflection by four. Figures 80 and 81 at upper left are electrometers for measuring the electrification of rain. Figure 119 in this plate from the second edition (1785) represents a medical applicator combined with a small Leyden jar, for giving small shocks to specific areas; this was not in the first edition (1784). (*Author's collection*)

did not mention any activities of his own in this area. Unlike most writers on medical electricity, he gave no specific examples of cures effected by this means. This part of the book was considerably extended in later editions, especially the fourth (1792).

A separate twenty-eight-page 'Essay on Magnetism' concluded the book. This, he said, was extracted from a larger work, which he had laid aside for the present, as he was expecting Cavallo to produce a work on this subject shortly.

In his second edition (1785), published only a year after the first, Adams paid more attention to the conflicting theories of the nature of electricity then current. Personally, he was inclined to favour the two-fluid theory ('in the course of this Essay many observations will occur, which tend to confirm this'), but not being sufficiently sure of his ground to demolish the one-fluid theory, he outlined the significant points of both, so that readers could make up their own minds.

The enlarged chapter on medical electricity in the second and later editions was reinforced by the addition of an engraved frontispiece (replacing plate I), showing a cylinder electrical machine in use for this purpose. This plate went through at least three versions. At first the human figures were sketched in outline only, the most prominent feature being the electrical machine. For the fourth edition the figures were filled in (see figure 7.3), and the opportunity was taken to dramatically increase the size of the lady's hat, presumably reflecting the latest fashion. The Jones brothers, in their edition of 1799, added some accessories in the otherwise blank part of the plate.

Expansion of the coverage of magnetism in the second and third editions was largely due, Adams said, to 'the ingenious and kind hints of Dr.Lorimer'. The additional material included remarks on the magnetic variation and compasses designed for its measurement, and Dr Lorimer's dipping needle; these were illustrated in an extra plate, which occurs only in the second and third editions. The long association between George Adams senior and Dr Gowin Knight in the production of compasses for the Navy, however, was not mentioned, perhaps because by that time both parties had been dead for thirteen years.

The make-up of the third edition (1787) differs markedly from that of its predecessors by the inclusion of a long supplement of fifty pages at the front of the book, 'Containing an Account of the PRINCIPAL DISCOVERIES made in that Science [electricity] since the Publication of the first Edition', illustrated by an extra plate.[5] About half of this extra material consists of extracts from De Saussure's *Inquiries and*

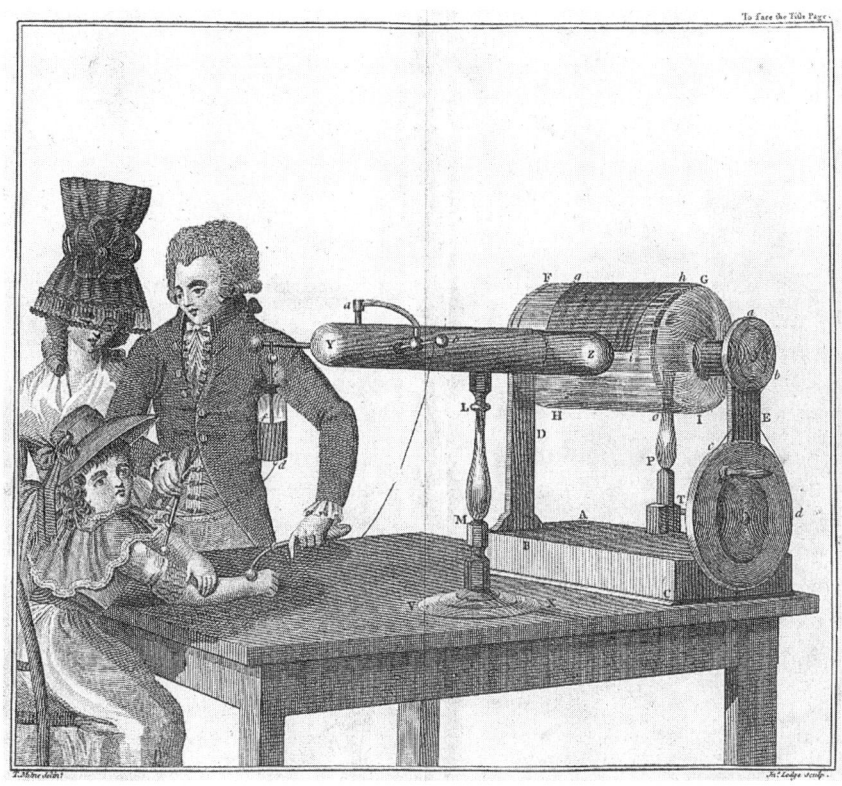

7.3 Frontispiece of the fourth edition (1792) of *Essay on Electricity*. For this edition the human figures, previously shown in outline, were filled in. (*Author's collection*)

Observations on Atmospherical Electricity, with copious notes and comments by Adams. The other additions describe Bennett's atmospherical electrometer; Cuthbert's large electrical machine for Teyler's Foundation at Haarlem; experiments on electric light, and the non-conducting power of a perfect vacuum, by William Morgan; and experiments on air by Henry Cavendish.

One enthusiastic electrical experimenter at this time was John Hill, of Wells-next-the-Sea, Norfolk, who signed himself 'John Hill Terts.' (that is, the third).[6] In 1787 he sent a copy of Adams's second edition (1785) to his former teacher, the Revd W. Steggall, of Wyverston, Suffolk, annotated with numerous manuscript comments, and bound with a manuscript appendix of about twenty pages containing further practical hints.[7] The annotations indicate that Adams and Hill were in contact, though possibly only through correspondence. For example, at the point where Adams cautions readers against rubbing the glass insulating pillars too hard when cleaning them, as they are varnished, Hill inserted a footnote:

> To Varnish Glass Pillars. First Rub the glass with spirits of Wine, then set the spirits on fire; and when burnt out, lay on (with a camels hair brush) Copal Varnish. – G.Adams to J.H.

Hill was also interested in microscopy: he purchased from Adams one of the latter's improved lucernal microscopes designed to project an image on to a translucent screen. This led to some correspondence in the *Gentleman's Magazine* in 1796, as will be seen later.

In his chapter on medical electricity in his second edition (1785), Adams mentioned in passing a short course of lectures on this subject read recently by Mr John Birch, an assistant surgeon at St Thomas's Hospital, London. Adams hoped that Birch's position would result in electricity becoming accepted by the public as a recognized method of healing (leading, Adams presumably hoped, to increased sales of electrical machines). In 1779 Birch had published a tract on the use of electricity for removing female obstructions, which was, said Adams, 'the only Treatise we have yet had from the Faculty' on medical electricity. Adams was so impressed by the favourable results obtained by Birch in apparently hopeless cases that for his fourth edition (1792) he dropped magnetism completely, to make room for a long 'letter' from Birch occupying fifty-five printed pages. Birch and Adams evidently worked closely together, for the 'letter' opens with praise for the way Adams had 'remedied the defects I found in the machine, as they presented themselves to me in the course of my experience'. The latter extended over more than twelve years, during which Birch had gradually come to the conclusion that electricity operated in three distinct modes, depending on whether it was applied as a discharge from a point, from a ball, or from a Leyden jar. The first mode acted as a sedative, the second as a stimulant, and the third as a deobstruent. Guided by his knowledge of anatomy (which few, if any, other writers on medical electricity possessed), he applied whichever technique seemed appropriate in specific cases, about forty of which he proceeded to describe. Many of these had been referred to him in his capacity as a surgeon, but by using electrical treatment the necessity for surgery had in most cases been avoided – no doubt to the immense relief of the patients concerned, anaesthetics being then unknown.

When the Jones brothers produced a fifth edition of Adams's book in 1799 they added a section on animal electricity, which by then was beginning to supersede static electricity as the main area of interest to electrical experimenters and the general public. William Jones was rather critical of Adams's text, which he said contained too many 'referential errors'; the chief reason for reprinting the book, he said, was its popularity with readers due to the large number of experiments described.

Essays on the Microscope

While he was engaged on revising his *Essay on Electricity* for the second and third editions, Adams was also compiling a major work on microscopy. Advance publicity for this was provided by an announcement in his third edition (1787): 'Speedily will be published ... Microscopical Essays'. The title eventually adopted was *Essays on the Microscope*, the plural form presumably being chosen because of the multiplicity of objects (and types of instruments) described, though the text was not divided into separate essays. The thirty-two illustrative plates in this work all bear the copyright date 20 May 1787, while the allegorical frontispiece entitled 'Truth discovering to Time, Science instructing her Children in the Improvements on the Microscope' is dated 1 July 1787. The earliest newspaper advertisements that have been noticed were published in February 1788 (see figure 7.4), but it is possible that Adams relied at first on selling to callers at his shop, and only resorted to advertising when sales began to flag. In addition to selling books and instruments, Adams also attracted some potential customers to his shop at this time by acting as an agent for Adam Walker's annual course of lectures on natural and experimental philosophy at Founders' Hall, Lothbury: tickets for this, price 1 guinea, were available at 60 Fleet Street and at Nairne & Blunt's, Cornhill.[8]

Essays on the Microscope was dedicated (by permission) to the king; moreover, the dedication implies that the king had personally encouraged Adams to undertake the task of writing works on mathematical and philosophical instruments. As Dr Johnson had died in 1784, and George junior was appreciably more literate than his father, there is no reason to suppose that in this instance the dedication was written by anyone other than Adams himself.

In his preface, Adams said that he had first thought of simply republishing his father's *Micrographia Illustrata*, last issued sixteen years

This Day was published,
Illustrated with Thirty-two Plates, and an elegant
Frontispiece,
In One Volume Quarto, Price 1l. 6s. in boards,
The Cuts separate,
Dedicated, by Permission, to his Majesty,
ESSAYS on the MICROSCOPE. Containing a description of the most improved Microscopes, with particular instructions for using them, and preparing objects for examination; a general history of insects, their transformations, peculiar habits, and œconomy; an account of the various species and singular properties of the Hydræ and Vorticellæ; a description of 379 animalcula; with a concise catalogue of interesting objects; a view of the organization of timber, and the configuration of salts when under the microscope.
By GEORGE ADAMS,
Mathematical Instrument-maker to his Majesty, and Optician to his Royal Highness the Prince of Wales.
Printed for and sold by the Author, at his shop, Tycho Brahe's Head, No. 60, Fleet-street.

7.4 An advertisement for *Essays on the Microscope* published in the *London Chronicle* for 21/23 February 1788. (*Author's collection*)

earlier in 1771, 'but I soon found that both his and Mr. Baker's tracts on the microscope were very imperfect. Natural History had not been so much cultivated at the period when they wrote, as it is in the present day'. He had tried to improve on previous works, he said, by arranging subjects in systematic order, and introducing the reader to the system of Linnaeus, as far as it related to insects.

How many of the microscopic objects described had actually been examined by Adams himself is uncertain. Almost all of his descriptions were taken from published writings by other named authors, as he was careful to point out, to avoid charges of plagiarism. His list of works consulted runs to sixty-three titles by fifty-one authors, from Adams (his father's *Micrographia Illustrata*) to Valmont de Bomare (*Dictionnaire Raisonné universal d'Histoire Naturelle*). Fifteen of the titles were published after 1772, and represent his sources of information on researches carried out since his father's death. For the Linnaean names of insects described he was, he said, indebted to 'Mr.Marsham', meaning Thomas Marsham (1743–1827), who in 1788 was one of the founders and first secretary of the Linnean Society and later its treasurer.[9] Unless there was an amazing coincidence of names, this was almost certainly Adams's brother-in-law (Hannah's brother Thomas).[10]

In chapters I–III, dealing with the evolution of the microscope as an optical instrument, Adams could write from personal experience, as these chapters were based on instruments sold in his shop or by his competitors. All of the principal types then current, simple and compound, are covered, but no mention is made of the 'variable' pattern introduced in 1770. This evidently had a short life. A possible explanation for this is that the design was due in part to the Dowager Princess of Wales, who allowed John Hill the free run of her gardens at Kew while he was writing *The Construction of Timber*. Her death in 1772 would have absolved George Adams junior from any obligation to continue making this model. Instead, the principal type of universal compound microscope described and illustrated in Adams's book had a compass joint between the top of the base pillar and the limb, where the 'variable' had a toothed wheel.

Extant examples of Adams's universal compound microscopes made in the 1780s and 1790s differ in detail. None exactly matching his own illustration is known, but several examples exist of a broadly similar model with the compass joint attached to a solid rectangular bar forming the limb. No provision was apparently made for fine focusing in any of George junior's compound microscopes, probably because (judging by extant examples) the objective lenses normally provided were of fairly low power, the minimum focal length being about 1/3rd of an inch. For large magnifications, Adams seems to have preferred to use low-power objectives with high-power (up to ×30) eyepieces, a combination which is not to be recommended on optical grounds. Dr S. Bradbury, who carried out some measurements on the performance of eighteenth- and early nineteenth-century microscopes in the 1960s, found that the particular Adams microscope that he examined was optically poor in comparison with contemporary instruments by other

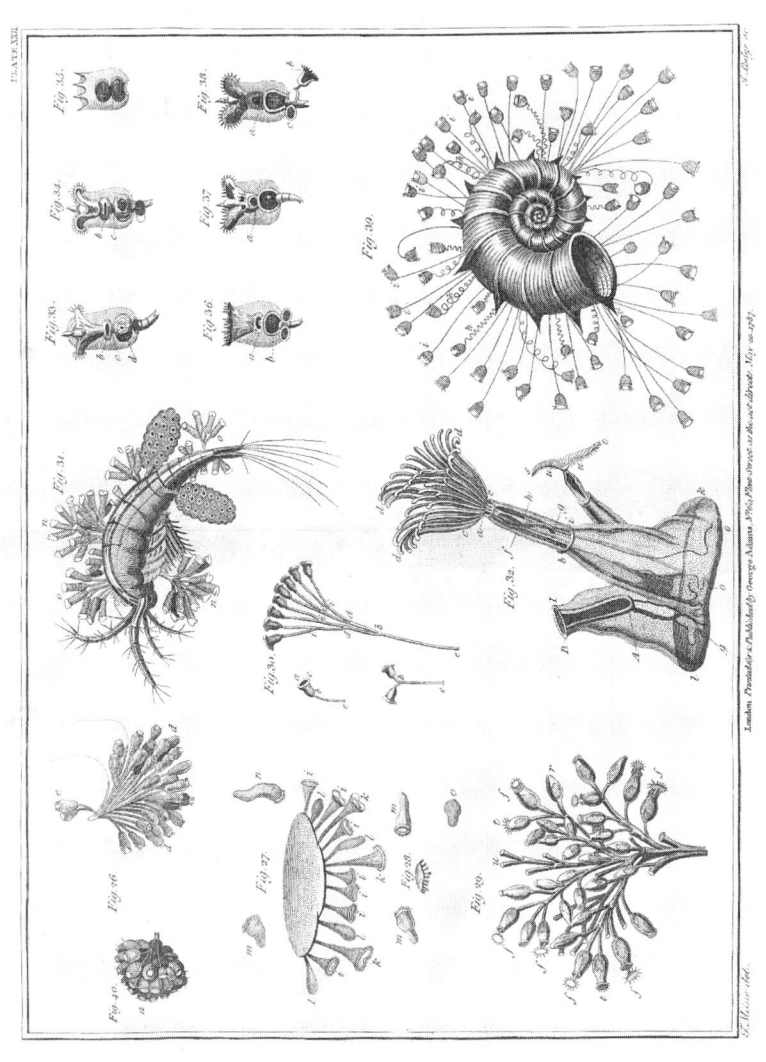

7.5 Plate XXII of *Essays on the Microscope*, drawn by T. Milne from illustrations of microscopic objects (various forms of Vorticellae) previously published elsewhere; for example, Figure 39 represents *Vorticella simplex campanulata*, 'as found by M. Roesel, fixed to a curious cornu ammonis'. Figures 33–8 are two species of *Vorticella urceolaris* discovered by Henry Baker. (*Author's collection*)

makers: the empirical addition of extra lenses in the body tube and eyepiece (to increase the power of the latter) reduced distortion and chromatic aberration, but increased spherical aberration to such an extent that the overall performance was worse than that of instruments in which no attempt had been made to overcome their optical shortcomings.[11] The image also suffered badly from glare, which had the effect of reducing contrast.

If these comments are difficult to reconcile with the fine detail shown in Adams's plates of microscopic objects, such as that reproduced in figure 7.5, several qualifying factors should be borne in mind (quite apart from the fact that the particular microscope examined may not be typical of Adams's output). In the first place the magnification employed to produce his illustrations was in most cases quite low: the range obtainable with Adams's numbered objectives was from ×15 to ×150 only. The plates in *Essays on the Microscope* are large in format because they were drawn to accompany a quarto text, not because a large size was necessary to show fine detail. In the second place, the reproduction technique (copper engraving) involved the engraver necessarily working with a sharp point: he had no means of showing areas of tone, or fuzzy edges, except by individual lines of varying thickness and spacing. Thirdly, the engraver was not working directly from the image produced by the microscope: in the case of a drawing signed by an artist, he was converting into a line engraving a drawing done by a professional artist based on sketches produced by an observer whose skill with the pencil is unknown. What the reader of *Essays on the Microscope* sees today is several stages removed from the actual image produced by the microscope.

The engraving of the sixty or so small plates for the first edition of *Micrographia Illustrata* (1746) was shared between J. Wrigley and T. Bowles, and possibly others as a few plates are unsigned. For about half of the much larger plates in *Essays on the Microscope*, George junior employed the artist T. Milne, who appears to have copied many of the individual figures on the plates from illustrations previously published elsewhere. Most of Milne's drawings were engraved by John Lodge, though one of instruments is signed by R. Laurie. Of the remaining large plates, ten are not signed at all. A medium-sized plate showing various forms of micrometer, signed 'Goodnight sc.', had previously been used (folded) in the fourth edition of *Micrographia Illustrata* (1771). Several of the large plates were produced by assembling two of the much smaller octavo plates from the first edition of

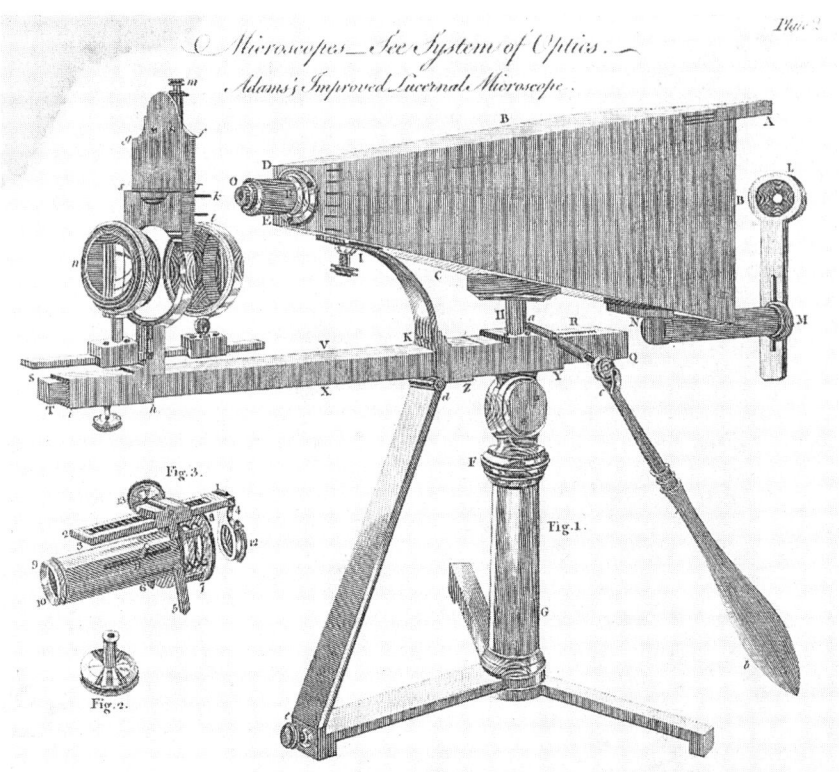

7.6 Adams's lucernal microscope for transparent or opaque objects, seen here in a plate from *Hall's Encyclopaedia* faithfully copied (but laterally inverted) from Adams's own plate III. In this form the stage assembly is mounted at the end of a substantial brass bar which can be moved longitudinally by a rack and pinion for focusing. To illuminate opaque objects, a lamp was placed in line with the bull's-eye condensing lens and the concentrated light thrown on to the front of the specimen by the adjustable mirror. (*Author's collection*)

Micrographia Illustrata (1746) side by side: these were more than forty years old when reused by George junior, presumably with his mother's permission, since all of his father's copperplates belonged to her.

Another new instrument described and illustrated in *Essays on the Microscope*, besides the universal compound microscope, was the 'lucernal' (projection) microscope. In the fourth edition of *Micrographia Illustrata* (1771) George Adams senior had shown a camera obscura fitted with a microscope in place of the usual single projection lens, to produce a real image of microscopic objects on a screen, using either the sun or a lamp as the light source. This made a rather unwieldy device. The version produced by George junior (see figure 7.6) consisted of a microscope permanently mounted with its optical axis horizontal, the space between the objective lens and a translucent viewing screen being enclosed by a pyramidical wooden box (or occasionally in some instruments a large brass tube) to exclude extraneous light. The box was mounted on a brass tripod foot, incorporating a substantial horizontal brass bar carrying the stage and its accessories. This bar could be moved horizontally relative to the box by means of a rack and pinion for focusing. Illumination was provided by an Argand lamp: this invention of a lamp with a hollow tubular wick and controlled draught, greatly increasing the light output, was what transformed the lucernal microscope from an interesting idea into a practical proposition. If the illumination from the lamp was nevertheless too weak to produce an image on the translucent screen (which might be the case with opaque objects), the screen could be taken out and the aerial image observed directly. A removable eye-location aperture was provided to assist the observer in placing his or her eye at precisely the right point. In this configuration, of course, only one observer could use the instrument at a time.

In the winter of 1792/3 the Revd John Prince of Salem, Massachusetts, wrote to Adams suggesting that it would be better to keep the stage stationary and mount the objective/box on a universal joint, to facilitate observation of living specimens without disturbing them. Prince (1751–1836) was an experimental philosopher in his own right who also acted as agent for British manufacturers' products in America; he was responsible for several improvements in scientific instruments, especially the air-pump.[12] John Hill, of Wells, Norfolk, the electrical experimenter mentioned earlier in this chapter, heard of this suggestion (though he did not know who had made it) and asked Adams to make him a lucernal microscope incorporating both this suggestion and a small improvement of his own. Adams duly complied, and Hill's instrument was delivered, with instructions for its use, not long before Adams died. In October 1796 Hill wrote to the *Gentleman's Magazine* saying that as he had not seen any mention of this improved model, he thought it ought to be described in that journal:

> I was the first person for whom he [Adams] made one on the new principle, and Mr. Adams's health declining soon after, makes it very probable that mine may be the only one that ever came from his hands; and, not having seen since his death any account given of the alteration alluded to, I deem the Gentleman's Magazine no improper vehicle through which to make it

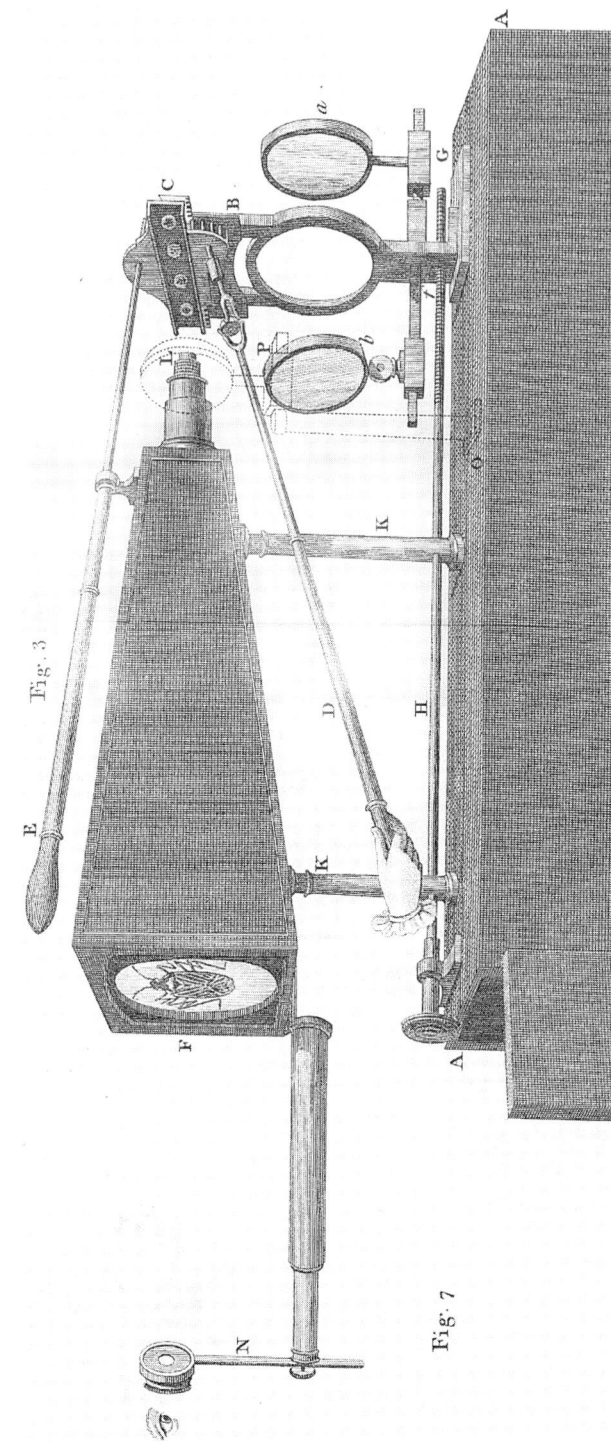

7.7 For the second edition W. & S. Jones deleted various small accessories from Adams's plate IX (some were transferred to other plates) to make room for this illustration, showing the lucernal microscope made in a more portable form with a wooden box foot. In this form it is still a stage-focusing instrument, but the body and stage assembly can be dismounted and packed inside the box for transport. Long-handled remote controls are provided for moving the specimen around relative to the objective lens. (*Author's collection*)

214

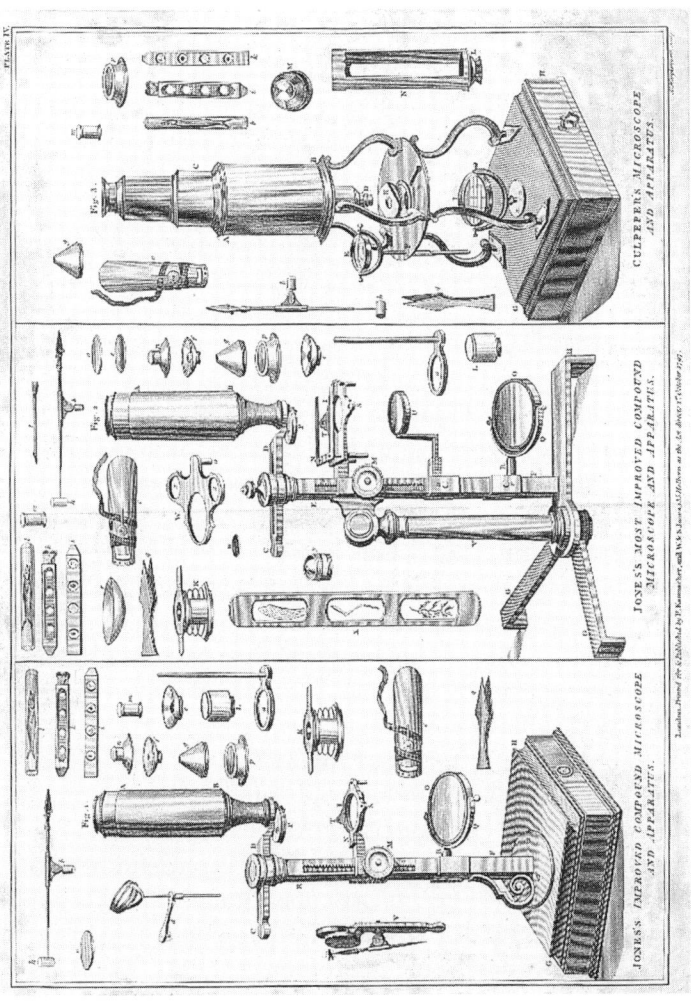

7.8 W. & S. Jones completely replaced Adams's plate IV, depicting the various forms of microscopes then made, to bring it up to date (1798). The most advanced compound microscope is shown in the middle illustration; it has a compass joint to provide adjustable inclination, a substantial rectangular bar limb incorporating a rack for stage focusing, and full aquatic movements to the body. Microscopes basically of this form had actually been made by George Adams junior for some time after publishing his first edition in 1787, but largely because of the legend on this plate they are known to collectors today as Jones's 'most improved'. (*Author's collection*)

known, together with a farther improvement of my own in the manner of applying the lamp, and which I had the honour of communicating to, and receiving Mr. Adams's entire approbation of.[13]

Hill's letter was accompanied by a sketch showing the box resting on a universal joint, and controlled in lateral and vertical movements by two screws at right angles at the screen end. It was also movable longitudinally for focusing. His own 'improvement' consisted only of mounting the lamp on an adjustable bracket attached to the stage assembly instead of having it free-standing. The whole self-contained instrument was mounted on a wooden baseboard, rather than the brass pillar and tripod shown in Adams's plate. An instrument basically of this form is illustrated in Clay and Court (1932);[14] it is said to bear a manuscript label (of unknown date) stating that the lamp was attached to it at the suggestion of Mr John Hill of Wells, Norfolk. It may, therefore, be the actual instrument made by Adams at Hill's request.

It is debatable whether Prince's suggestion constituted a real improvement or not. When the Jones brothers brought out a second edition of Adams's book in 1798, three years after his death, they included (as well as Adams's original plate) an illustration of the lucernal microscope as then made by them (see figure 7.7). This reverted to stage focusing, abandoning the universal joint suggested by Prince but retaining the idea of mounting the whole instrument on a wooden baseboard; this made it easier to pack into a wooden carrying case, the top of the case forming a convenient stand for the instrument when in use. (It was probably also cheaper to produce than Adams's brass tripod foot.) Several instruments of the Jones form are extant in museums,[15] and examples occasionally come on to the market.

Though they left almost all of the rest of Adams's plates unchanged, the Jones brothers replaced his plate showing universal compound microscopes completely. Their version is shown in figure 7.8. The instrument in the middle of this plate is what has become known to collectors as 'Jones's most improved' microscope, though in reality it had its origins in instruments made by Adams in the 1790s.

For updating Adams's text describing microscopic objects, the Jones brothers called in Frederick Kanmacher, MD, a fellow of the Linnean Society, so (unlike Adams's other works, which were edited by William Jones personally) the second edition of *Essays on the Microscope* appeared as 'with considerable additions and improvements, by Frederick Kanmacher, F.L.S.'. His name also occurs in the copyright notice on the redrawn plates. Kanmacher took the opportunity to add an 'advert' (that is, preface) of his own, which is signed 'Apothecaries Hall, Jan.1.1798'.

This second edition received a favourable eleven-page review in the *British Critic* for July 1798, which must have boosted sales of this work despite its considerable price of £1 8s. The book, said the anonymous reviewer in conclusion, 'stands unrivalled as a general and particular history of the microscope and its principal objects; and reflects the highest credit both on the author and editor'.[16] It remained a standard work on the subject until the introduction in the 1830s of objectives corrected for both chromatic and spherical aberration on Lister's principle, which opened a new chapter in the history of the microscope as a scientific instrument.

Astronomical and Geographical Essays

After *Essays on the Microscope*, Adams turned his attention to a replacement for his father's *Treatise on the Globes*, last reprinted (fifth edition) in 1782. Rather than reprint it again, he decided to rewrite the book completely, partly because he wished to arrange the 'problems' in a more methodical manner than his father had done, and partly to make the book more comprehensive. To this end he added an introduction to astronomy in general, before turning to the use of the globes and other instruments in particular; he also added a section on the use of observatory instruments. The resultant volume was published in 1789 with the omnibus title *Astronomical and Geographical Essays*. The introduction to astronomy formed essay I, the use of the globes essay II, the use of the planetarium and so on essay III, and an introduction to 'practical astronomy' essay IV. With an overall total of 665 octavo text pages (in the first edition) and twenty-one plates plus a frontispiece, this made a much more substantial volume than *Treatise on the Globes*. By having entirely new plates engraved, this approach made George junior independent of both his father's text and plates, now the property of his mother. The new plates, bearing the copyright date 1 September 1788 (except plate XX: 10 October 1788), were drawn by T. Milne and engraved by J. Lodge. A pictorial frontispiece depicting three ladies looking at a shooting star through a reflecting telescope, drawn by Burney and engraved by Laurie, bears the copyright date 10 October 1788.

The book went into a second edition in 1790, and a third in 1795, the year of George junior's death. The text was completely reset for these editions, resulting in changes to the numbers of pages, and a few sections were rewritten or otherwise amended, chiefly to incorporate the latest ideas on the structure of the universe arising from observations with Herschel's large telescopes. Before the third edition was published several of the plates were used to illustrate one of the volumes of *Lectures* (1794), for which purpose they were renumbered; consequently, when they were required again for the third edition of *Astronomical and Geographical Essays* in 1795, the order of the plates and their numbering was altered from previous editions. Also, some plates were omitted from this edition, and one new one was added. This needs to be borne in mind when the relevant plates are cited.

The globes and the problems illustrating their use in essay II were mounted in the manner introduced by George senior in 1766, though George junior's plate shows them in somewhat plainer stands than those originally provided for the king. In his 1766 *Treatise* George senior described how to use the terrestrial globe to demonstrate the seasons and related phenomena by setting the axis at an angle of 23½ degrees to the vertical. George junior retained this section, but added essay III describing the construction and use of a geared planetarium/tellurian/lunarium of the form now generally called a 'compound orrery'.[17] Separate plates showed it set up as a planetarium and as a tellurian, with the lunarium attachment inset (see figures 7.9 and 7.10). These plates show the 'drum' containing the wheelwork supported on three short metal legs to raise it off the table, but this is just artistic licence: all extant examples (of which there are many) have the standard brass pillar and folding tripod which was also used for small telescopes and other table-top instruments.

Models of this type, with teeth round the edge of the top plate which mesh with wheelwork on the attachments, evolved from an experimental version supplied by Benjamin Martin to Harvard College in 1766.[18] Martin's mechanical arrangement was not very satisfactory, however. In all later models the attachments incorporate a radial shaft with a contrate wheel and pinion at its inner end, driven from one of the coaxial tubes of the planetarium mechanism, the outer end having wheelwork to connect it to the 'rolling pinion' and the tellurian or lunarium gears. When the planetarium wheelwork is actuated by turning the operating handle, the 'rolling pinion' is caused to rotate and move the attachment round the central sun.

There are two versions of the tellurian attachment in extant instruments: with or without an extra dial on the arm. Examples of the two

Essays and Lectures 219

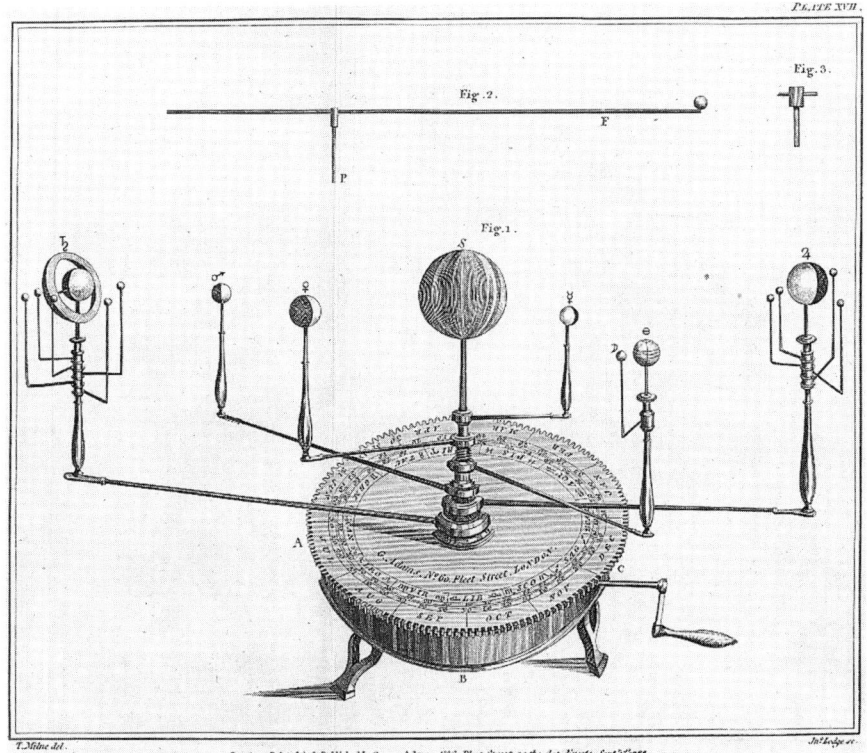

7.9 Plate XVII from *Astronomical and Geographical Essays*, showing a compound orrery assembled as a planetarium. In this configuration the teeth round the edge of the top plate serve no purpose, all the wheelwork being contained within the brass drum. The illustration shows the drum mounted on three short metal legs, but this is just artistic licence: all extant models of this type have the drum supported by a brass pillar on a tripod foot, the legs of which normally fold to enable the instrument to be packed into a flat case. (*Author's collection*)

versions are shown in figures 7.11 and 7.12. As one extant model with the extra dial is signed by Martin (and is therefore pre-1782), this is probably the earlier version, the extra dial having been found unnecessary by the time Adams's plate was engraved in 1788.

Broadly similar models by W. & S. Jones, and other makers or retailers of the early nineteenth century, are fairly common. Between *c.* 1790 and *c.* 1830 the all-brass compound orrery with two or more attachments, costing £30–£40, was the most expensive solar system model in commercial production, grand orreries being virtually obsolete by then.

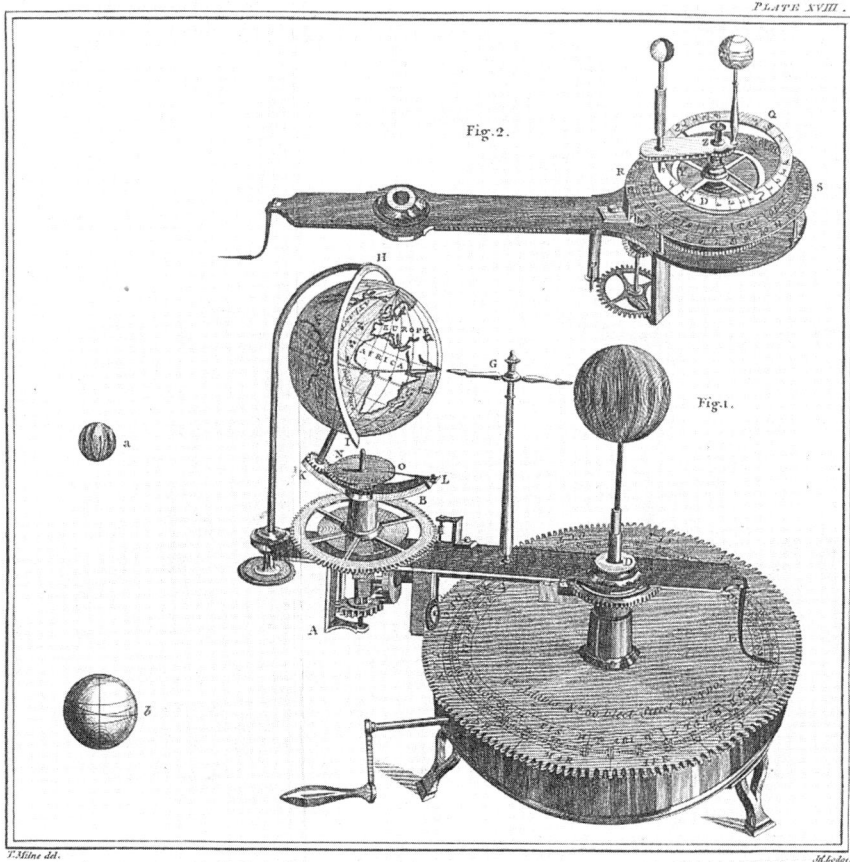

7.10 The compound orrery of figure 7.9 depicted with the planet arms removed and the tellurian attachment fitted in their place, demonstrating in detail the phenomena associated with the rotation of the Earth on its inclined axis, while maintaining its parallism as it moves round the sun. The alternative lunarium attachment, demonstrating the phenomena of the Earth/moon system (including the regression of the nodes of the moon's orbit), is shown separately above. Both attachments incorporate a wheel or pinion which meshes with the teeth on the top plate when fitted. (*Author's collection*)

The models shown in Adams's plates included several cheaper designs, such as the 'hybrid' form shown in figure 7.14, in which the outer planets are attached to coaxial tubes, with wheelwork hidden beneath the wooden baseboard, while the inner planets are actuated by exposed wheelwork on the combined Earth/moon arm. Such models

7.11 A compound orrery assembled as a tellurian. The top plate, which is 8¾ inches in diameter, is signed 'G*ADAMS, MATHEMATICAL Instrument-Maker, To His Majesty, Fleet Str^t LONDON'. For a full description see the Whipple Museum's *Catalogue 4: Spheres, Globes & Orreries*, item 33. (*Whipple Museum, Cambridge, Inv.Wh.1273*)

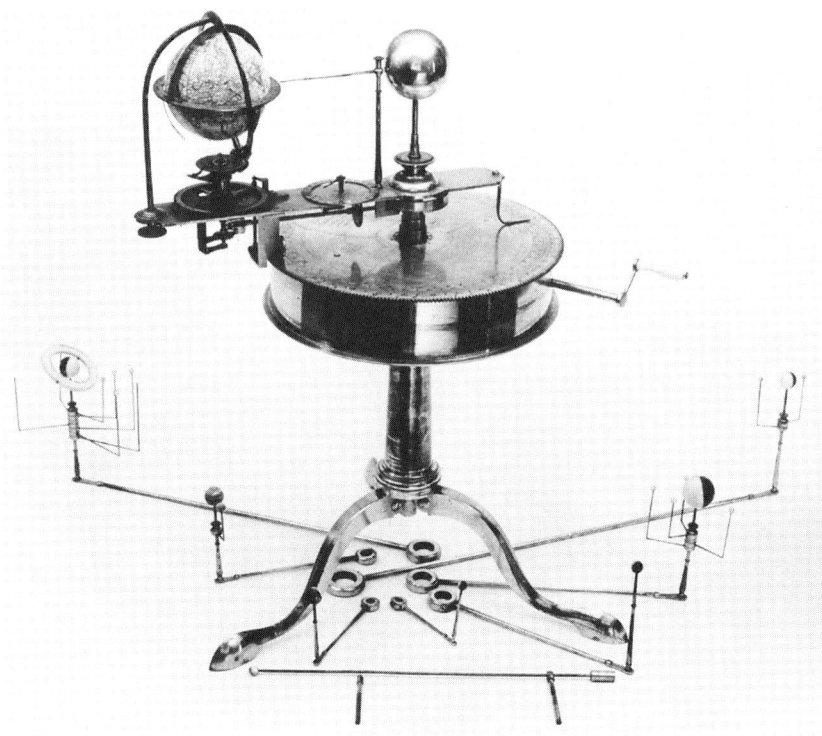

7.12   A similar instrument to that shown in figure 7.11, but with an extra 2 × 12-hour dial on the tellurian attachment. This instrument is fully described in F.R. Maddison, *A Supplement to a Catalogue of Scientific Instruments in the Collection of J.A. Billmeir, Esq.* (Oxford & London, 1957), pp. 94–5. (*Museum of the History of Science, Oxford, inventory no. 58762*)

cost less than half as much as the all-brass compound orrery, the actual price depending on how many motions were included (diurnal rotation of the Earth by wheelwork, for example, was an optional extra). Another of Adams's plates (see figure 7.15) showed an all-brass armillary sphere on an adjustable-inclination stand, copied from a plate in James Ferguson's *Lectures* first published in 1760.[19]

The last part of Adam's book, essay IV, 'An Introduction to Practical Astronomy', was also sold separately in 1795 as a self-contained work. It included an illustration and description of an equatorially-mounted refracting telescope, incorporating means for rigidly supporting each axis of rotation together with adjustments for correcting any deviation of the optical axis of the telescope from its desired position. (See figure

Essays and Lectures 223

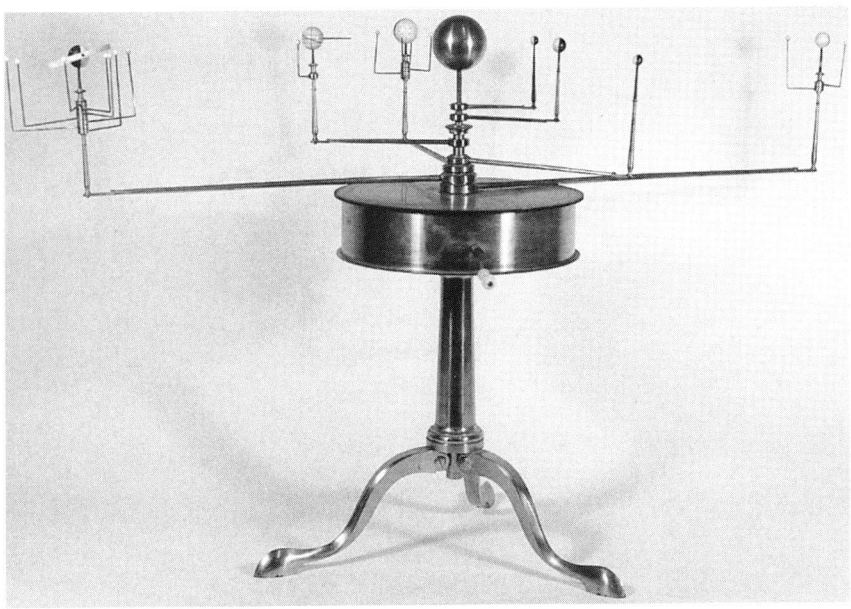

7.13 The compound orrery of figure 7.11 assembled as a planetarium. In this configuration the instrument shown in figure 7.12 is virtually identical. (*Whipple Museum, Cambridge, Inv.Wh.1273*)

7.16.) Full instructions for using the adjustments to set up the telescope formed the last section of Adams's book.

It was at about this time that Adams lent the proprietors of Hall's folio *Encyclopaedia* some of his globes, astronomical instruments, and microscopes, so that they could be drawn by the illustrators of that work (with his name prominently displayed, of course). A full-page engraving of an 'improved equatorial', dated 20 September 1788, almost exactly contemporary with figure 7.16, shows a different mechanical arrangement (figure 7.17), with the telescope and a large calibrated circle at opposite ends of the declination axis. Evidently Adams was experimenting with different designs at this time: in his correspondence with van Marum (chapter 8) he offered to make an equatorial mount with circles several feet in diameter on request. None by Adams as large as that is known, but figure 7.18 shows an extant equatorially-mounted telescope broadly matching the encyclopaedia plate, possibly the actual instrument that was copied by the draughtsman for the latter.

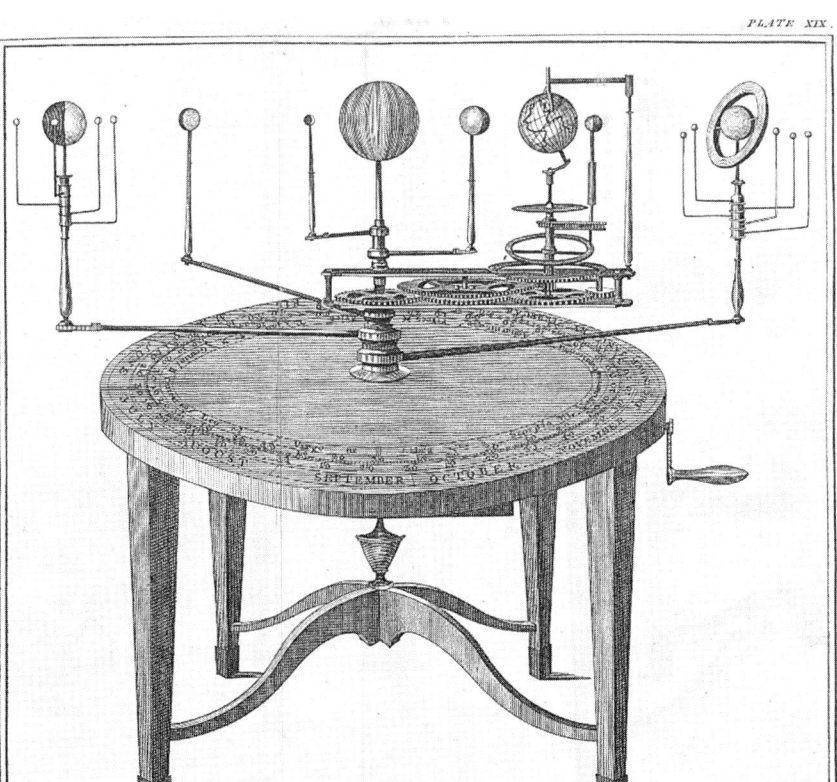

7.14 A hybrid orrery (combined planetarium and orrery), depicted here in Adams's plate XIX as a table model made mainly of wood with brass fittings. Similar models were also made with a wooden baseboard supported on short wooden legs. In the latter form they continued in production up to the mid-twentieth century. (*Author's collection*)

An Essay on Vision

From his remarks in the preface to this 153-page work (somewhat smaller than the foregoing *Essays*), it seems that Adams wrote it primarily because many people were using spectacles who did not really need them. By giving 'proper rules for ascertaining when spectacles are necessary, and how to choose them without injuring the sight' (part of his subtitle), no doubt he hoped to dissociate himself from pedlars and

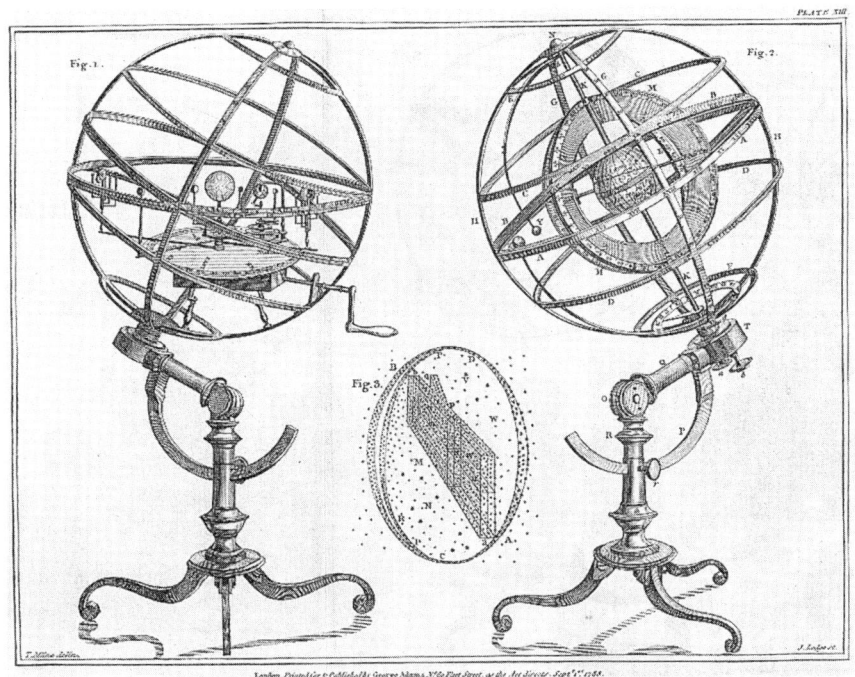

7.15 The basic armillary sphere with adjustable inclination depicted in this plate was copied from plate XX in James Ferguson's *Lectures*, first published in 1760. One matching the right-hand figure was supplied by George Adams junior to van Marum and is now in Teyler's Museum, Haarlem. None by Adams with a complete planetarium inside, as depicted in the left-hand figure, has been found. (*Author's collection*)

hawkers who sold unsuitable glasses with little regard for the harm they might do.

Like most of his writings, *An Essay on Vision* includes a list of authors consulted, in this case numbering fifteen, from Ayscough (1757) to Warner (1773). The most recent was Nicholson's *Introduction to Natural Philosophy* (1787), but in his preface Adams referred also to a paper by O'Halloran just published in the *Transactions of the Royal Irish Academy* (1788), which contained new information on the nature of the iris diaphragm. Though Benjamin Martin had died seven years earlier, Adams could not resist commenting unfavourably on the latter's 'Visual Glasses', even though, he said, 'they are [now] worn by few but those who, from long habit, have accustomed their eyes to these pernicious shades'.[20] Martin's 'Visual Glasses' not only had wide

7.16 This illustration in *Astronomical and Geographical Essays* (plate XXI, dated 1 September 1788) apparently depicts an actual instrument made by Adams (perhaps for experimental purposes) incorporating means for setting the axes truly at right angles. Full instructions on how to achieve this were given by Adams in his text. (*Author's collection*)

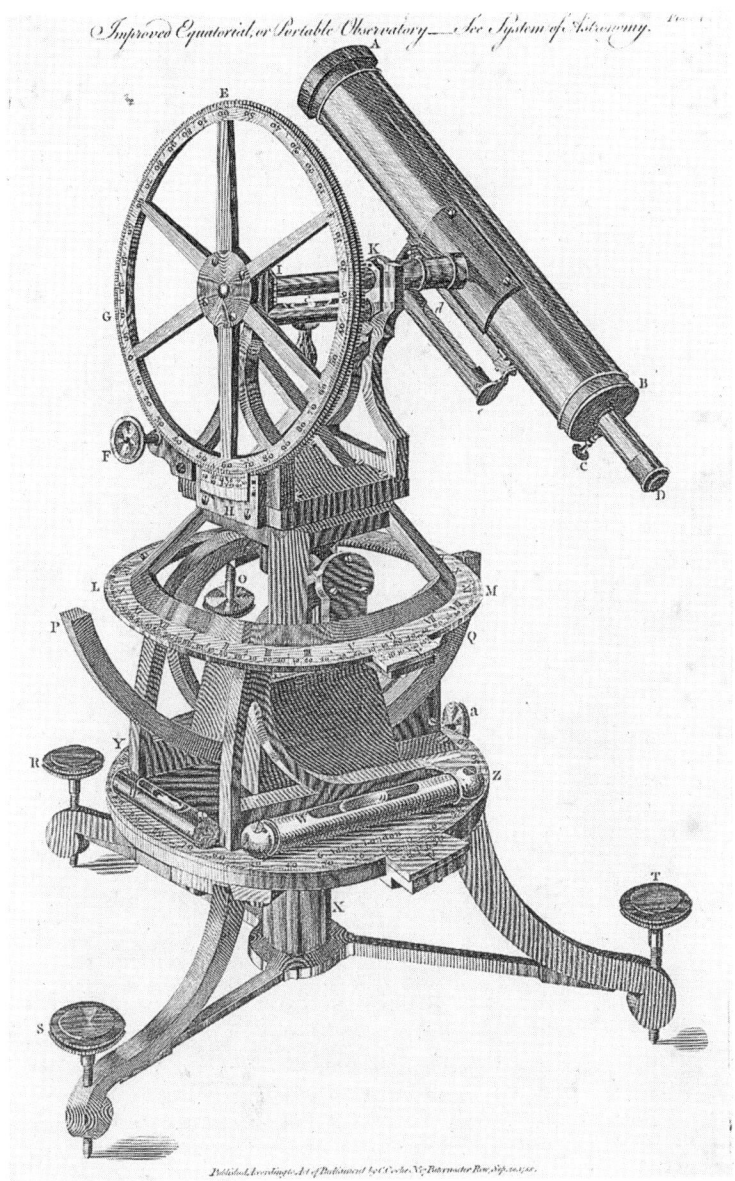

7.17 This engraving of an equatorially-mounted telescope does not appear in any of Adams's own books, but was published in *Hall's Encyclopaedia* with an 'Act of Parliament' copyright date of 20 September 1788, only three weeks later than that of figure 7.16. The legend 'G.Adams London' is just visible by one of the spirit levels. (*Author's collection*)

7.18 A universal equatorial mounting and telescope, signed 'Adams, London', with a long declination axis and circular declination scale, similar to the design depicted in figure 7.17. The diameter of the base is 7 inches, a size commonly used by the Adams firm for Ordnance theodolites. (*Private collection*) Earlier equatorial mountings by George Adams, based on the rather less stable mid-eighteenth century design by James Short, are located at the University of Bologna (formerly the property of Lord George Cowper), and the Harvard Collection, Massachusetts (acquired second-hand in 1803: see Wheatland, *Apparatus of Science* pp. 27–9).

horn rims to restrict the aperture, they also were generally made of coloured glass, violet being Martin's favourite colour, though some opticians considered green to be the least hurtful to the eyes (green being the most common colour in nature). Adams pointed out that 'though green is a pleasant colour to look at, it is by no means so to look through', as green-tinted glass distorted all other colours.

First published in 1789, *An Essay on Vision* went to a second edition three years later in 1792, and was translated into Dutch and German, but it does not appear to have been reissued by the Jones brothers after Adams's death.

Description, Use, and Method of Adjusting Hadley's Quadrant and Sextant

Hadley's quadrant, though clearly superior in performance to its predecessors and competitors (such as Caleb Smith's instrument made by George Adams senior at the start of his career), required careful manufacture and frequent adjustment if its capabilities were to be fully realized. It was also more costly than the relatively simple instruments, such as the backstaff, to which seamen had become accustomed. Many years elapsed after its first trials in the early 1730s before it became the standard instrument for navigators. Moreover, for obtaining the longitude from lunar distances a sextant (measuring angles up to 120 degrees), rather than an octant, was required, and if the instrument was not to become too unwieldy for use at sea its dimensions had to be reduced. This in turn necessitated greater accuracy in scale division, and it was really only when this became possible through the use of mechanical dividing engines that Hadley's instrument achieved lasting superiority. By the 1780s this situation had been reached, and Adams evidently felt that he should draw attention to his products in this field by including them in his literary coverage. By this time, of course, numerous other firms were producing or selling quadrants and sextants – it was a highly competitive field – and innumerable descriptions of the instrument and how to use it, by writers of varying competence, had been published.

The tract that he issued in 1789 was an octavo of seventy pages, printed by Hindmarsh, plus one folding plate. In a short preface, Adams said that although prior publications by Ludlam and Magellan had covered this subject in considerable detail, one was out of print and the other was written in a foreign language, so he felt justified in producing his own work, especially as 'the descriptions usually given by the instrument-maker with the quadrant are so imperfect'. His aim was to explain the principles of the instrument, and its adjustments, not merely to describe its parts, and also to demonstrate how it should be used to obtain the best results.

For numerical examples Adams relied on Nicholson's *Navigator's Assistant*, but he added practical hints throughout the tract which seem to indicate that he had acquired some experience in the use of the instrument at sea. So far as is known, he made no long sea voyages, but he certainly crossed the English Channel at least once (in each direction), which would have enabled him to try taking observations from the deck of a moving ship.

In a two-page catalogue of optical, navigational, and meteorological instruments (and globes) at the end of his tract, Adams listed:

	£	s	d
Hadley's quadrants in mahogany frames	2	2	0
Ditto, in black ebony frames	3	3	0
Ditto, on the best construction	4	14	6
Hadley's sextant in wood	6	16	6
Ditto in brass, on the most improved plan, from £11 11s. 0d. to	15	15	0
Circular instruments, to answer the purpose of the sextant	14	14	0

Bearing in mind that Hadley's quadrants were made in a variety of sizes as well as materials, it is not surprising that one seldom finds two examples exactly alike today.

A Short Dissertation on the Barometer, Thermometer, and other Meteorological Instruments

Issued in 1790, this, Adams admitted, was a 'hastily written tract' (of sixty pages), but he hoped it would prove useful to his customers. Other commitments had prevented him devoting the time to the subject that he would have liked.

It was only by assiduous observation of atmospheric changes, he said, that we could ever hope to accumulate sufficient knowledge of atmospheric phenomena to enable us to predict 'the changes of the elements, with as much certainty as we now do those of the planetary bodies'. He urged everybody who possessed meteorological instruments to keep a daily diary, and 'transmit the result of his observations to the public'. By meteorological instruments he meant: barometer, thermometer, hygrometer, rain gauge (Adams consistently spells this word 'guage'), electrometer, anemometer, and evaporation gauge.

With regard to the 'common barometer', he pointed out that one frequent source of error was the variable height of the mercury in the cistern:

> To remedy this evil, it is necesary that the lower surface [i.e. the mercury in the cistern] should always be kept at the same height from the divisions on the scale fixed to the instrument. This is effected by means of a floating guage, which was first applied to the barometer by MY FATHER, though others since his time, assumed the merit to themselves ... This guage is never applied to the common portable barometer, but only to those of the best kind.[21]

For accurate measurement of the height of the mercury column, Adams's 'best' barometers were provided with a nonius reading to 1/100th of an inch, and in order to ensure that the index was accurately aligned with the surface of the mercury it was made double, one piece being behind the tube, thereby eliminating parallax errors. This improvement Adams attributed to Ramsden. These 'best' barometers had a temperature correction scale, for use if the ambient temperature differed appreciably from 55 degrees Fahrenheit. Presumably this was thought to be the long-term mean for London, though Adams did not say so.

On the subject of hygrometers, Adams pointed out that a great deal could be deduced about the state of the atmosphere even without instruments, from the appearance of the sun and moon. For measurement of the moisture content, the hygrometer of De Saussure, using a human hair, was preferable to that of De Luc, though it was difficult to make. It would be considered in detail, Adams said, in his 'Meteorological Essays'. This never materialized as a separate work, though the subject was eventually covered in his five-volume *Lectures*.

Adams clearly regarded his best meteorological products as instruments for scientific research, rather than domestic ornaments, though the range of prices quoted in a short catalogue at the end of his tract indicates that he supplied the domestic market as well. His cheapest 'plain portable barometer' cost only 2 guineas, while a barometer with a long cylindric thermometer and a hygrometer cost 7 guineas. The most expensive instrument listed was a 'barometer for measuring the heights of mountains', at 9 guineas. Adams devoted several pages to the method of calculating heights by barometric observations, with corrections for temperature derived from tables by Sir George Shuckburgh.

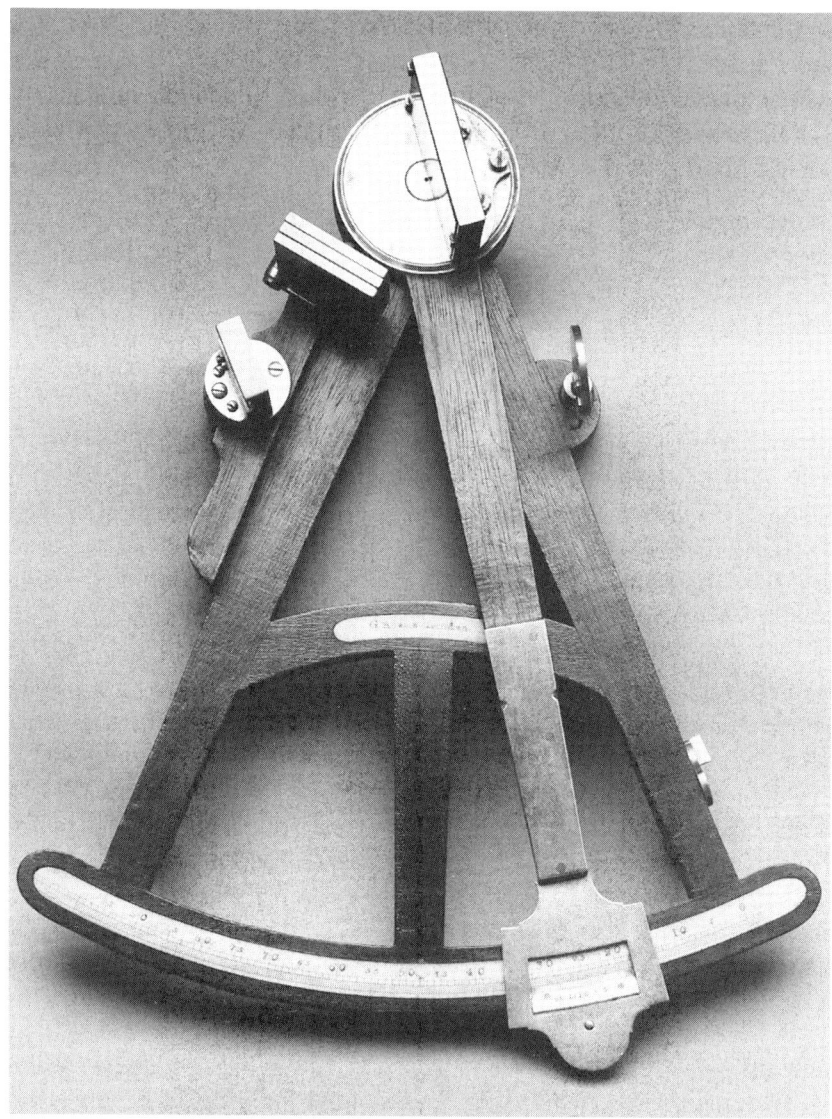

7.19 A small wooden octant signed 'George Adams, London' with an index arm of radius 8¾ inches. The ivory scale has the 'fouled anchor + I.R.' mark, indicating that it was divided on Ramsden's engine, with divisions every 20 minutes of arc (see figure 7.20). A vernier enables readings to be taken to one minute of arc. The combination of small size, wooden construction, and engine division is unusual, and suggests that this was made to a special order for someone to whom portability and accuracy were equally important. (*Private collection*)

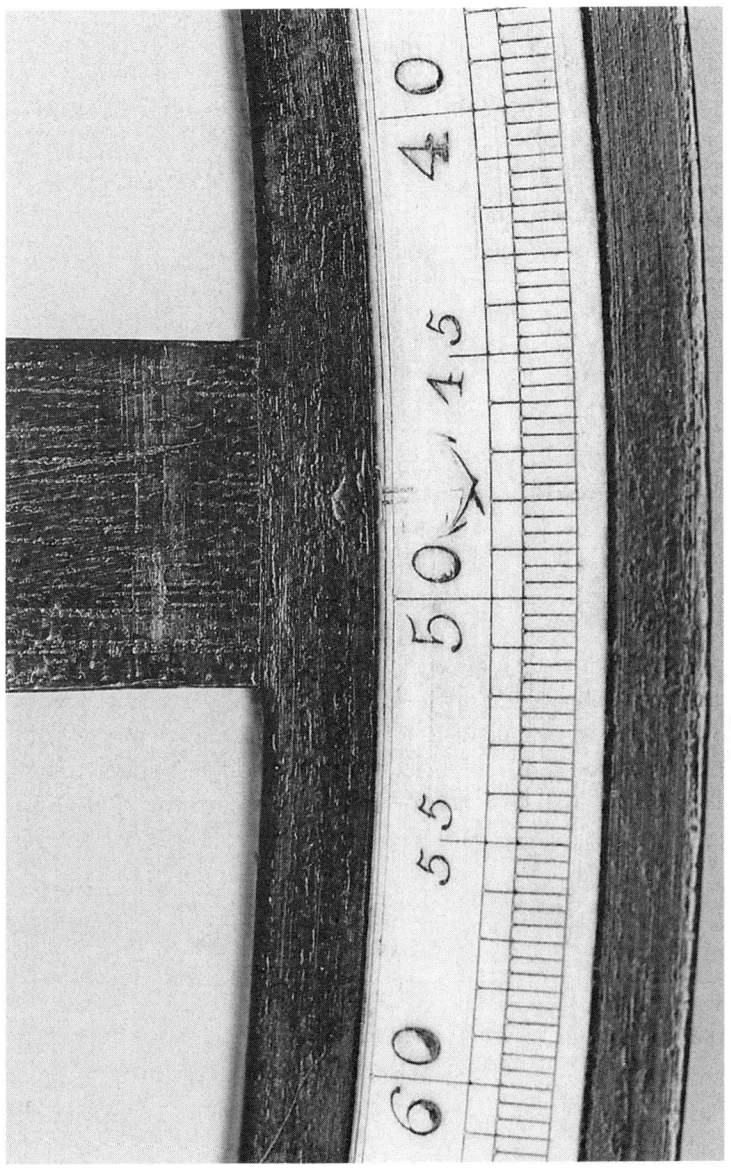

7.20 Part of the scale of the small octant shown in figure 7.19, engraved on ivory. Though not clear in this reproduction, on the instrument itself the fouled anchor and 'I.R.', indicating that the scale was divided on Ramsden's engine, can be made out. For an example of a degree scale engraved on brass, see figure 7.28. (*Private collection*)

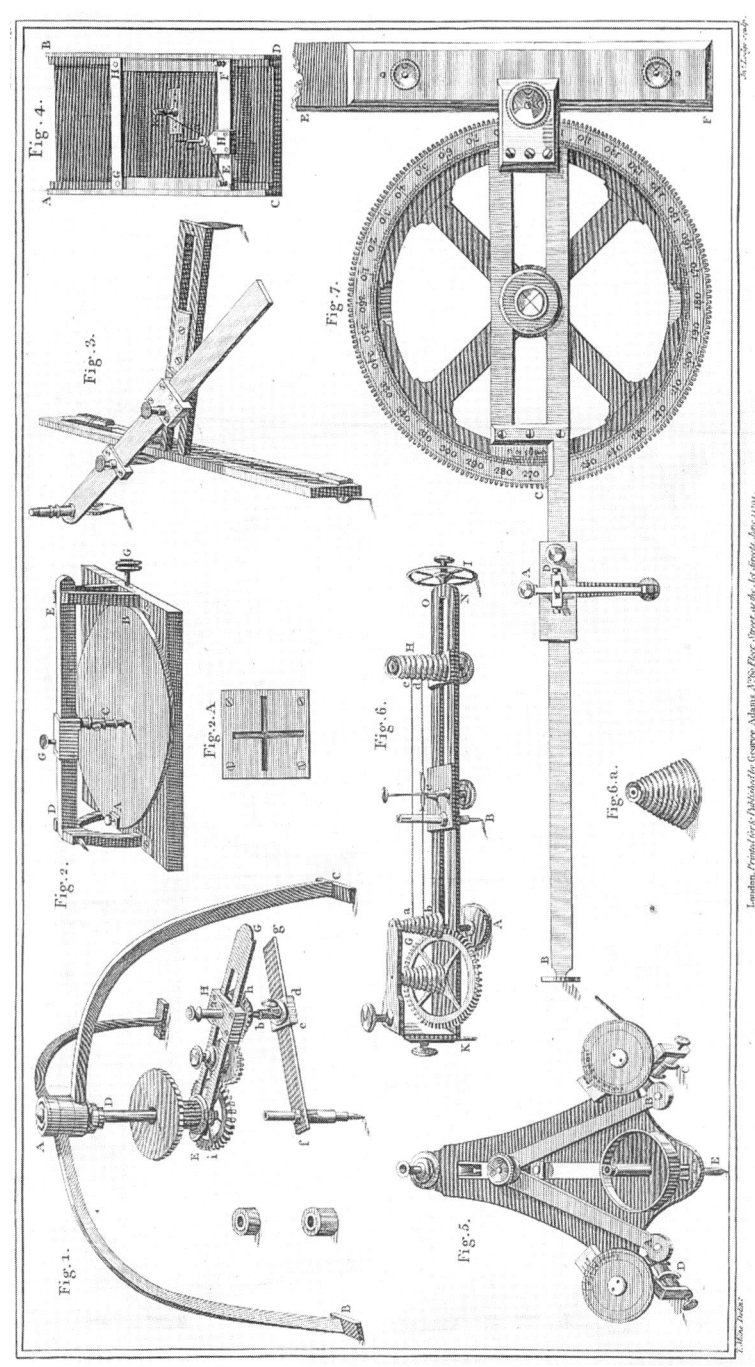

7.21 Some complex drawing instruments for geometrical curves, illustrated in plate XI of *Geometrical and Graphical Essays*. Heywood's instrument for drawing arcs of large radius is shown at figure 5. Suardi's geometric pen, which is capable of drawing a wide range of curves by changing the gear ratios, is shown at figure 1. Figure 6 is a device for drawing spirals. Figure 3 is a trammel for drawing ellipses; this device, unlike the others shown in this plate, is often found in magazine cases of drawing instruments. (*Author's collection*)

Geometrical and Graphical Essays

The subtitle of this work, published in 1791, states that it contains a description of the mathematical instruments used in geometry, civil and military surveying, levelling, and perspective; but Adams could with justification have added 'and their use', as his text went well beyond a mere description, explaining in detail how each instrument was used in practice. The drawing instruments covered were not restricted to those commonly found in pocket cases: they included specialized devices such as ellipse-drawing and spiral-drawing instruments, architectural sectors, and aids to drawing in perspective. (See figure 7.21 for some examples.) The surveying instruments included all types used in land, maritime, and military operations, from a simple magnetic compass to the most advanced theodolites. Such extensive coverage, Adams claimed in his preface, had not been attempted since Stone translated Bion's French work on the subject in 1723.[22] He appended a list of twenty-eight authors whose works (in Latin and French as well as English) he had consulted, ranging in date from Clavius' *Astrolabium Tribus Libris Explicatum* (1611) to Hutton's *Treatise on Mensuration* (1788).

Much of the contents of this 500-page work (with thirty-two plates, many of large size, plus a frontispiece) was relevant to the activities of the draughtsmen and engineers employed by the Office of Ordnance, and no doubt with that market in mind Adams dedicated his book to the Duke of Richmond, master-general at the time. In his preface he particularly thanked Mr Landmann, Professor of Fortification and Artillery at the Royal Military Academy, Woolwich, for supplying him with details of the 'course of practical geometry on the ground' conducted there (see figure 7.22), which was not only useful to the military officer but also 'would make a useful and entertaining part of every gentleman's education'. Other contributors personally mentioned included Mr Gale, whose plotting tables were printed in an appendix, available separately for the benefit of users of that method.

The terminology of surveying instruments remains confused despite attempts in recent years, particularly by staff at the Whipple Museum, to define the various types more precisely. In increasing order of complexity and cost, the following four categories of angle-measuring instruments (other than those for fixed angles, such as optical squares) were distinguished by Adams and shown in his plates:

a) The common circumferentor, consisting of a compass mounted at the centre of a long bar with plain sights at each end, the only angular scale being on the compass itself (figure 7.23). This, Adams said, was not much used in England, where land is valuable, but was useful in America for surveying large tracts of ground. In an attempt to improve the accuracy of reading some instruments were provided with an index and nonius inside the box.

b) The common theodolite, consisting of a large horizontal circular scale and two pairs of plain sights: one pair on a fixed diameter, as in a circumferentor, and the other on a movable diameter (figure 7.24). With this instrument readings are taken off the circular scale, the compass in the centre serving only to relate a set of measurements at a station to the magnetic meridian.

c) The second-best theodolite, consisting of a telescopic sight movable about a horizontal axis carried on a turntable on a vertical axis. As in the common theodolite, a compass at the centre of the instrument serves only to relate a set of readings to the local magnetic meridian, angular measurements in both the horizontal and the vertical planes being taken from degree scales centred on their respective axes.

d) The theodolite as improved by Ramsden, consisting of an instrument with telescopic sights broadly similar to the second-best theodolite, but with refinements of construction to improve its accuracy, and having a second telescopic sight underneath the horizontal circle (figure 7.25). The second telescope could be set to observe a reference object while measurements were taken with the main instrument. Theodolites of this pattern, said Adams, were normally made with horizontal circles of 6, 7, and 8 inches diameter, but generally of 7 inches.

These were not the only types of surveying instruments made, of course. As Adams himself pointed out, there were many variations designed to suit individual purposes (and pockets); consequently extant examples seldom match his illustrations exactly. For customers who wanted a telescopic sight but could not afford the second-best pattern, a slightly simplified version was produced which had a pivot instead of a long axle for motion in the vertical plane. The provision of rack-and-pinion controls for all adjustments was taken for granted on the best instruments, but the price could be reduced substantially if they were omitted (figure 7.26). Smaller 'portable' models with a limited range of vertical movement were also made (see figures 7.27 and 7.28).

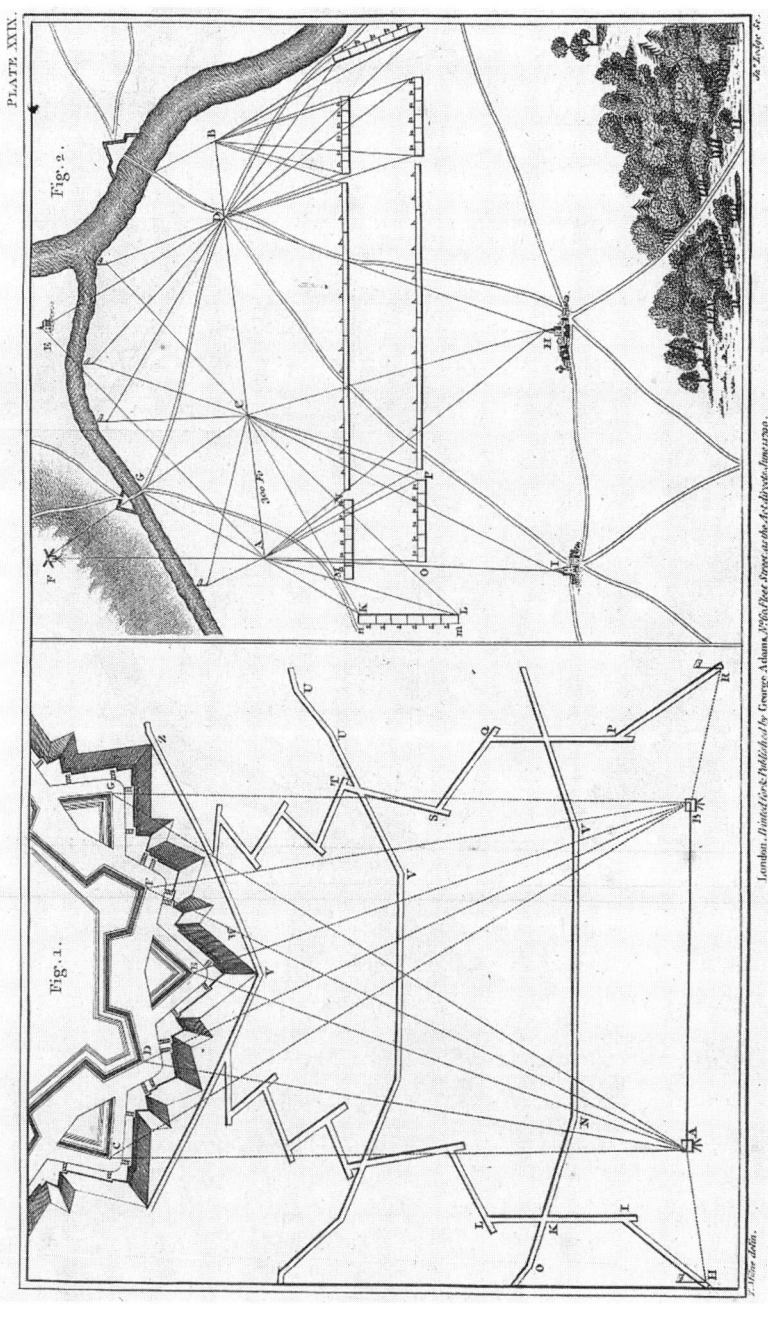

7.22 Two examples of 'The use of the Plain Table in Military Operations', communicated to Adams by Professor Landmann of the Royal Military Academy. The angles of a fortification are carefully planned to provide covering fire, and the angles of an approaching system of trenches have to take this into account. The right-hand diagram shows how to take the plan of an encampment. (*Author's collection*)

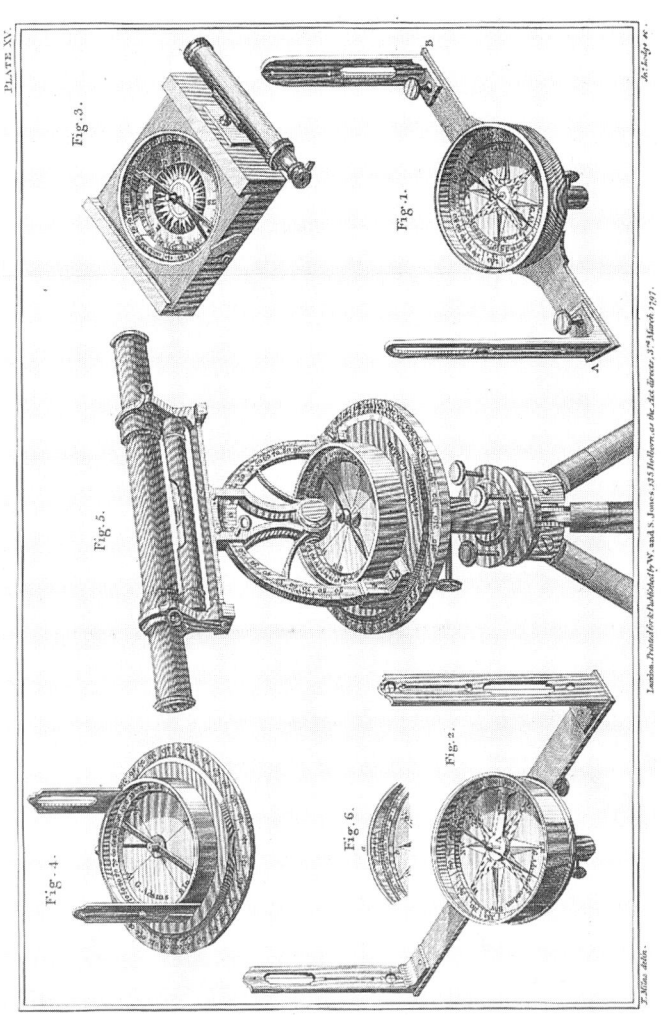

7.23 Plate XV from *Geometrical and Graphical Essays*, showing (at left and right) circumferentors; (centre) an inexpensive telescopic theodolite with a vertical arch but no rackwork, and a pivot rather than an axle for elevation movement. The circumferentor shown at the left (figure 2) incorporates an attempt to improve the accuracy of angular readings by providing an index AB and degree scale within the box, independent of the compass. This effectively turns the instrument into a 'simple theodolite' but with an angular scale limited in diameter to that of the compass box. In the second edition W. Jones said in a footnote that the index AB had been found to interfere too much with the compass needle; he provided instead an external nonius, shown at figure 6, added by him to Adams's plate. This reproduction is from Jones's edition (1797). The other figures are unchanged, retaining Adams's name on the instruments depicted. (*Author's collection*)

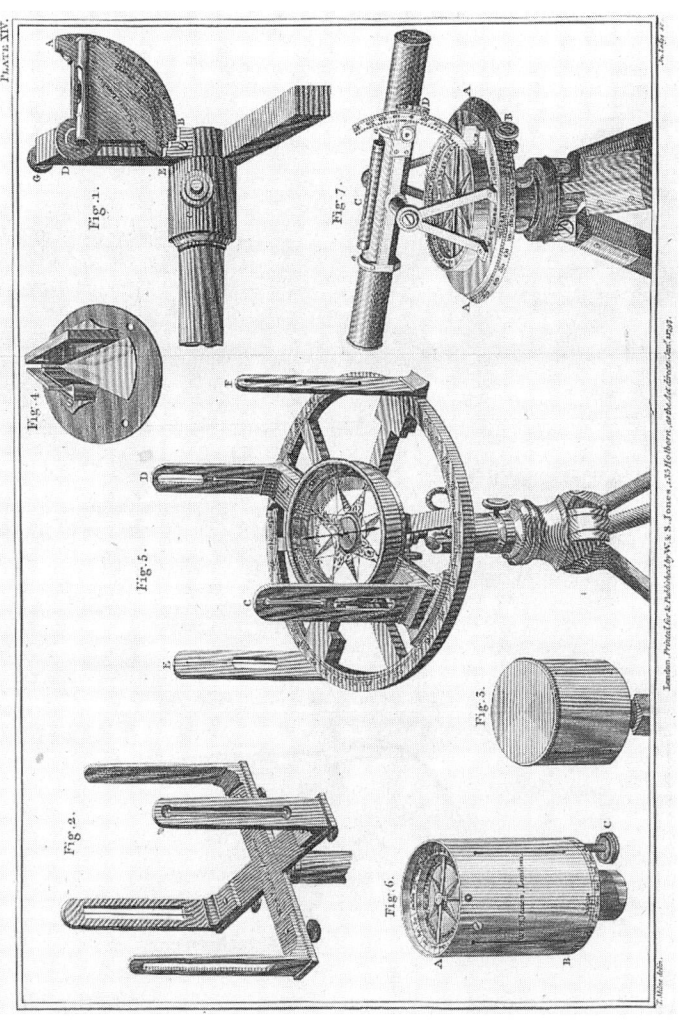

7.24 Plate XIV, from the second edition (Jones, 1797), showing in the middle a 'Ten inch common theodolite' with 'Jones London' on the compass card in place of Adams. In this instrument horizontal angles are measured on the large graduated circle (or vertical angles, if the instrument is suspended like an astrolabe), while the magnetic compass serves to relate the set of measurements to the magnetic meridian. Instruments of this type are sometimes catalogued as circumferentors, but the eighteenth-century terms were 'simple theodolite' or 'common theodolite'. The figure in the lower right-hand corner (figure 7) shows a small telescopic theodolite with rackwork but a limited range of elevation movement. Adams referred to a 'miniature theodolite' in a footnote but did not describe or illustrate it, so Jones added this illustration to Adams's plate, saying that it was about 4 inches in diameter, with a telescope 6 inches long. (*Author's collection*)

239

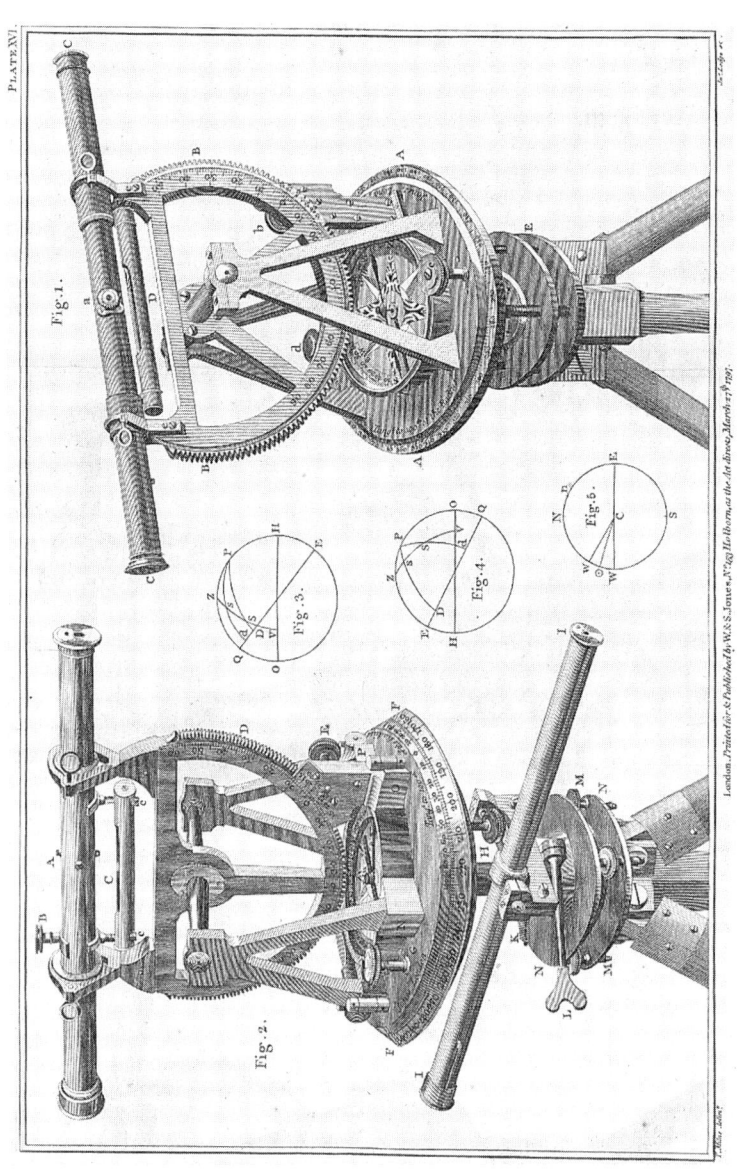

7.25 Plate XVI, showing theodolites of what Adams called the 'best' form of construction, with geared controls for both horizontal and vertical movements, and a reversible telescopic sight. That on the left has an auxiliary telescope (which Adams attributed to Ramsden) for checking the back station while measurements are taken with the main telescope. Both instruments are shown with 'parallel plates' between the top of the tripod and the base, to enable the latter to be set accurately horizontal. This reproduction is from the Jones edition (1797), with 'Jones London' on the instruments depicted. (*Author's collection*)

7.26 A 4-inch theodolite with limited elevation movement signed 'Geo. Adams London' matching a description by William Jones in the 1797 edition, and the illustration (see figure 7.24) which he added to Adams's plate XIV. (*Private collection*)

For the frontispiece in his first edition Adams provided an illustration of a more complex instrument based on Ramsden's two-telescope design, but with the whole instrument perched on what amounted to an equatorial mounting, in addition to having the usual horizontal and vertical axes. In his preface he said that this (experimental) instrument was not finished in time for it to be inserted in its proper place, so he devoted several pages to describing the method of adjusting it there. However, the number of axes incorporated in the mounting seems excessive, and it is doubtful whether this design was ever used in practice; it was omitted entirely from later editions by W. & S. Jones.

Chronologically, this was the first of Adams's publications to be revised and reissued by the Jones brothers after they had acquired the stocks and copyrights in 1796. A second edition 'corrected and enlarged by William Jones' appeared in 1797. Jones listed his principal

7.27 The 4-inch theodolite shown in figure 7.26, packed in its carrying case. (*Private collection*)

7.28 Detail of the signature and main scale divisions of the 4-inch theodolite. A nonius (just visible at the right of figure 7.26) enables readings to be taken to 1/20th of a degree (3 minutes of arc). Jones said in his description that if the observer did not object to very fine divisions, a nonius reading to 2 minutes of arc could be provided instead. (*Private collection*)

244 George Adams Junior

7.29 The dial of a perambulator by George Adams indicating miles, furlongs, poles, and yards. The smaller divisions on the outer scale each represent one yard, and the larger ones one pole (5½ yards); those on the inner scale represent one furlong (40 poles or 220 yards). A complete circuit of the inner scale represents 10 miles. This form of scale division corresponds with that described by George Adams junior in *Geometrical and Graphical Essays*. The associated wheel had a circumference of half a pole, or 8 feet 3 inches. Adams distinguished between a pedestrian 'perambulator', constructed as above, and a 'waywiser' fitted to a carriage, with a dial indicating the number of turns made by the wheels. The latter instrument had to be calibrated to suit the particular carriage in which it was mounted. (*Christie's, London, 14 December 1989, lot 151*)

additions as: a description of a new pair of pocket compasses, containing the ink and pencil points in its two legs; improved perambulator; way wiser; improved surveying cross; improved circumferentor; complete portable theodolite; great theodolite by Ramsden; pocket box sextant; artificial horizon; pocket spirit levels; a pair of perspective compasses; Keith's improved parallel scale; new method of surveying and keeping a field book; gunner's callipers; gunner's quadrant; gunner's level, and so on.

The last three items, which Adams (surprisingly, in view of his Ordnance connections) did not mention at all, were amongst those illustrated in two new plates added by the Jones brothers, bringing the total of the numbered plates up to thirty-four. Adams's frontispiece was scrapped and replaced by one showing Ramsden's 'Great Theodolite' for the Ordnance survey of England. Many of the plates depicting instruments were partially re-engraved to accommodate the new ones described by Jones; these revised plates (which have Adams's name on the instruments erased and replaced by W. & S. Jones) bear copyright dates from December 1796 to April 1797. Jones's additions to the text, and altered plates, make the second and subsequent editions appreciably different from the first. Two further editions, dated 1803 and 1813 respectively, were issued by the Jones brothers with only minor alterations.

Lectures on Natural and Experimental Philosophy

Adams's last and most ambitious venture, issued in five volumes (four of text and one containing the plates and a comprehensive index) in 1794, must have occupied much of his leisure time during the last two or three years of his life, when he was in poor health. Uncertain, perhaps, whether this multi-volume work would meet with the same success as his individual essays, he chose to publish it by subscription, a method which enables the probable sales to be determined before printing starts. Advance notices of his intention appeared in several of his publications during the early 1790s. Three booksellers as well as Adams himself took in subscriptions, at a pre-publication price of £1 4s, though the title-page simply states that it was 'sold by the author'. He need not have worried about the possibility of failure to find buyers: the list of subscribers printed in the first volume shows that over a thousand copies were sold in advance of publication, which must have brought in a subscription income of at least £1,200.

Dedicated by permission to the Princess Royal, *Lectures* was intended to be not merely a statement of physical principles, but also 'a display of the divine goodness, wisdom, power, and order, manifested in the works of creation'. In a thinly-veiled reference to the French Revolution, which had just resulted in the execution of Louis XVI, Adams assured the princess that 'true philosophy' does not undermine or assault 'the fair structure and mutual dependence of civil society, by infusing the spirit of ambitious discontent, and introducing the principles of levelling authority'. It would be his earnest wish that her father's realms should flourish in prosperity and peace, 'neither seduced by false philosophy, nor convulsed by democratic violence'.

A strong religious theme pervades Adams's approach to his subject, coupled with a horror of what was happening in the name of natural philosophy on the Continent, especially though not exclusively in France. The opening remarks in his preface reveal that he had visited France and Switzerland about twenty-five years earlier – long before the French Revolution – and even then had been alarmed by the atheistic approach to philosophy that he had found to be prevalent on the Continent. He had there and then resolved to compile a comprehensive work on the subject based on religious principles, to counter the continental atheistic approach, but the pressures of business life had intervened and it was only in recent years that he had managed to find the time to realize his ambition.

If 'twenty-five years' was accurate, George Adams junior must have visited France and Switzerland around 1768/9, when he was aged eighteen or nineteen and about two-thirds of the way through his apprenticeship. This period coincides with the time when George Adams senior was apparently absent from London and unable to take up his appointment to the court of the Grocers' Company (chapter 5). Perhaps George senior personally escorted an important order to a continental customer, and took his son (and perhaps Dudley as well) with him as part of his education. Elaborate grand orreries by Adams are currently located in Brussels and Geneva;[23] they are undated, but are thought more likely to have been supplied by George senior than his son. One of these could well have required his personal attention to ensure that it was correctly set up.

With over 2,200 pages of text and thirty-nine folding plates, obviously the contents of *Lectures* cannot be summarized in a few paragraphs. The contents of volume IV (astronomy, electricity, and meteorology), and parts of volume II (vision, and optical instruments) duplicate to some extent Adams's previously published essays on these subjects, and reuse some of his plates in these earlier works, but he also drew upon (with acknowledgements) publications by numerous other authors to expand his coverage. In his preface he particularly acknowledged his indebtedness to De Luc, the Revd William Jones, FRS (not William Jones the optician), and 'my friend the Rev.Mr.Agutter, MA'. His list of cited sources runs to about fifty titles by over forty authors. All but a handful (for example Desaguliers, 1744; Cotes, 1738) were published after 1772 and could not have been in his father's library: he must have purchased and studied these himself.

While the sections on optics, mechanics, hydraulics and pneumatics, astronomy, and electricity were directly relevant to his instrument-making business, and in modern terminology would be classified as 'pure and applied physics', in volume I Adams ventured into a different field which today would come under the heading 'chemistry'. His writings on this subject have an extraordinarily old-fashioned flavour, treating fire as an elemental substance and denying most of the recent continental discoveries on chemical composition and reactions. He was evidently a firm believer in phlogiston, and regarded Lavoisier's discovery of oxygen as a misinterpretation of his experimental results, which would soon be recognized as erroneous by the rest of the scientific world. Whether Adams's dismissal of the chemical experiments performed in France was based on genuine disbelief, or was simply part of his general antipathy towards everything emanating from an atheistic regime, is difficult to say. In this respect his attitude had much in common with King George III's on pointed lightning conductors (they could not be of any use, as they had been invented by an American).

The plates in *Lectures* bear the copyright date 4 December 1793, except for the allegorical frontispiece, which is dated 1 January 1794.[24] A news item under the heading 'Domestic Literature' in the *British Critic* for May 1794 (p. 599) announced that Adams would publish his lectures in the course of the next month, with the comment (possibly provided by Adams himself): 'One great Object with this Author has been to oppose the Phaenomena of Nature to the false Philosophy of the French Atheistical Writers.'

Analysis of the subscribers list shows that over half of those who gave addresses were located in what may be loosely called the 'London area', stretching from Hampstead to Dulwich and from Shepherd's Bush to Greenwich. Surprisingly, the next largest concentration was in Manchester. Examination of the Manchester names reveals that amongst them was the Revd Mr Clowes, a prominent member of the Swedenborgian movement there, so it is likely that the number of Manchester subscribers was swelled by local Swedenborgians aware of both Adams's and Hindmarsh's connections with the movement. A total of forty-five copies went to America, three subscribers there taking twelve copies each, including the Revd John Prince of Salem.

Hindmarsh's *New Magazine of Knowledge* had ceased publication in October 1791, too soon for *Lectures* to receive a mention therein. A long and highly complimentary review appeared in the *British Critic*, but not until August 1795, over a year after publication. As Adams died (at Southampton) on the 14th of that month, it is doubtful whether he ever saw these published comments on his *magnum opus*. In a preface added to the 1795 volume of this magazine at the end of the year, the editor mentioned Adams's untimely death and added that: 'Our commendations ... are now of no importance to him; but we renew them, because they are gratifying to ourselves, and may be useful to others. The praise of one is often the incitement to many.'

The Jones brothers issued a second edition of *Lectures* in the same five-volume format in 1799, with some minor editorial amendments by William, plus longer additions inserted in the form of appendices to lectures XIV, XVI, XXII, XLII, and LII. These additions, Jones said, were mainly of an experimental kind, Adams's theoretical text being left untouched. They included, for example, an enlarged description of the compound orrery with tellurian and lunarium attachments, and of Ferguson's 'whirling table' or centrifugal force demonstration apparatus.

This 1799 edition attracted the attention of Professor Robert Patterson of Philadelphia, who had already produced American editions of several of James Ferguson's popular scientific works. In 1806 he edited the Jones version of Adams's *Lectures* for a four-volume American edition, apparently published partly by subscription as a list of 192 names is included in the fourth volume, representing 247 copies with multiple orders. Nearly a quarter of these were students at Dartmouth College, New Hampshire, where the Professor of Mathematics and Natural Philosophy, John M'Cormick, was also a subscriber, so it is quite likely that he was the prime mover in getting the American edition published.

During the next twenty years or so all of Adams's major titles were republished by the Jones brothers, as indicated above. Bibliographic details of these, and the American editions derived from them, are included in the short-title list (appendix IV). These Jones editions usually have the current W. & S. Jones catalogue of instruments and books inserted at the end when the volume was bound. Some of the Jones editions may have been reprinted, without a new edition number, more often than the entries suggest: catalogues dated as late as 1835 have been found in volumes nominally dated 1812/13. An advertisement in a sixth edition (1812) copy of *Astronomical and Geographical Essays*, which has an 1825 Jones catalogue bound in,[25] mentions that both the second edition of *Essays on the Microscope* and the Jones edition of *Lectures* were then 'reprinting'. These late reprints can only be identified by evidence such as watermarks or positive dating by the original purchaser. None of Adams's works has so far been found in a publisher's cloth binding of the type introduced in early Victorian times, so it is probable that no further reprints were issued after the 1830s.

8 Instruments for van Marum

In the 1760s George Adams senior, in his capacity as mathematical instrument maker to his majesty, supplied numerous philosophical instruments and apparatus for what is now known as the King George III Collection, as outlined in chapter 3. A generation later, George Adams junior was fortunate in obtaining several orders for instruments and apparatus in the fields of astronomy, mechanics, optics, and electricity for the similar collection then being assembled in Haarlem by Martinus van Marum for Teyler's Foundation. This collection survives today largely intact in its purpose-built home (Teyler's Museum), which also houses much of van Marum's extensive correspondence relating to it.

The instruments and apparatus in Teyler's Museum at Haarlem have been fully described and illustrated in volume IV of *Martinus Van Marum, Life and Work* (1973); the descriptive catalogue itself, by G.L'E. Turner, included in that volume as pp. 129–401, is also available as a separate item. A selection of letters to and from van Marum, published as volume VI of the series (1976), includes transcripts of six of the fourteen written by George Adams junior, one of which is also reproduced in facsimile. Adams's contributions to the instruments in the collection are therefore well known; but a summary of what he provided, and when, may be of interest in connection with the Adams story as a whole. This is given in table 8.1, derived not only from the above-mentioned descriptive catalogue but also from comments and invoices in van Marum's correspondence file (the whole of which is available to researchers on microfiche),[1] plus information contained in van Marum's diary of his visit to London in the summer of 1790.[2]

When van Marum started ordering instruments and books from England he made use of Magellan's services as an agent in London, leaving it to him to arrange delivery of the goods and pay for them in

Table 8.1
Instruments supplied by George Adams junior to Martinus van Marum for Teyler's Foundation, Haarlem

Cat. No.	Description	Date Ordered	Date Sent	Price £	Price s	Price d
312	Brooks' electrometer		Dec.'85	6	6	0
269	Lucernal microscope	Nov.'88	Sep.'89	25	10	0
272	Microtome	Nov.'88	Sep.'89	6	0	0
47	Atwood's machine		1790	27	6	0
320	Nicholson's doubler		Jun.'90	3	3	0
321	Nicholson's spinner		Dec.'90	2	10	0
94	Armillary sphere	Jun.'90	Dec.'90	31	10	0
95	Planetarium	Jun.'90	Dec.'90)			
96	Tellurian	Jun.'90	Dec.'90)	48	0	0
97	Lunarium	Jun.'90	Dec.'90)			
92	Terrestrial globe	Jul.'90	Nov.'91	40	0	0
93	Celestial globe	Jul.'90	Mar.'92	20	0	0
98	Jovarium	Oct.'91	Mar.'92)			
99	Saturnium	Oct.'91	Mar.'92)	33	12	0

The following are lost or unidentifiable:

	Description	Date Ordered	Date Sent	£	s	d
	Bennet's electrometer		Jun.'90	1	1	0
	Eudiometer		Jun.'90	4	4	0
	Various magnets	Jun.'90	Dec.'90	3	4	0
	Magnetometer etc	Jun.'90	Dec.'90	3	3	0
			Total	£255	9	0

sterling. Sixteen letters from Magellan to van Marum, dated between 1783 and 1789, are in the file, several of them containing references to instruments by Adams. However, in 1789 Magellan fell ill and was unable to continue acting as an agent for overseas buyers. In February 1790 he died, leaving numerous orders only partially completed. From the middle of 1789 Adams bypassed Magellan and communicated directly with van Marum, so the bulk of the information on goods that he supplied is contained in Adams's own letters, rather than Magellan's.

It seems that Adams first came to van Marum's notice through his *Essay on Electricity*, published in 1784. On 7 October 1785 Magellan wrote to van Marum saying that he had just despatched a Brooks' electrometer made by Adams, together with the second edition of his book on electricity, amongst other items.[3] At this time, as shown in chapter 6, Adams and Magellan were both members of the Chapter House Philosophical Society, and must have been personally known to each other. Three years elapsed before there were any further business dealings between van Marum and Adams, mainly because the disturbed political situation on the Continent had resulted in a temporary cessation of purchases by Teyler's Foundation. On 9 November 1788 van Marum wrote to Magellan saying that the directors were now proceeding with their plan to assemble a 'cabinet of choice physical instruments', and wished to purchase a lucernal microscope as described in Adams's *Essays on the Microscope* (1787), complete with all the relevant accessories.[4] In replying to this letter, Magellan mentioned that he had had an accident but hoped it would not affect his work.

In this he was sadly mistaken. On 19 June 1789 he wrote a brief reply to three letters received from van Marum in the past six months, in which he said that the lucernal microscope had been completed by Adams some time ago but illness had prevented him (Magellan) attending to it.[5] On 30 June, Adams wrote to van Marum himself, confirming that the microscope, lamp, and microtome were ready, and asking how they should be delivered.[6] By October it had become obvious that Magellan was not going to recover, so Adams wrote to van Marum saying that they had been put on a ship bound for Amsterdam.[7] In the course of this correspondence Adams told van Marum that hitherto he had made all of 'Mr.Atwood's instruments', both for Magellan and for Atwood himself, at 25 guineas, and central force machines at 30 guineas.

Meanwhile Dr Jan Ingenhouz, FRS (1730–99) was trying to find out what was happening about Magellan's agency. Primarily a botanist, he had only recently become acquainted with van Marum, but he had lived intermittently in London for about twenty years and was familiar with the principal English instrument makers. On 7 December 1789 he wrote to van Marum with an assessment of the situation.[8] He said that Magellan was now unable to do any work at all, as a result of his fall. The two executors of his will were already 'arranging his affairs', but they had been unable to sort out which of the numerous orders being processed by Magellan had been completed. Finished and unfinished instruments abounded, with no indication of their intended destina-

Instruments for van Marum 253

8.1 The time-measuring part of an Atwood's machine at Teyler's Museum, Haarlem, consisting of an anchor escapement and seconds pendulum. Separated from the machine, such a piece of equipment may be mistaken for a 'journeyman clock' as used in observatories for timing short intervals. (*Photograph by G.L'E. Turner*)

8.2 The terrestrial globe at Teyler's Museum supplied by Adams to van Marum in 1791, specially mounted with its axis at a fixed inclination of 23½ degrees to the vertical. Two pillars at opposite ends of a diametrical bar support terminator and twilight rings, and (on the left) a sub-sun pointer. The globe itself is signed by Dudley Adams and dated 1789 (see figure 8.3), while the brass stand is signed by George Adams (figure 8.4). (*Photograph by G.L'E. Turner*)

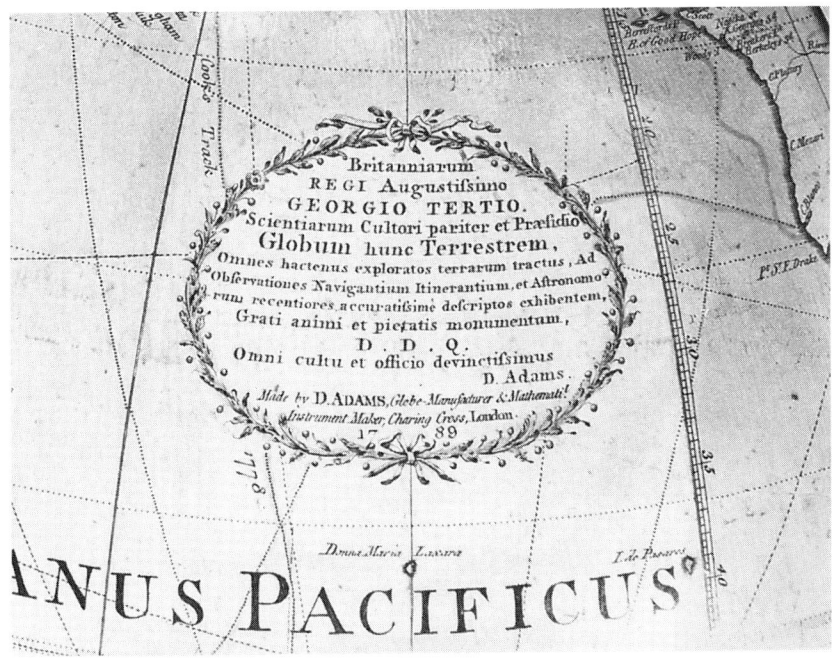

8.3 The Latin dedication on van Marum's 18-inch terrestrial globe is the same as on those launched by George Adams senior in 1766, except that it ends 'D' rather than 'G' Adams. Below the dedication, the original Latin imprint of G.Adams in Fleet Street has been erased and replaced by: 'Made by D.ADAMS, Globe Manufacturer & Mathemati¹ Instrument Maker, Charing Cross, London. 1789'. (*Photograph by G.L'E. Turner*)

tions, so it was up to the persons who had ordered them to communicate with the executors to resolve the confusion.

At almost the same time, on 8 December, Adams wrote to van Marum himself with details of various astronomical models, and of his equatorial mounts for telescopes. That described in his *Astronomical and Geographical Essays*, he said, was about 14 inches high with circles 6 inches in diameter, but he could supply a larger and stronger version for 50 guineas. Even larger ones, with circles up to 3 or 4 feet in diameter, could be supplied on request. However, after examining one by Dollond priced at 150 guineas, which he considered too expensive, van Marum eventually decided to buy a second-hand one by Ramsden from the Ebeling Collection.[9]

On 4 June 1790 Adams wrote to van Marum to inform him that a eudiometer had been completed and despatched, after some delay.[10]

8.4 Though the terrestrial globe at Teyler's Museum is signed by Dudley Adams, George Adams signed the brass stand himself as Mathematical Instrument Maker to his Majesty. (*Photograph by G.L'E. Turner*)

Made under the supervision of Dr Ingenhouz, this had presented some difficulties as the workmen had not made one before and did not understand how it worked. A month later van Marum decided to visit England himself. His diary of this visit describes the actual journey in some detail, and the events of his first week in London, but after that it peters out, though he did not return to Holland for another month at least. Fortunately the first week included two visits to Adams's shop, on 14 and 17 July. The first occasion was an evening appointment, arranged the previous day, when van Marum was shown Adams's latest form of globe mounting with a fixed inclined axis, his new pantograph, work in progress on a planetarium and tellurian, and other instruments. On 17 July van Marum ordered a planetarium, tellurian, and lunarium (specially made for him on individual stands, not combined). In a long letter to van Zeeburgh at Haarlem written on 26 July,[11] van Marum said that he had also ordered from Adams a 12-inch armillary sphere which would cost 30 guineas; this is not mentioned in his diary, so it is quite likely that he visited Adams at least once more

8.5 Van Marum's celestial globe at Teyler's Museum is mounted like its terrestrial companion with its axis fixed at 23½ degrees to the vertical. A loose-fitting meridian and horizon assembly can be positioned anywhere on the globe to show what portion of the celestial sphere is visible at a given place and time. (*Photograph by G.L'E. Turner*)

258 *George Adams Junior*

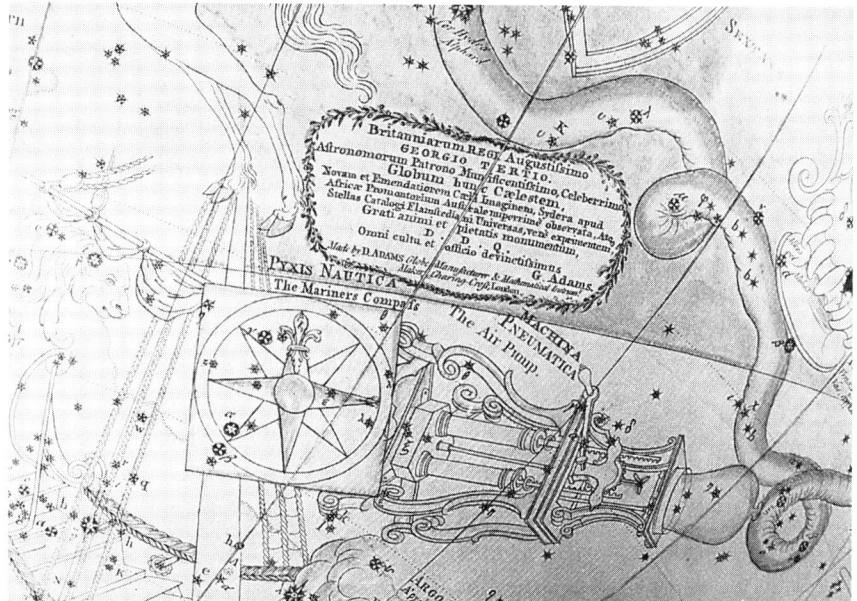

8.6 The Latin dedication on van Marum's 18-inch celestial globe, like that on the terrestrial, is the same as on George Adams senior's original 1766 version, including in this case the signature 'G.Adams'. The Latin imprint has been changed to: 'Made by D.ADAMS Globe Manufacturer & Mathematical Instrum^t Maker, Charing Cross, London'. This globe is not dated. (*Photograph by G.L'E. Turner*)

after 17 July. On one of these unrecorded visits he must have ordered an Atwood's machine also.

Adjacent to Adams's letter of 4 June in van Marum's correspondence file is an undated sheet headed 'Estimates for Mr. van Marum'.[12] Rather roughly written in Adams's hand, with numerous corrections and alterations, it could have been included with the letter, but more likely was given personally to van Marum when he visited London. On this sheet Adams quoted £48 for the planetarium, tellurian, and lunarium as three self-contained instruments; £31 10s for the armillary sphere; £37 for an 18-inch globe with the new form of mounting on a wooden foot, or £43 on a brass foot; and 9 or 10 guineas for an 18-inch celestial globe. Various pieces of magnetic apparatus were also mentioned, mostly unpriced.

Some, but not all, of the items ordered by van Marum from Adams in July 1790 were delivered in December that year. On 17 December Adams wrote detailing the items sent, and apologizing for the delay, which he blamed on the illness of the relevant workman.[13] It would appear from this comment that one man was assigned to a particular customer's order, presumably making all the instruments concerned – or any rate, all the astronomical models which formed the principal part of this consignment. At the foot of his letter Adams wrote an invoice covering both this shipment and the previous one (eudiometer and so on), totalling £99 2s. (There are no headed invoices by Adams in the file: all the prices quoted are written in manuscript on the same sheets as the relevant correspondence.)

According to the catalogue of the van Marum Collection the Atwood's machine was also ordered and delivered in 1790, but although this device figures several times in the correspondence between van Marum, Magellan, and Adams, it is not clear when it was actually sent. As this was a standard piece of apparatus that Adams had made before, it is possible that he had one in stock. The timing part of this apparatus, consisting only of a seconds pendulum, escapement, and weight drive, is very similar to the 'journeyman clocks' used in observatories for measuring short intervals of time, leading to uncertainty about the use of such devices when they have become separated from the rest of the apparatus (see figure 8.1).[14]

In a letter of 18 January 1791 acknowledging receipt of these items, van Marum asked Adams to quote for making two similar models showing the satellite systems of Jupiter and Saturn. Adams had mentioned the possibility of such models in his *Astronomical and Geographical Essays*.[15] In his reply a month later, Adams said that he had not been able to write earlier due to a 'fever and violent cold'.[16] This may have been brought on by disruption around then caused by building operations in Fleet Street. The ratebooks reveal that in the third quarter of 1790 Adams's neighbours on the east, the silversmiths Whipham & North, vacated their premises (number 61) temporarily while they were being rebuilt. This seems to have started a chain reaction, perhaps due to inadequate party walls (though as this part of Fleet Street had been rebuilt after the Great Fire they should have been properly constructed). In the last quarter of 1790 Adams's property (number 60) is shown as 'empty' also, followed in subsequent quarters by numbers 62 and 63. By the middle of 1791 all four properties, numbers 60–63 inclusive, were entered as 'empty, rebuilding'. When he next wrote to van Marum, on 12 September, he mentioned that he had only just moved back into his 'rebuilt house'.[17] The ratebooks show that Whipham & North managed to find a temporary home just across the street in number 171 (Benjamin Martin's old shop), but how Adams coped during this period has not been discovered. He most probably had workshop premises somewhere in London in addition to his shop and residence at 60 Fleet Street, but no clues to their possible location have emerged. (In the early nineteenth century Dudley Adams had a 'factory' in Gough Square, as will be seen later, which was only revealed by his bankruptcy and consequent defaulting on the rates.)

The terrestrial and celestial globes that van Marum had ordered in July 1790 still had not been delivered when Adams wrote in September 1791. The delay, he explained, was due to a strike by the specialist workmen involved, who were demanding an increase in their wages. Quite probably (though Adams did not say so) this demand was prompted by the unusual nature of the globe mountings, which necessitated changes in the men's long-established working practices – a likely catalyst for an industrial dispute. In the normal form of construction the globe core of papier mâché, wood, and plaster is built up on a solid iron axle, longer than the ball diameter, which is used to support the globe in a frame during the manufacturing process, the ends of this axle being pivoted in the brass meridian ring in the finished article. The globes for van Marum, by contrast, were built up on a tubular axis, the north end of which was flush with the globe surface while the south end projected a few inches to enable it to be fitted with a time dial; the completed globe was supported by a substantial iron rod, fixed to a baseplate at an angle of 23½ degrees to the vertical, instead of a meridian ring. This form of mounting, which in effect foreshadowed that in almost universal use for schoolroom globes today, was not illustrated in *Astronomical and Geographical Essays* but was subsequently shown in a plate in Adams's five-volume *Lectures* (1794).[18] The general idea was that it enabled the globe to be used to demonstrate the seasons and associated phenomena, without the obscuration produced by the horizon frame and meridian ring in the normal form of mounting (see figures 8.2 to 8.6).

The terrestrial globe was eventually despatched on 4 November 1791, and the celestial on 10 March 1792, much to Adams's evident relief: the excuses in his multi-page letters of this period were wearing rather thin by then. He subsequently claimed that he had made no profit on these special globes, which he said were better adapted to explain the phenomena of the Earth and heavens than any other pair in Europe, as it was impossible to estimate accurately in advance the cost of one-off products involving some development work.

As the uses of globes mounted in this manner were not described in his *Essays*, he gave instructions (for the celestial in particular) in a three-page letter dated 16 March 1792, in which he told van Marum that the celestial globe had just been completed and shipped.[19] At the end of this letter Adams listed his charges for the models of Jupiter's and Saturn's satellite systems as well as the globes, making a total (with packing cases and so on) of £97 17s.

There is one more instrument by Adams in Teyler's Museum, not listed in the table, namely a brass universal compound microscope (catalogue no. 267) of a type current around 1790. No documentation relating to this has been found; possibly van Marum purchased it for his own use when he visited London in 1790.

Two further letters from Adams are in the correspondence file,[20] but no more orders from van Marum were forthcoming. Perhaps he was more annoyed than the correspondence reveals by the repeated delays in obtaining the terrestrial and celestial globes that he wanted, even though he must have known that non-standard products could not be produced in a hurry.

Throughout this correspondence Adams endeavoured to establish a relationship with van Marum on the basis of one philosopher communicating with another, rather than a mere tradesman writing to a valued customer. In a letter of 8 December 1789, Adams drew van Marum's attention to a new book on electricity by Mr Bennet. Following their meeting in London in the summer of 1790, in his next letter (December 1790) Adams referred to Dr Blair's experimental improvements in telescopes, which he said promised to form a new epoch in the science of optics.[21] His lengthy apologetic letter of 12 February 1791, explaining the delay in delivering the terrestrial and celestial globes, contained a further reference to Dr Blair's work. In his letter of 11 November that year, saying that the terrestrial globe had at last been despatched, he asked van Marum if he would forward details of his latest electrical experiments, as he (Adams) was preparing a new edition of his book on electricity and would like to include them. van Marum responded favourably to this request.

Adams's penultimate letter of 29 June 1792 was almost entirely concerned with comments on various scientific topics not directly relevant to the instruments supplied by him, such as remarks on an 'electric picture' which he and Dr Ingenhouz had examined together. It had been made by (or for) The Revd Mr Bennett, author of the book on electricity previously mentioned. Apparently in reply to enquiries by van Marum, Adams said that he did not know what Nicholson was doing at the moment, as he had some reason (not specified) to be displeased with Nicholson and did not see him very often. Steam engines equal in power to one man could be obtained to go in a box 3 feet square: Adams suggested that van Marum might find one useful for working some of his machines, such as air-pumps. Adams himself had made an experimental pedometer, but it still needed some development to make it a practical instrument. In conclusion he thanked van Marum for supplying details of his gasometer, which would be inserted in his (Adams's) *Lectures*.

The steam engine figures again in Adams's last letter (18 September 1792). The workman (not named) now said that it would be contained in a box only 18 inches square, and would cost £40. When it was finished, Adams said, he would examine it and send van Marum his opinion of it. However, this apparently did not materialize, as no further communication from Adams is in the file. When his five-volume *Lectures* appeared in 1794, van Marum was not amongst the 950 pre-publication subscribers listed in the first volume, though it is possible that he subsequently acquired a set through a bookseller.

9 Hannah Adams and the Succession

In the preface to his five-volume *Lectures* (1794), Adams referred to being afflicted with 'much weakness and languor' during the compilation of that work. A year after its publication he went to Southampton in an effort to recover his health, but without success: he died there on Friday 14 August 1795, a few weeks after his forty-fifth birthday. His place of burial has not been located.[1] His death received the usual brief mention in London daily and tri-weekly newspapers on the following Monday: 'On Friday at Southampton, Mr.George Adams, Mathematical Instrument Maker in Fleet Street.' The local newspaper, the weekly *Hampshire Chronicle*, missed the event in its issue of 17 August but noticed it under 'Southampton News' a week later: 'Yesterday se'night died, Mr.George Adams, Mathematical instrumentmaker, of Fleet Street, London; author of the Essay on Vision.' Selection of this one example of his numerous scientific works for particular mention suggests that the information came from a local Swedenborgian. A longer obituary was published in the *Gentleman's Magazine*, running to eighty-five lines. 'He was a man most attentive and industrious in his business', the writer said, and had also produced 'several literary works highly useful to promote the cause of Natural Philosophy'. The obituarist added that Adams was personally known to the king as a staunch Tory of the old school.

In his will dated 22 July 1794 – just over a year before his death – George Adams junior left the instrument business, and the stocks and copyrights of all his writings, to his widow Hannah Adams.[2] However, the fact that his mother was still alive when he died at the early age of forty-five, coupled with the terms of his father's will, led to uncertainty about the future of the Fleet Street business. (Ann's date of death has not been positively determined, but there is strong and consistent evidence that she died in 1809 at the advanced age of eighty-six, outliving most of her children.)[3] That being so, the position after George junior's death in 1795 was as follows:

a) The premises of 60 Fleet Street, and the copyright of books written by George senior, together with the copperplates of their illustrations, and any unpublished manuscripts by him, belonged to Ann for her lifetime, and at her death were to be divided between her surviving children.
b) The copyrights and printed stocks of books written by George junior belonged to his widow Hannah absolutely.
c) The ownership of the goodwill, stock-in-trade, tools and so on of the instrument business at 60 Fleet Street was debatable, as the part attributable to George senior should have devolved on his widow Ann, but this would probably be disputed by Hannah, who would naturally expect to receive the benefit of her late husband's contribution over the past twenty-three years.
d) Similar considerations applied to the substantial library built up over many years by both George senior and junior, as books dated before 1773 could have been bought by either.
e) The globe business at Charing Cross, on the other hand, which Dudley Adams had been running since 1788 (see chapter 10), seems to have been his property by mutual agreement, so it was not affected by George junior's death.

Hannah moved quickly to establish her position: the lord chamberlain's books show that she was appointed mathematical instrument maker to his majesty sometime in August (the precise date is not entered), in other words within a fortnight of the death of her husband. The latter's will was proved by the four executors, one of whom was Hannah, on 8 September. Apart from leaving all his household goods, and the disputed stock-in-trade and so on at 60 Fleet Street, to Hannah, together with the stocks and copyrights of his own writings, its main provisions were:

a) Legacies of £50 or £100 to each of thirteen beneficiaries, including his brother Dudley and surviving sisters Lucy Adams, Isabella Adams, and (Mrs) Hannah Robinson.
b) His quarter share in a freehold estate at Linfield in Surrey to Hannah for life.[4]

c) His copyhold estate at Galwood Common near Great Baddow in Essex to Hannah absolutely; this was originally hers anyway, inherited from her grandmother at the age of seven, and had only come to George junior as part of a marriage settlement.[5]
d) £3,000 in trust, the interest to be paid to Hannah for life, and then the capital to Dudley and any surviving sisters equally.
e) £100 to each of three trustees.
f) The residue to Hannah absolutely.

No mention was made of the property in Buckinghamshire, as this belonged to his mother Ann for life. The total value of the monetary bequests was £4,300, but the residue could have been substantially greater than this figure. As George junior died just before death duties were introduced, there is no record of the total value of his estate. When Hannah died fifteen years later she left a total estate, including land and investments, of nearly £20,000, which means that in today's money George junior would have been a millionaire.

Hannah evidently thought she could carry on the Adams instrument and bookselling business indefinitely: it was not uncommon for widows to continue to run their late husbands' businesses, especially if the manual work involved had been done by employees or subcontractors.[6] However, she would undoubtedly have encountered considerable opposition – to put it mildly – from her brother-in-law Dudley, who had been trained as a general mathematical instrument maker though for the past six or seven years he had concentrated on making and selling globes. There was also the matter of the premises to be considered, for George senior's widow Ann may well have wanted to see her younger son Dudley installed there instead of Hannah, especially as the latter had no children.

At least one instrument (a barometer, untraced) has been reported with the signature Hannah Adams, but little evidence has survived of her activities in the instrument-making field. If she had counted on retaining the Ordnance trade, she was unlucky – or not quite alert enough. No detailed board minutes have survived for the period around the time of George junior's death, but the index volumes to the 1795 Ordnance correspondence show that on 16 August, only two days after George junior died at Southampton, Dudley Adams wrote to the board requesting permission to succeed his late brother as a supplier of mathematical instruments; he must have put pen to paper as soon as news of his brother's death reached London.[7] Rather surprisingly, according to the register his letter was not received at the Office of Ordnance until 25 August. It was considered at a meeting on the same day. Perhaps Dudley deliberately backdated his letter in case anyone else had applied for the appointment in the interim. On 26 August, the day after Dudley's letter was received and considered by the board, Hannah Adams wrote making the same request to succeed her late husband; her letter was received the day it was written (the 26th), and considered on the 27th. It is quite possible that she intended to continue trading under the name 'George Adams', and did not think it necessary to inform the board of her husband's death at first. If so, she must have been furious when she discovered what Dudley had done, but by then it was too late to retrieve the situation. The board evidently agreed to Dudley's request to succeed his late brother, for according to the register he wrote on 28 August 'returning acknowledgement for his appointment'. (The actual letters written by Dudley and Hannah have not survived, so the story has to be pieced together from the register or index entries.)

A few weeks elapsed before Dudley's Ordnance appointment took effect. In the meantime there were a number of orders to George awaiting completion. Hannah presumably dealt with these on behalf of her late husband: no entries have been found in the bill books in her own name. The sequence of Adams bills dated in the second half of 1795, listed in table 9.1 in chronological order of bill date, shows that from the beginning of October 1795 onwards, six weeks after George died, all orders went to Dudley.

Hannah continued to trade at 60 Fleet Street until early 1796. Though she lost the Ordnance business, she managed to continue the link with Christ's Hospital: V.K. Chew's (unpublished) researches show that the ledgers record one payment of £11 7s 6d to her late husband's executors in 1795, and then one in her own name in 1796. Possibly Dudley thought that in comparison with the Ordnance trade, that with Christ's Hospital was not worth pursuing.

It must have been very difficult for Hannah to maintain her position in the face of the combined opposition of her brother-in-law and mother-in-law. In February 1796 she decided to dispose of the library collected over a period of sixty years or more by her late husband and his father. This naturally upset Dudley, who, writing fourteen years later within a few days of Hannah's death, claimed that he could have stopped the sale if he had so wished, as he had an exclusive right to 'all his father's works'.[8] However, he had no right to books purchased by his brother after 1772, so his intervention might have delayed the sale and necessitated sorting

Table 9.1
Ordnance bills in the second half of 1795 in the name of George (G) or Dudley (D) Adams, extracted from WO52/78–93 and rearranged in chronological order of bill date

Bill date	G/D	Order date	£	s	d	Allowed date	WO52/ No.	Page
28 Jul.	G	*31 Aug.	8	15	0	17 Dec.95	84	350
28 Jul.	G	3 Jul.	15	10	0	10 Nov.95	88	134
28 Jul.	G	30 Jun.	4	7	6	5 Nov.95	79	287
5 Aug.	G	30 Jun.	3	8	6	10 Nov.95	83	328
(posthumous bill dates):								
21 Aug.	G	24 Jun.	50	8	0	17 Dec.95	86	323
26 Aug.	G	30 Jun.	17	14	0	12 Nov.95	87	308
31 Aug.	G	28 Jul.	10	3	0	15 Feb.96	93	87
22 Sep.	G	*30 Sep.	6	15	0	18 Dec.95	84	352
1 Oct.	G	30 Sep.	63	0	0	17 Dec.95	78	314
Total (George)			£180	1	0			
31 Oct.	D	29 Oct.	2	12	6	4 May 96	92	271
31 Oct.	D	31 Oct.	3	10	0	4 May 96	93	259
7 Nov.	D	31 Oct.	209	8	0	4 May 96	82	93
7 Nov.	D	30 Sep.	25	4	0	4 May 96	90	299
10 Dec.	D	10 Dec.	99	12	6	3 May 96	91	116
12 Dec.	D	30 Nov.	120	0	0	4 May 96	93	252
31 Dec.	D	12 Dec.	3	8	8	4 May 96	91	134
Total (Dudley)			£463	15	8			

*Retrospective authorization

the books into three categories (his father's, his brother's, and those of undetermined provenance), but he could not have stopped Hannah selling those that were undoubtedly hers by inheritance.

The sale by Leigh & Sotheby took place over five days (Sunday excepted) from 24 February to 1 March 1796. The auctioneer's copy of the thirty-eight-page printed catalogue, annotated with prices fetched and buyers' names, is in the British Library.[9] An analysis undertaken as part of this study of the Adams family shows that there were around 2,000 volumes, including some long runs of periodicals such as the *Philosophical Transactions*, sold in a total of 1,151 lots. (For the benefit of other researchers in this field, the analysis was published in 1988 on microfiches.)[10] Though this was essentially a working library, its contents were not limited to books or periodicals on subjects directly relevant to instrument making: they also included anatomy and medicine, biographies, geography and travels, literature, and theology. About 16 per cent of the books were in French, and a few (2.3 per cent) in Latin. A breakdown by year of publication shows that at least 40 per cent were published after 1772 and could not, therefore, have been purchased by George senior; these were undoubtedly Hannah's by right. A further breakdown of the post-1772 volumes shows that twenty-seven items were published in 1795, the year of George junior's death. As he died in August, in the thirty-third week, this means that he was then buying new books at an average rate of one per eight days.

The sale was evidently well attended. About eighty people made purchases, ranging from a single lot to as many as seventy-eight (by 'Bruce', probably William Bruce, bookseller). Prices realized covered a wide range, from a minimum of sixpence to 13 guineas, the latter figure being reached by two lots, namely a run of thirty-eight volumes of the *Philosophical Transactions* and two volumes of Lavater's *Physiogonomy* respectively. The total for the five-day sale came to a round figure of exactly £400 (hammer price).

Only a few weeks later, Hannah decided to abandon her attempt to continue the instrument business. A hint that she may have been forced out by her mother-in-law occurs in the wording of advertisements which Hannah inserted in several London newspapers early in April 1796, in which she said that 'being obliged to quit her House', she was giving up the business and disposing of 'her valuable Stock at reduced Prices, for Ready Money'. All persons indebted to her late husband's estate were requested to settle their accounts by 1 May 1796, and any that had left instruments to be repaired were asked to take them away immediately, 'as the Shop will be cleared in a short time'.

These advertisements included a list of books written by George junior that were still available, but Hannah could hardly have expected to continue selling these herself once she had vacated the Fleet Street shop. Having failed to dispose of the stocks of these to booksellers by auction or other means, she sold them to William & Samuel Jones, of Holborn, together with the copyrights.[11] During the next few years, catalogues of instruments by W. & S. Jones usually had a page or two listing books by the late George Adams, sold by them. When stocks of the original editions ran out, William Jones revised and updated the text and plates, and issued new editions under the W. & S. Jones imprint. The Jones editions are included in the short-title list in appendix IV, together with American editions based on them.

The stock-in-trade, tools and so on remaining at 60 Fleet Street after the business had closed were sold by auction by Mr Christie, at his Great Room in Pall Mall, on Wednesday and Thursday, 29–30 June, and Saturday 2 July 1796. According to the advertisements for this in London newspapers, the sale was 'by Order of the Executors' of the late George Adams, indicating that Hannah had failed in her attempt to convince the rest of the Adams family that she was the sole and rightful owner of the business. The instruments specifically listed in the advertisements included 'a portable Transit Instrument, a Universal Equatorial, a capital Variation Compass, large and pocket compasses, marine and portable Barometers, Thermometers and Hygrometers, Hydrostatic Balances, Apparatus for Electrical Machines and Air-Pumps, Artificial Magnets, Orreries, Spheres, a Set of Mechanical Powers, Camera Obscuras, Achromatick Telescopes, Pocket Microscopes, Spectacles, Reading Glasses, Quadrants, Cases of Mathematical Instruments, Brass and Wood Scales, Mechanical and Optical Models, Theodolites and other Surveying Instruments'.

Thus ended Hannah's brief career as proprietor of an instrument-making business. Financially she was quite comfortably situated, with a secure income from her Essex estate, the trust fund of £3,000 provided in her late husband's will, and all the residue from the latter. At the time of her death fifteen years later in 1810 she was living in Clapham, Surrey, with her unmarried elder sister Mary Marsham (who, like Hannah, had inherited property in Essex from their grandmother) and her niece Mary Driver. Her will, signed in 1807 with a codicil added on 17 September 1810, is a lengthy and complicated document listing over thirty separate bequests.[12] As she and George junior had no children, the bulk of her £20,000 estate was left to her own blood relations in the Marsham family: her sister, brother, and nieces and nephews. Her brother-in-law, Dudley Adams, received a token bequest of £20, and his unmarried sister Isabella Adams £100 in consols. Death duties at 10 per cent (as they were not her blood relations) reduced these figures to £18 and £60 2s 6d respectively, consols at that time standing well below par.

Part III:

Dudley Adams

10   Globe Maker and Instrument Maker

The last child – a boy – of George Adams senior and his second wife Ann née Dudley was born on 27 November 1762, when George was aged fifty-three and his wife Ann forty. He was baptized 'Dudley' at St Dunstan's in the West on 5 December,[1] thereby perpetuating his mother's maiden name.

When he was not quite ten years old his father died (October 1772). Presumably he was then living at 60 Fleet Street with his mother and elder brother George junior, who jointly continued the instrument business there, and also his unmarried sisters. Two years later, as related in chapter 6, George junior married, creating a difficult domestic situation. On 6 February 1777, aged fourteen and a quarter, Dudley was formally apprenticed to his brother in the Grocers' Company, to be instructed in mathematical instrument making.[2] That George junior charged his mother £49 for this, a sum roughly equal to a year's wages for a journeyman, suggests (as mentioned earlier) that there may have been some friction in the household by this time. Perhaps Ann, as owner for life of 60 Fleet Street, was charging George rent for its use, and he seized the opportunity to retaliate by charging her a premium for educating his younger brother. Be that as it may, the normal practice would have been for the apprentice (Dudley) to live with and be maintained by his master (George) for the duration of the apprenticeship, so presumably he continued to live at 60 Fleet Street. Whether Ann and her daughters also continued to live there, or had to move elsewhere, is not known.

If the apprenticeship lasted the normal seven years, it should have been completed by 1784. Dudley probably continued to work for his

274 Dudley Adams

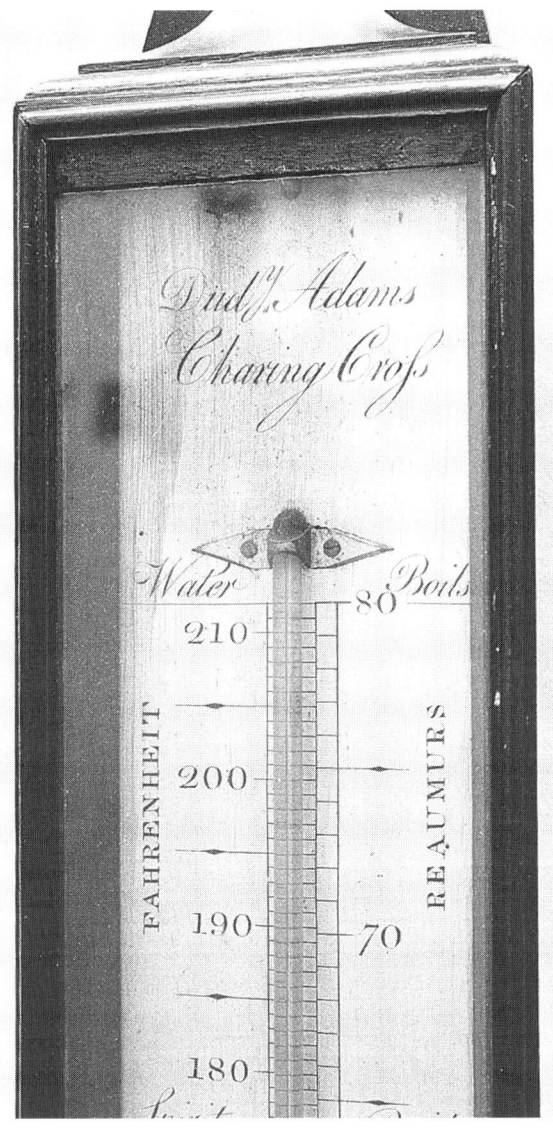

10.1 Detail of the top of a mercury-in-glass wall thermometer with Dudley Adams's Charing Cross address, dating it between 1788 and 1796. The silvered brass register plate is mounted in a mahogany frame with a glazed door, exposing the whole of the 20-inch-long tube to view. It is calibrated from 5 to 212 degrees Fahrenheit on one side of the tube and from minus 13 to plus 80 degrees Réamur on the other. It is not clear what this instrument would have been used for. (*Museum of the History of Science, Oxford, inventory no. 30776*)

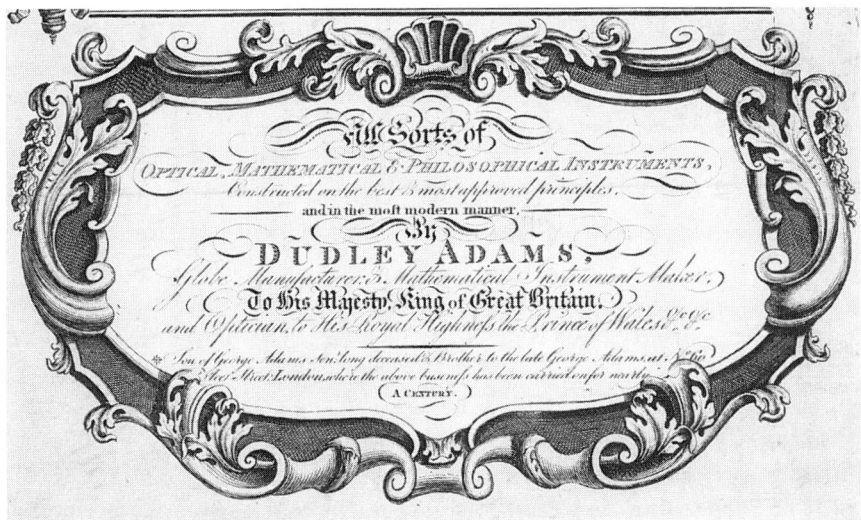

10.2 Detail of the inscription on Dudley Adams's trade card (see the frontispiece for a reproduction of the whole card). His appointment to King George III, and the reference to the Prince of Wales (rather than the Prince Regent), indicate that it was engraved after November 1796 and before 1811. 'Nearly a century' for the age of the business is an exaggeration: it was founded in 1734. (*British Museum*)

brother as a fully-trained journeyman for a while; however, circumstantial evidence suggests that he may have spent some time abroad at this point (see chapter 12). Then on 11 February 1787, when he was aged twenty-four, he married a widow, Margaret Sophia de Langlade, at St Marylebone parish church.[3] This fact was revealed by the International Genealogy Index (IGI); except for her death in 1801, nothing has been discovered about Margaret's life or background, or how she and Dudley met. In the following year, perhaps finding that he could not possibly live with his brother and sister-in-law now that he was married (but this is just conjecture), Dudley decided to go into business on his own account. For this he needed separate premises, and his freedom, which was granted by the Grocers' Company on 5 June 1788, his address on that occasion being given as Spring Gardens, Charing Cross.[4]

This address suggests a possible link with his somewhat older half-sister Sarah Blunt and her linen-draper husband Robert, whose son Robert Blunt junior had been apprenticed to George Adams junior in 1782, two years before Dudley finished his own apprenticeship in the same business. Dudley and Robert junior must have worked alongside each other for about two years. Shortly before Dudley obtained his freedom, Robert's father died (April 1788), leaving the linen-drapery business to Sarah on condition that she admitted Robert junior as an equal partner when he came of age in 1789.[5] Though no documentary evidence of a business link between Dudley and the Blunt family has been discovered, it surely cannot have been just coincidence that Dudley chose to establish his own business only a few doors from that of the Blunts.

The ratebooks of St Martin in the Fields show that Dudley began paying the quarterly Poor Rate at Michaelmas 1788 on premises which, from the sequence of entries, were next-but-one to the eastern entrance to Spring Gardens,[6] opposite King Charles's statue, identifiable as 53 Charing Cross on Horwood's large-scale map of 1799.[7] The Blunt linen-drapery business was at number 64, opposite the Mews Gate. All the buildings in this area were swept away in the early twentieth century, when The Mall was made into a processional carriage route from Buckingham Palace to Trafalgar Square: Dudley's shop would have been roughly in the middle of the modern roadway, a little to the north of the present Royal Bank of Scotland (formerly Drummond's Bank).

Almost immediately after opening his shop, Dudley took an apprentice, Joseph Morris, son of a victualler in the Strand, for a premium of £49.[8] Dudley gave his occupation in the company's books as 'optician', and indeed this is how he described himself in later years despite having been trained as a mathematical instrument maker. For the six or seven years that he traded at Charing Cross he dealt in all kinds of instruments, but specialized in the production and sale of terrestrial and celestial globes. He said later that his mother, owner for life of her late husband's globe plates and tools, had given these to him to enable him to establish his own distinct business; perhaps she hoped that by this means, friction between the two brothers (and their wives) at 60 Fleet Street would be avoided. The globes supplied to van Marum in 1791–92 have gores signed by Dudley Adams and dated 1789, so they must have been amongst his first productions. Other examples of Dudley's 1789 globes, in more conventional stands, are known.[9]

An undated fifteen-page catalogue issued by Dudley Adams, 'West Side of Charing Cross', is at the Museum of the History of Science, Oxford. On the title-page Dudley called himself 'Globe Manufacturer and Mathematical Instrument Maker'. It is almost identical in layout, content, and prices to George junior's fifteen-page catalogue of 1790, except that where George's refers to plates in his books, these references are omitted in Dudley's. Both catalogues include globes of 18, 12, 9, 6, and 3 inches diameter, but only Dudley's mentions the larger size, 28-inch. The inference is that Dudley supplied the usual sizes from 18-inch downwards to George (and other firms) for retail sale, and also sold them retail himself, but reserved the manufacture and sale of the 28-inch size for himself alone.

The plates for these, the largest globes made commercially in England at that time, were originally Senex's, which had been updated by James Ferguson and then Benjamin Martin in the mid-eighteenth century, and then sold by auction after the latter's death in 1782. Precisely how they passed into the possession of the Adams family is not clear. A pair of globes of this size, made by Dudley Adams, was chosen as one of the gifts from Britain to the Emperor of China sent with Lord Macartney's mission in 1792.[10] On 21 July 1792 Dudley wrote a brief note to Sir Joseph Banks, President of the Royal Society, informing him that this 'pair of Magnificent Globes' was almost ready for despatch, should he or any of his friends wish to see them.[11]

A pair of 28-inch globes by Dudley Adams is reported to be in the Military Academy, Lisbon. No extant example has been located in Great Britain, but in 1797 a pair was at Stowe House, Buckinghamshire, in the ante-library. A description of the house and contents published that year said that 'The Coelestial globe differs from all others in being graduated in the same manner as the terrestrial globe, by lines of latitude and longitude, on a dark blue ground: the stars of seven different magnitudes are marked by foils of different colours, and the nebulae are marked in silver.'[12] These globes formed lot 2549 when the contents of the house were sold by auction in 1848.[13]

Possibly as a result of the commission to the Emperor of China, on 31 July 1794 Dudley was appointed globe maker to his majesty.[14] This is believed to have been the first appointment of this nature: no earlier ones have been found in the lord chamberlain's books.

Meanwhile, on 5 February 1791 Dudley took another apprentice, Charles Stewart, son of the late Deputy Governor of Florida, for a premium of £100, paid by the widow.[15] Soon after this, on 13 May 1791 the court of the Grocers' Company resolved to elect twelve new liverymen. Sixteen were nominated, including Dudley Adams and Robert Blunt junior, and the former was amongst those chosen.[16] (Robert Blunt junior had to wait until 1795 for election to the livery.) On 14 July 1793 Dudley took his third apprentice, John Blenkinsop, son of a

10.3 A 4-inch brass reflecting telescope signed on the eyepiece backplate 'D.ADAMS LONDON' without a street address, mounted on a pillar-and-tripod stand with inward curling feet. This shape was used in the 1750s by Benjamin Martin, amongst others, for small planetariums and other table-top instruments, and is usually associated with the eighteenth century rather than the nineteenth. However, Dudley continued to use it long after other makers had changed to more stable outward-turning 'pad' feet, as in (for example) figure 7.11. (*Christie's, London, 20 August 1987, lot 138*)

deceased surgeon of Reading, Berkshire. No premium was paid on this occasion; perhaps the boy's father had performed some medical services for Dudley or his wife in the past.

Extant instruments, other than globes, signed with Dudley's Charing Cross address (rather than just 'London'), and hence positively attributable to the period 1788–96, include several barometers, a wall-mounted thermometer (see figure 10.1), a brass Culpeper-type microscope, a magnetic compass, a surveyor's spirit level, and a gunner's quadrant.[17]

George junior's early death at the age of forty-five in 1795 provided Dudley with an opportunity to take over the long-established instrument business in Fleet Street. It seems likely that Hannah's decision to attempt to carry on the business there herself took Dudley by surprise. His immediate application to the Board of Ordnance for permission to succeed his brother may have been intended as much to forestall possible competitors as to upstage Hannah. The table in the previous chapter shows that the flow of Ordnance orders to Adams switched from George to Dudley without any obvious break, suggesting that Dudley must have been closely involved in this work before his brother's death. His first two bills in his own name, dated 31 October 1795, were quite small,[18] but they were followed a week later by a large one for over £200. The instruments detailed in this bill included three dozen quadrants with plummets, three dozen perpendiculars, three dozen diagonal scales, twelve pairs of brass calipers, and no fewer than sixty pairs of steel compasses with bows.[19] The warrant for this order was dated 31 October, only a week earlier, which suggests that there must have been a production line for these military items in continuous operation, so that quantities of a few dozen could be supplied virtually from stock. During the three months October to December 1795 Dudley's Ordnance bills amounted to a total of £463. As this was equivalent to around 260 man-weeks (taking one day's work by an instrument maker to be worth 5s), this implies at least twenty men working on Ordnance orders alone, quite apart from those engaged on production of instruments for the civilian market. Either most of this work must have been subcontracted, or else the Fleet Street business employed several dozen workmen at another undiscovered address, whom Dudley took on immediately after his brother's death. Whether this workforce – subcontracted or otherwise – was managed by Hannah or by Dudley during George junior's last illness has not been discovered.

The run-down of Hannah's business activities, culminating in the sale by auction of all the remaining stock of George junior's instruments at 60 Fleet Street, has been outlined in the previous chapter. Hannah having departed, and the shop having been cleared out, at the beginning of August 1796 Dudley announced in the newspapers that he, 'Globe-Maker to the King, Optician to their Royal Highnesses the Prince of Wales, Duke of York, &c. &c., and Mathematical Instrument Maker to the Honourable Board of Ordnance &c.', had removed from Charing Cross to the Fleet Street premises formerly occupied by his late brother, 'where the businesses of both houses will continue to be carried on in all its branches'. Answering insinuations that he was only a globe maker, not a mathematical instrument maker, he pointed out that he had been 'initiated into the working branch of the Mathematical Business at an early age', had served a regular seven-year apprenticeship in that trade, and was competent to construct 'every species of Mechanical Instruments, as well as Globes, whether Optical, Mathematical, Philosophical, or otherwise'.

This advertisement (which also mentioned several specific products) first appeared on the front page of *The Times* on 5 August 1796,[20] and was repeated no fewer than eighteen times in this newspaper up to the end of October. In the ratebooks of St Dunstan's, Dudley's occupation of 60 Fleet Street begins at midsummer; those of St Martin's indicate that he did not vacate his 53 Charing Cross address immediately: he continued to pay the rates there until the middle of the next year (1797), when Robert Blunt junior moved the Blunt linen-drapery business there from number 64.[21]

It will be noted that the appointments listed at the beginning of Dudley's advertisement do not include mathematical instrument maker to his majesty. Hannah retained this herself for the time being; the lord chamberlain's books show that it was not relinquished by her and assigned to Dudley Adams until 17 November 1796. On the other hand, the appointments include Optician to the Duke of York, as well as to the Prince of Wales. Dudley first mentioned his appointment to the Duke of York in the inscription on a 12-inch terrestrial globe dated 1795 bearing his Charing Cross address. Presumably he took over the appointment to the Prince of Wales on his brother's death.

It was probably soon after his move to 60 Fleet Street that Dudley had the elaborate trade card engraved which is reproduced on a reduced scale in the frontispiece (see also figure 10.2). The somewhat archaic design, incorporating obsolete instruments such as Davis's quadrant, and reviving the redundant 'Tycho Brahe's Head' shop sign, may have been a deliberate attempt to emphasize the ancient foundation of

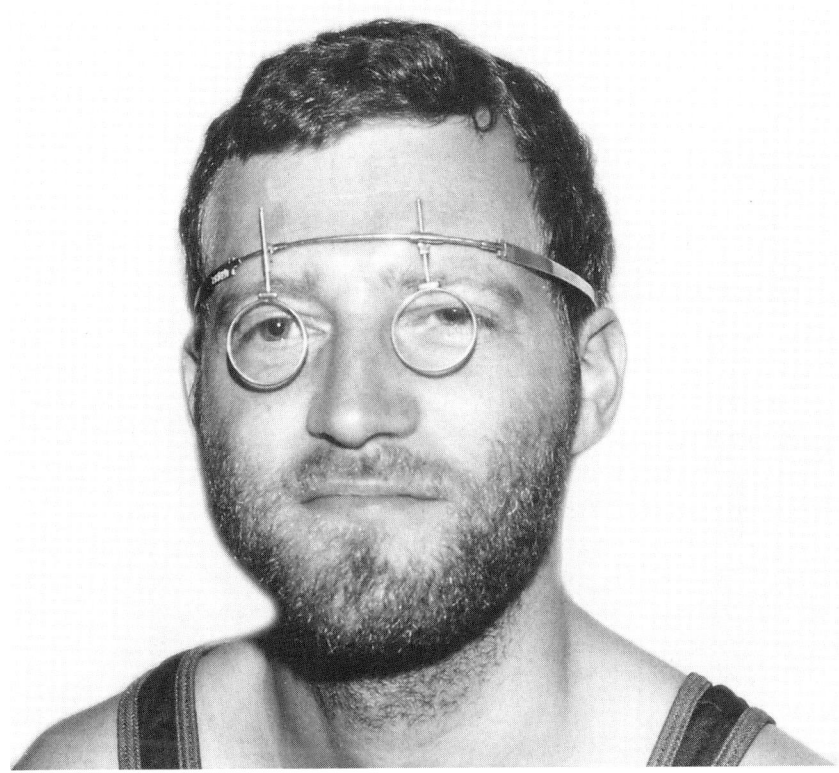

10.4 Dudley Adams's patent spectacles in use. The positions of the eye lenses can be adjusted individually to suit the wearer. (*Photograph: Hugh Orr*)

the Adams instrument business. But Dudley was exaggerating its age by thirty years or more when he claimed that it had been established in Fleet Street for nearly a century. As he was only nine when his father died, quite possibly neither he nor his mother (his father's second wife) knew much about the early years of the business – or that his grandfather was a cook.

The stocks and copyrights of George Adams junior's books having been sold by Hannah to W. & S. Jones, Dudley, in contrast to his late brother, had no books to sell when he took over the Fleet Street premises. Later in 1796 he became a bookseller indirectly by sharing the imprint on Benjamin Donne's *Essay on Mechanical Geometry* with Troughton and W. & S. Jones,[22] thereby renewing an association with Donne dating back to the 1750s. This seems to have been Dudley's only venture into bookselling before 1810.

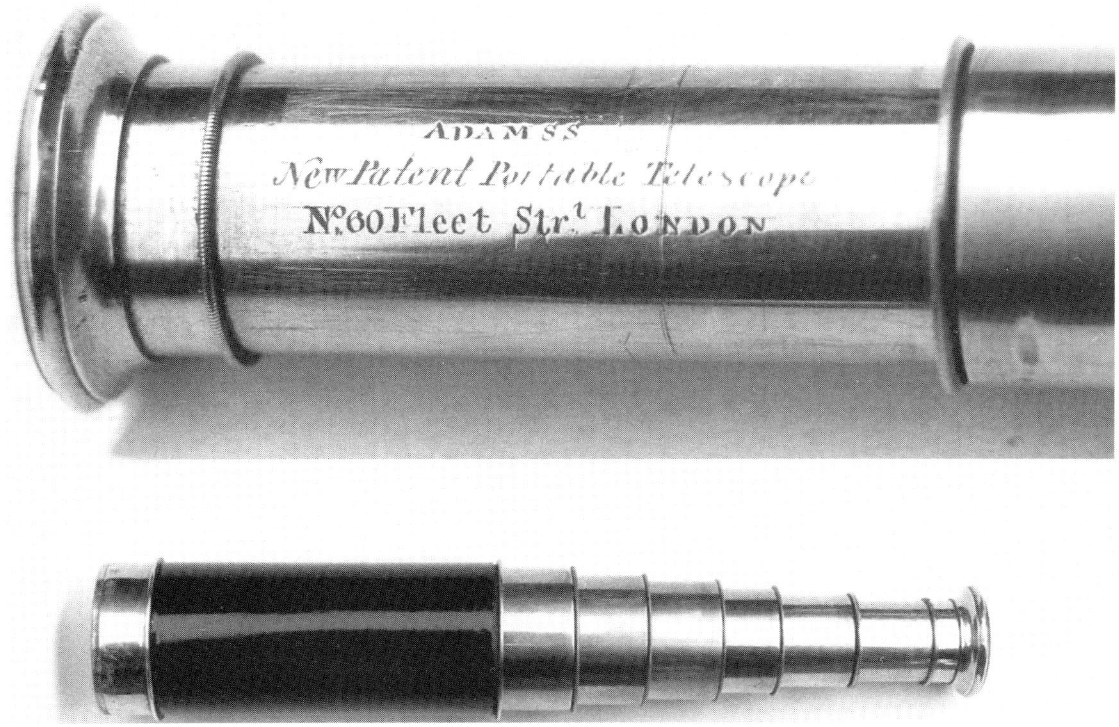

10.5 A portable telescope made in accordance with Dudley Adams's patent of 1800, with multiple draw-tubes. This was what he called a 'one-foot' telescope, shutting down to a closed length of only 4 inches. (*Tesseract, Cat.17 [1987], item 5*)

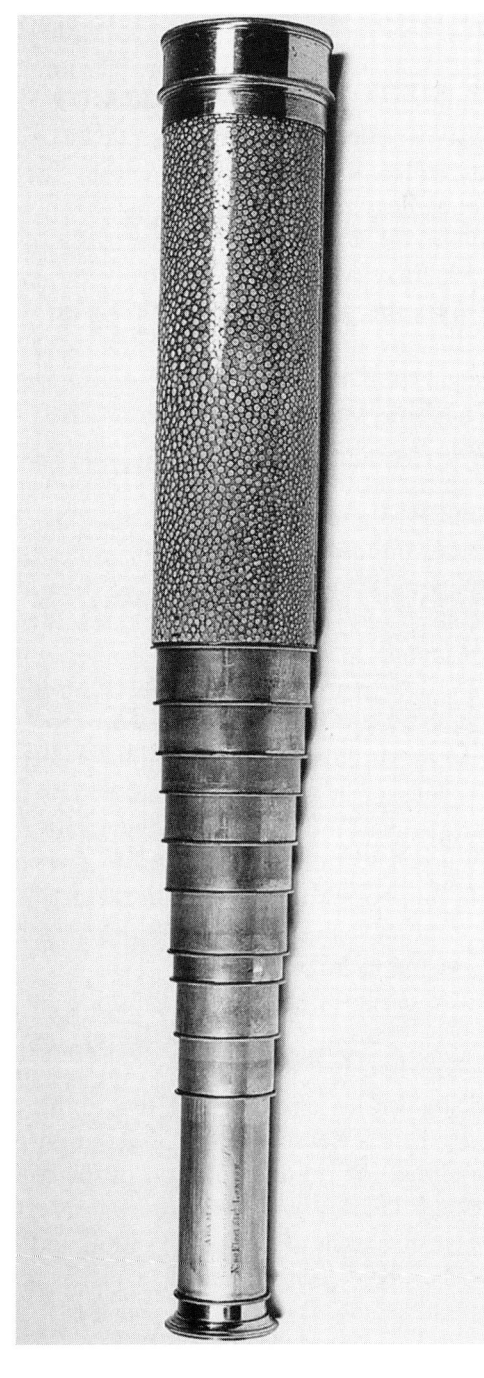

10.6 A portable telescope made, like that shown in figure 10.5, in accordance with Dudley's patent, and signed in the same manner, but of larger dimensions. The object glass is 2 inches in diameter, the closed length is 9¼ inches, and it has ten draw-tubes. (*Whipple Museum, Cambridge, Inv. Wh.897*)

In 1797 Dudley was granted the first of three patents that he secured between then and 1815. His father had applied for only one, in collaboration with Richard Jack in 1750, and his brother none at all, but Dudley evidently had more faith in both the patenting system and the commercial value of his own ideas.[23] In the preamble he called himself an 'optician', of Fleet Street, and gave his invention the short title 'Certain spectacles on an entire new Principle'. His specification (identified today as patent no. 2155, dated 23 January 1797) included a large folding plate with forty-two figures. Basically the construction consisted of a band round the head at the temples, with separate eye glasses suspended from it by hinged pieces in such a way that they could be adjusted in separation and also tilted up out of the way when not required. The headband was made of several pieces hinged together, so that the whole assembly folded into a small space (see figure 10.4). Examples turn up from time to time in the salerooms.

Three years later Dudley obtained another patent (no. 2407, dated 30 May 1800), entitled 'A mode of rendering Telescopes, Perspective, Prospect, and other Optical Glasses, more portable than has hitherto been executed'. At the heart of his invention was a new form of spring to be fitted to the inner ends of telescope draw-tubes to keep them steady when drawn out, and thereby enable many more than usual to be employed, reducing the overall length of the instrument when closed (see figures 10.5 and 10.6). As this is entirely hidden, telescopes made in accordance with this patent can only be distinguished by dismantling them, unless they bear a legend indicating that they were patented by Dudley Adams in Fleet Street. (Patent telescopes of this period will not, of course, bear a patent number, as these were allocated retrospectively in the mid-nineteenth century, when all the specifications recorded in the patent rolls were ordered to be printed.)

One of Dudley's trade cards in the Heal Collection at the British Museum has a manuscript table on the back listing his patent telescopes in various sizes.[24] It seems to be a draft (in carefully-written script) for an advertising leaflet or pamphlet describing a 'package deal', aimed particularly at overseas agents. A paragraph of explanatory text says it is 'Intended for the information of the Mercantile & Shipping Interests'. What Dudley was offering was a collection of forty-two instruments (twelve each of 1 foot, 1½ foot, and 2 foot, and six of 3 foot) packed in a box about 2 feet × 8 inches × 10 inches, weighing 60 pounds in all. The total retail value was £148, and Dudley was offering it to merchants at £111, that is, a discount of 25 per cent on the price to the public. He pointed out that despite the low prices and 'great discount', all the instruments were warranted to be 'Glass'd on the best and most perfect Principles, the Object Glasses not being single but Achromatic'. This is chiefly of interest for the information it gives on closed and open lengths of his standard sizes, and their prices:

One-foot	4 inches closed	£2 2s.0d. each (retail)
Eighteen-inch	4⅞ inches	£3 3s.0d.
Two-feet	5¾ inches	£4 4s.0d.
Three-feet	7⅛ inches	£5 15s.6d.

Meanwhile Dudley was not neglecting his globe-making business (see figures 10.7 and 10.8). Several sizes of Adams globes bearing the publication date 1798 are extant. The date is significant, for in 1799 a completely new set of 18-inch globes was launched by W. & S. Jones with the title 'New British Globes'. Preparations for these had been going on for at least five years, and should have been known to Dudley as they had been mentioned by several writers, such as Samuel Vince in his *Complete System of Astronomy* (1797). Vince commented that no new globes of a large size had been introduced since the time of Senex; this may have been what prompted Dudley to update his own at that point. 'New British Globes', made by Bardin under the direction of W. & S. Jones, were favourably received by the public, and must have captured an appreciable portion of the market from Dudley. A 12-inch size was added in 1800. Competition for Senex's 28-inch globes, however, did not arise (in the form of Addison's 3-foot globes) until after Dudley's death, so he continued to have a clear field in this size.

In 1801 Dudley suffered a domestic tragedy: his wife Margaret died, having been ill for some time. Her death was noticed briefly in the *Gentleman's Magazine*: 'Nov.11th. At Chelsea, after a long series of illness, the wife of Dudley Adams, Fleet Street.' Little has been discovered about Dudley's domestic affairs. It is known (from the records of the Grocers' Company) that he had a son, a third-generation George, apprenticed to him in 1811 and hence probably born about the time of his move from Charing Cross to Fleet Street. The death of his wife in 1801 therefore left Dudley with at least one small child to look after. His father, faced with a similar situation in 1747, soon remarried, but Dudley does not appear to have adopted this course. In 1801 his mother and elder unmarried sisters Lucy and Isabella Adams were still living, so perhaps they had to assume responsibility for his domestic arrangements. Dudley's son George was the only Adams grandchild of

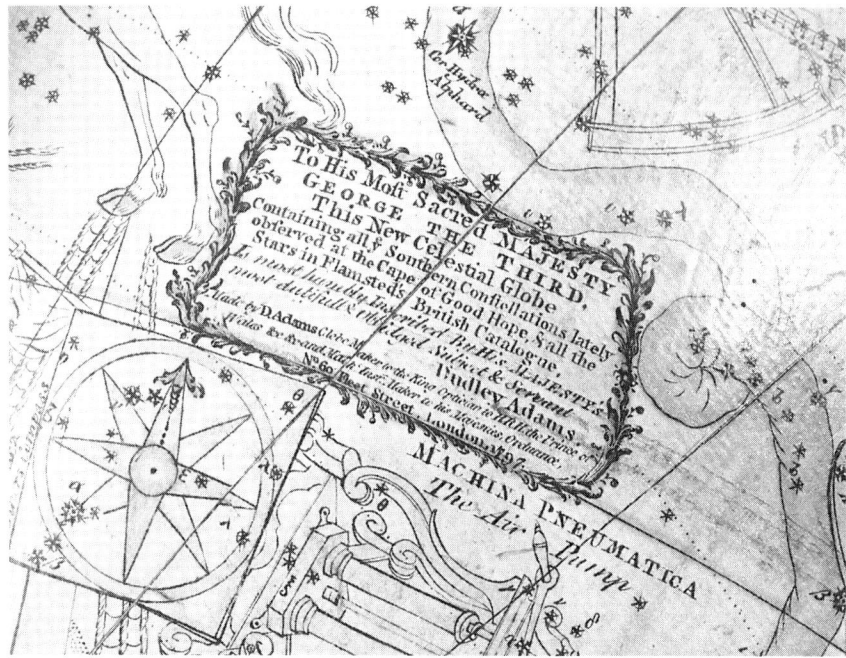

10.7 Detail of the inscription on a 12-inch Adams celestial globe dated 1797. Compared with that shown in figure 4.5, 'G.Adams' at the end of the dedication has been changed to 'Dudley Adams', and everything below this has been erased and replaced by Dudley's name, appointments, address, and the publication date. In order to accommodate the increased amount of information part of the inner edge of the border has been erased. For a later edition of this globe, in which the date has been changed and some of the appointments removed, see figure 10.20. (*Christie's, London, 27 September 1990, lot 22*)

George Adams senior, so Dudley's mother must have regarded his welfare as being of prime importance, since the continuance of both the family name and the business depended on his survival.

Scattered evidence of instruments supplied by Dudley during the early years of the nineteenth century survives in private archives, and of course in the files of the Office of Ordnance. For example, there are several Adams instruments with accompanying dated bills at Longleat, supplied to the then Marquis of Bath. In 1804 the latter bought from Dudley a large 'most improved' compound microscope, with a box of accessories, for thirty guineas. A copy of *Essays on the Microscope* at £1 12s 0d, and a packing case at half-a-crown, brought the invoice total to £33 4s 6d.[25] In 1807 he supplied three opera glasses to the

10.8 Detail of the inscription on an undated 12-inch terrestrial globe as modified by Dudley Adams after he had moved to Fleet Street in 1796. As well as Globemaker and Mathematical Instrument Maker to the King, and Optician to the Prince of Wales, Dudley claimed to be Instrument Maker to the Office of Ordnance. (*Christie's, London, 27 September 1990, lot 22*)

288 *Dudley Adams*

10.9   A gunner's perpendicular signed 'Adams London'. Though similar to the one shown in figure 10.10 there are minor differences in construction, and the style of the engraved signature is different. This instrument has a pocket case (not shown) of wood, pasteboard, and fish-skin, shaped to fit it. (*Museum of the History of Science, Oxford, inventory no. 51485*)

Globe Maker and Instrument Maker 289

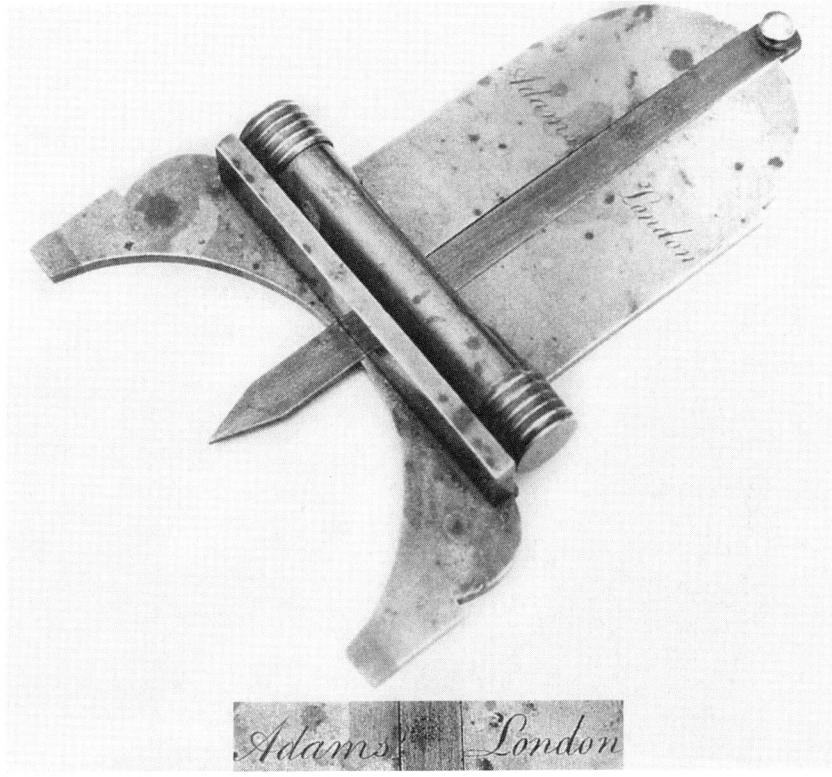

10.10 An example of an instrument signed 'Adams London' without an initial or street address, this is a perpendicular with a spirit level, similar in design to that shown in figure 5.10 signed 'G.Adams London' but with different proportions. (*Tesseract, Cat.16 [1987], item 47*)

Prince of Wales for £7 17s 6d.[26] A letter dated 20 October 1809 from Dudley to the Earl of Hardwicke, now in the British Library, refers to several instruments that the earl had sent in for repair, including a measuring wheel which Dudley said would cost £6–£7 and was not worth doing.[27]

Another noble customer was (presumably) the Duke of Bedford, for at Woburn Abbey there is a pair of extremely elaborate 'wheel instruments' signed (in a decorated Gothic style, as is all the lettering on these instruments) 'D.Adams, Fleet Street, London'. They are fully described and illustrated in Nicholas Goodison's *English Barometers 1680–1860*.[28] One instrument of this matching pair is a wheel barometer, the other is a circular-dial thermometer. The barometer also has a

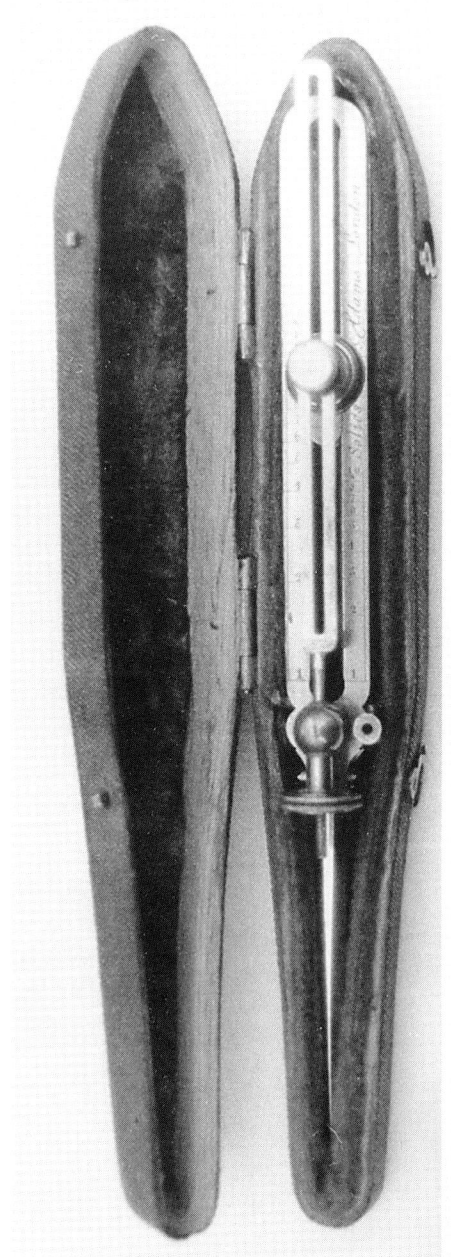

10.11 An example of a drawing instrument signed 'Adams London', namely a pair of proportional compasses in an individual case. *(Tesseract, Cat.28 [1990], item 42)*

10.12 A 360-degree brass protractor signed 'Adams London', with a vernier on the moving arm. The outside diameter is 6¾ inches, but this is probably what Adams would have called a '7-inch' instrument, matching his theodolites of that size. The fitted wooden case accompanying this particular instrument is inscribed on the outside 'R.Z. Mudge', that is, Lieutenant-Colonel Richard Zachariah Mudge, Royal Engineers, FRS (1790–1854). A later inscription inside the case states that after Mudge's death it was owned by Arthur H.D. Troyte, formerly Acland, who gave it to his daughter Frances in 1856; her son the Revd M.A. Bere presented it to the Lewis Evans Collection at Oxford in 1930. (*Museum of the History of Science, Oxford, inventory no. 45109*)

conventional thermometer with Fahrenheit and Reamur scales mounted on the stem of its case, while the thermometer has a Six's pattern max./min. thermometer in the corresponding position. Both instruments have an oat-beard hygrometer with a small circular dial at the top of the case. Goodison comments that the highly decorative cases belong stylistically to the period 1775–85, though Dudley's address suggests a date after 1795. A possible explanation for this anomaly is that they were made by Dudley as his 'masterpiece' at the completion of his apprenticeship, which would date them around 1784, but that they were not sold until he was in business in Fleet Street in his own right.

This is an appropriate point to mention that many late-eighteenth or early-nineteenth century instruments are extant (including some at Longleat) which have the abbreviated signature 'Adams London', without an initial or street address (see figures 10.11 to 10.13). Circumstantial evidence suggests that most – perhaps all – of these were made or sold by Dudley Adams after his brother's death, when the goodwill in the name 'Adams' was more important than Dudley's own forename. Dudley may have continued to trade under the name 'George Adams' for some time, at least so far as the shop premises in Fleet Street were concerned, if not for bookkeeping and invoicing. (Some garbled entries appeared in London directories in the next few years, including 'Adams & Dudley' at 60 Fleet Street; Hannah Adams was listed under her own name at that address in Lowndes's 1796 only; Kent's still had George Adams at that address until 1802.) There are other possible explanations for the 'Adams London' signature; for example, instruments so signed may have been supplied by subcontractors to both retail outlets during the period 1788–95, when George junior's and Dudley's businesses overlapped, but the post-1796 theory seems most likely.

Around 1806 Dudley suffered another blow: he lost the Ordnance trade. No reason for this has been found so far (the Ordnance records of this period are extremely voluminous). He was still supplying gunnery and other instruments in 1805, but the 'Extracts of Minutes' for 1806[29] show that in April that year the board considered tenders for supplying mathematical instruments by Fraser & Son, Watkins, and (Matthew) Berge. From these, the board selected Berge to be their sole supplier in future; he had the advantage of being already known to them, having taken over Ramsden's Ordnance work on the latter's death in 1800. The official price book of the Office of Ordnance for the early nineteenth century,[30] a volume about four inches thick, lists all types of mathematical instruments then purchased by the board on pages 142–6, with their prices, and gives Berge as their sole supplier, citing 7 April 1806 as his appointment date.

Several examples of shop bills with an engraved heading, incorporating the coats of arms of the king and the Prince of Wales, used by Dudley in 1807–1808, are extant. Significantly, the heading does not mention the Ordnance appointment. The latter has also been erased from the inscription on Dudley's globes dated 1813. Previously he had always drawn attention to his Ordnance appointment in his trade cards, shop bills, and newspaper advertisements.

The Buckinghamshire estate called Nutting Grove, revealed by George Adams senior's will, was mentioned in passing at the beginning of chapter 6. In 1809 the parish of Langley Marish, in which it was situated, was the subject of an Inclosure Act. This provides the first reliable information on the nature, size, and location of the estate. Coincidentally, Dudley's mother Ann died in that year. Under the terms of George senior's will the estate should have gone to all of Ann's surviving children in equal shares, but only Dudley and Isabella were still alive then (Lucy had died a spinster in 1805). The Inclosure documents[31] show that by the time the process had been completed, which took two years, Dudley Adams was the relevant landowner. Presumably he came to some private arrangement with his sister Isabella to provide her with her share in cash. The ratebooks of the parish, and the land tax records from 1781, show 'Mrs Adams' (that is, Ann) as the owner up to 1810, with a tenant in occupation. As no change in the rate and tax payments during the period of Mrs Adams's ownership is recorded, it is reasonable to assume that the following description, derived from the Inclosure documents (1809–11), can be applied retrospectively to the estate as it stood in 1772, when George senior died.

The homestead, Nutting Grove, was situated on a country lane now called Fulmer Lane, just south of the Turnpike Road from London to Gerrards Cross (now the A40). On the plan drawn by the surveyor to the Inclosure commissioners it is shown as a hollow square of buildings (a house, barns, stables and so on) adjoining the lane, with a piece of garden ground alongside, together with a 'home field' which formed a separate enclosure. A short distance along the lane towards the Turnpike Road were two fields and two meadows, on either side of the lane; these, together with the 'home field', formed the farmland belonging to the property. Their total area (including the ¾ acre occupied by the homestead and its garden) was just under 14 acres. When the estate

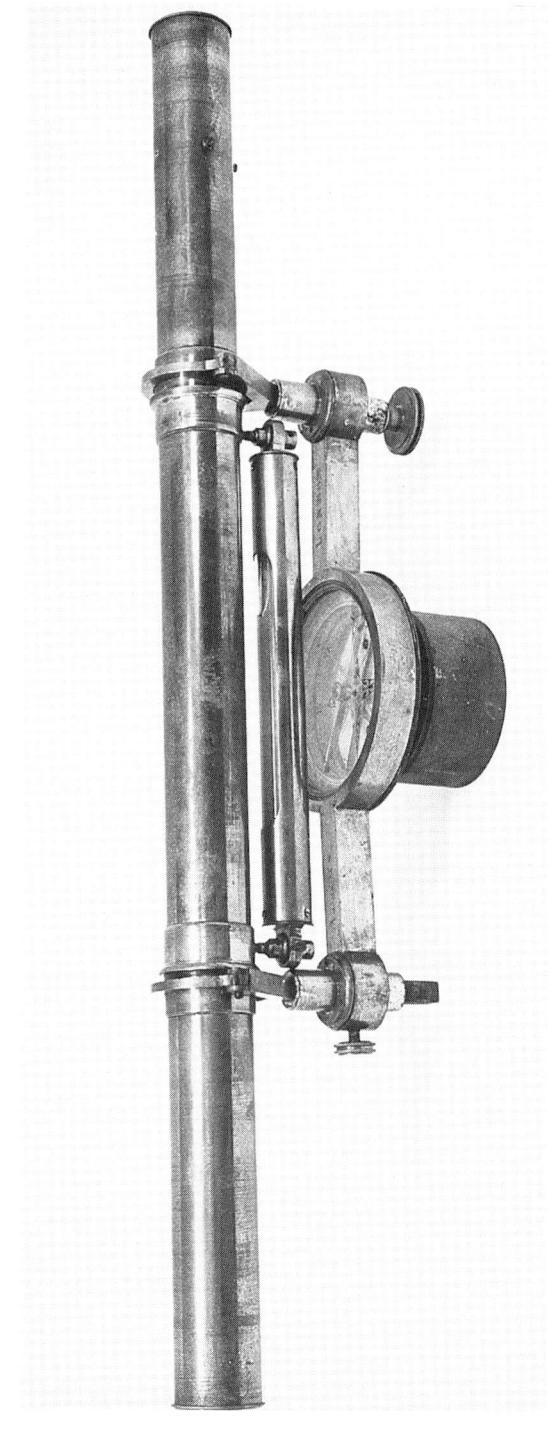

10.13 A surveyor's level, signed 'ADAMS LONDON' on the upper face of the horizontal bar. The front screw serves for coarse adjustment, and the back for fine. The wooden base on which the level currently rests is only for museum display purposes: it would normally be mounted as shown in figure 10.14(3). (Museum of the History of Science, Oxford, inventory no. 40612)

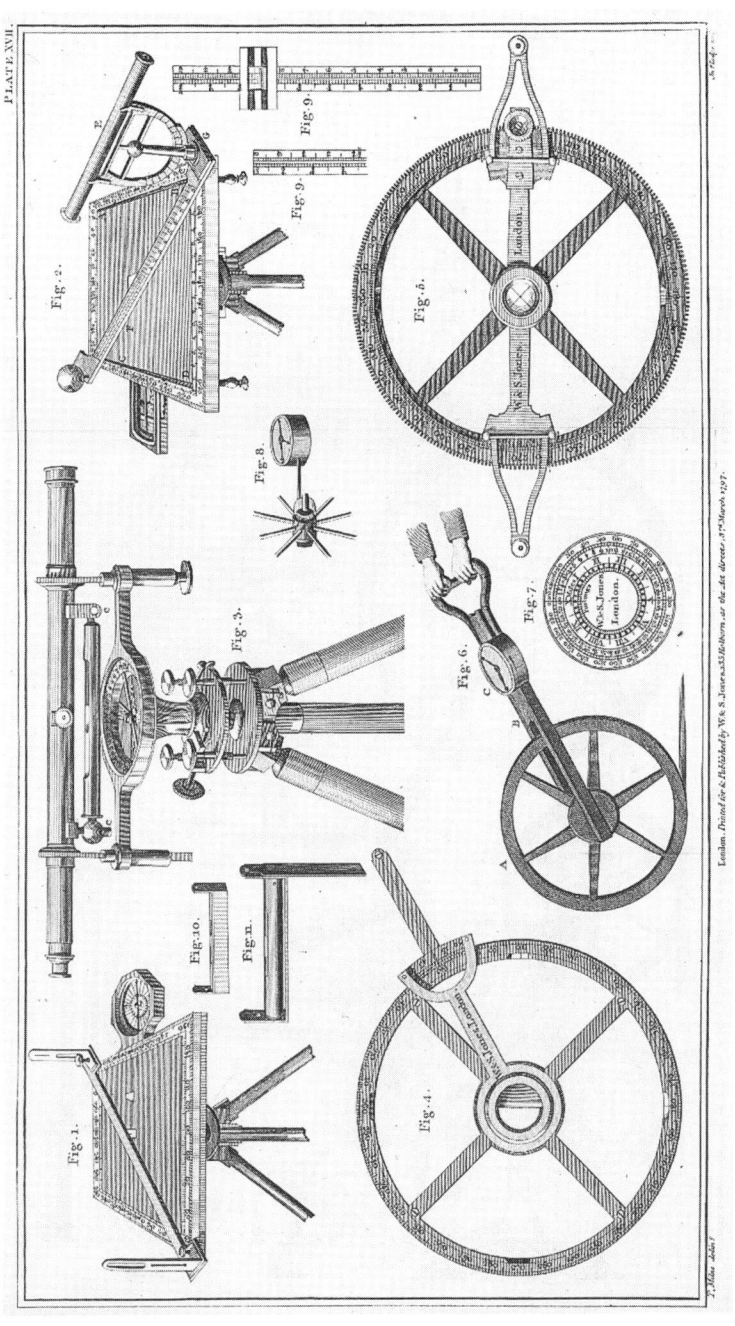

10.14 While Dudley Adams was running the Fleet Street business (post-1796), W. & S. Jones reissued George Adams junior's publications with their name on most of the instruments depicted instead of his. Figures 4 and 3 in this plate from *Geometrical and Graphical Essays*, second edition (Jones, 1797), showing miscellaneous surveying instruments, including level with telescope, match the contemporary instruments signed 'Adams London' shown in figures 10.12 and 10.13. (*Author's collection*)

295

10.15 Detail of the head of a mahogany stick barometer signed within a scroll at the top of the register plate 'ADAMS N°.60 Fleet Street LONDON'. (*Christie's, London, 18 July 1985, lot 20*)

10.16 A universal inclining dial signed 'ADAMS LONDON'. This type of instrument is sometimes mistaken for an equinoctial dial, but whereas the latter has a gnomon at right angles to the plane of the dial, an inclining dial has a gnomon and unequally-spaced hour divisions designed for a specific latitude, in this case 60 degrees north. For use in latitudes between 60 degrees and the equator the dial can be tilted to make the gnomon parallel to the Earth's axis, and hence the plane of the dial parallel to the horizon at latitude 60 degrees north, so that it still reads correctly. (*Whipple Museum, Cambridge, Inv.Wh.275*)

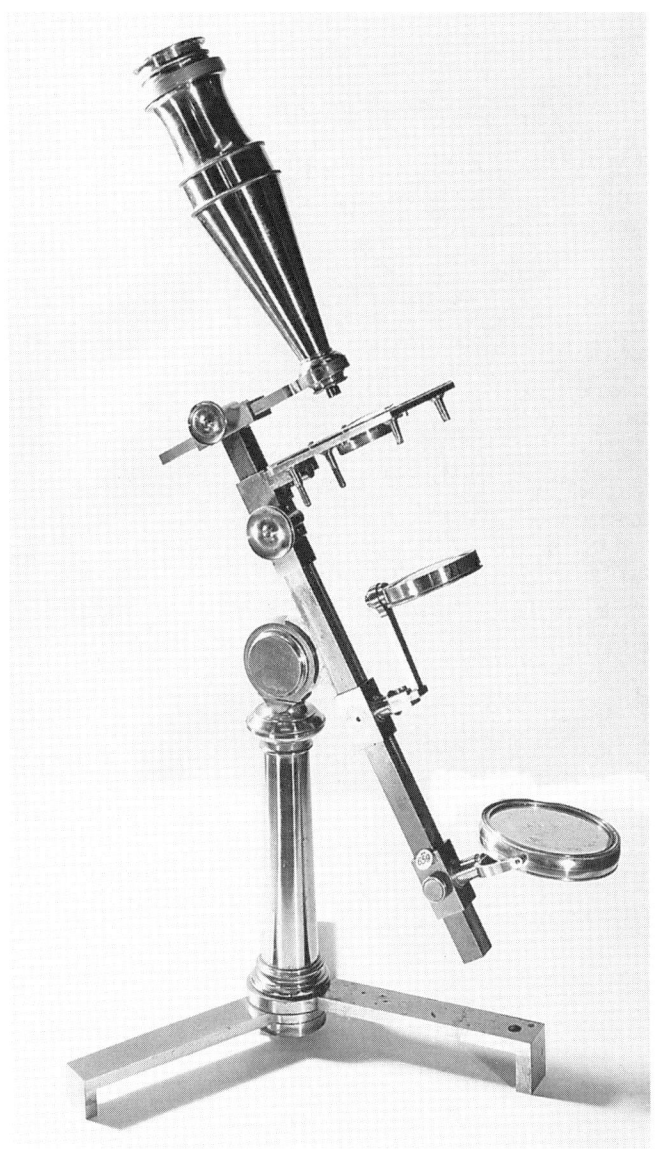

10.17 A compound microscope of the 'Jones's most improved' pattern (see figure 7.8), signed 'ADAMS LONDON'. Formerly in the Clay Collection, this large instrument, which stands about two feet high, is depicted with its numerous accessories on p. 200 of Clay and Court's *History of the Microscope* (1932). (*Museum of the History of Science, Oxford, inventory no. 45664*)

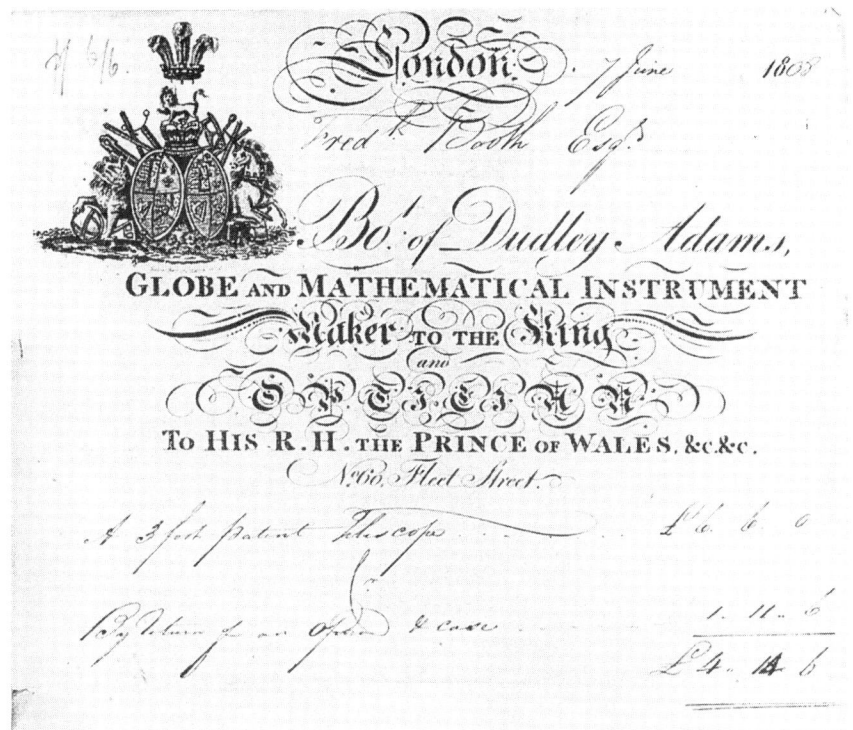

10.18 The engraved heading of this bill incorporates: (a) the shield of the royal arms as used from 1714 to 1800, which included the arms of France in one quarter as well as those of England, Scotland, and Ireland; (b) to the right of the above, the same shield with a 'label' of three points at the top, for the Prince of Wales; (c) the sovereign's crown; and (d) above that, the 'badge' of the Prince of Wales (the three feathers), all set against a background of mathematical instruments. By the date of the bill (1808) these arms were obsolete, the ancient English claim to the kingdom of France having been formally abandoned in 1800, but they were correct when Dudley received his royal appointments in the 1790s. (*Whipple Museum, Cambridge*)

was sold in 1817 the homestead was described as a pleasantly situated 'cottage residence' and outbuildings, with a verandah, lawn, paddock, gardens, vineyard, and orchard, 'surrounded by an interesting shrubbery and nut-walk'. (Some of these features may have been added by Dudley after 1809.)

An interesting point revealed by the Inclosure schedule is that Dudley's property was completely surrounded by land owned by the Duke of Somerset, of nearby Bulstrode Park (in another parish). The duke had

A
TREATISE,
DESCRIBING THE CONSTRUCTION AND EXPLAINING THE USE OF
NEW CELESTIAL AND TERRESTRIAL
GLOBES;
Designed to illustrate,
IN THE MOST EASY AND NATURAL MANNER,
THE PHENOMENA OF THE EARTH AND HEAVENS,
AND TO SHEW THE
CORRESPONDENCE OF THE TWO SPHERES;
WITH A GREAT VARIETY OF
ASTRONOMICAL AND GEOGRAPHICAL
PROBLEMS.

By GEORGE ADAMS, Sen.
LONG DECEASED. FATHER TO THE LATE GEORGE ADAMS.

THE THIRTIETH EDITION.
In which a Comprehensive View of the Solar System is given; and the Use of the Globes is farther shewn, in the explanation of Spherical Triangles.

Now Published by
DUDLEY ADAMS,
GLOBE AND MATHEMATICAL INSTRUMENT MAKER TO HIS MAJESTY; OPTICIAN
TO HIS R. H. THE PRINCE OF WALES, &c.; AND BROTHER
TO THE LATE GEORGE ADAMS.

LONDON:
Printed for, and sold by the Publisher, No. 60, Fleet-street.
1810.

10.19 Title-page of Dudley's reissue of his father's book on the globes, published in 1810 thirty-eight years after George senior's death. The wording follows the second and later editions of this title. Dudley's description of this edition as the 'thirtieth', when in fact only five editions had been published before it was superseded by George junior's *Astronomical and Geographical Essays* in 1789, led to some caustic comments by the instrument maker William Jones. (*Author's collection*)

10.20 A 12-inch celestial globe by Dudley Adams, probably produced in the first decade of the nineteenth century (see following page). (*Museum of the History of Science, Oxford, inventory no. 53095*)

10.21 Detail of the inscription on the globe shown in figure 10.20. Though apparently dated 1813, the last digit is a correction, perhaps from '0'. As the Prince of Wales became the Prince Regent in 1811, the basic engraving was probably last altered before then, to remove Dudley's appointment as instrument maker to the Office of Ordnance (*c.* 1806). Comparison with figure 10.7 reveals that this part of the inscription has been erased, leaving a blank space. (*Museum of the History of Science, Oxford, inventory no. 53095*)

purchased Bulstrode Park and its associated estate only just before the Inclosure of 1809, from the Duke of Portland. If no change had taken place since 1772, this means that when George senior owned Nutting Grove it was completely surrounded by land owned by the third Duke of Portland, William Henry Cavendish Bentinck, FRS (1738–1809), who succeeded to the title in 1762 and was Lord Chamberlain in 1765. The Bentinck family had lived at Bulstrode Park since the reign of William III. It seems likely, therefore, that Nutting Grove was originally a farm on the Bulstrode Park estate, and that George senior either bought it from the then Duke of Portland or received it in payment for some work done for the royal family.

At any rate, at the time of the 1809 Inclosure Dudley Adams was the relevant landowner, and as such was entitled to share in the distribution of the newly-enclosed common lands. He was allotted a piece with an area of 1¼ acres a little to the south of Nutting Grove. He also took the opportunity to purchase about 9 acres from another landowner who did not wish to take up his allotment, bringing his total holding from 1811 onwards to just over 24 acres.

The ratebooks and land tax records indicate that from 1811 until his bankruptcy six years later, Dudley both owned and occupied Nutting Grove himself. Situated about 18 miles from central London, adjacent to a major turnpike road (the eighteenth-century equivalent of a motorway), on a good day the travelling time to Fleet Street by coach or post-chaise would have been between two and three hours. While just feasible, it is unlikely that Dudley commuted daily; most probably he used his country estate as a weekend retreat. The auctioneer's sale particulars in 1817 reveal that it was then well stocked with sheep, pigs, poultry, seven milch-cows, and seven 'capital or useful blood horses'. The estate as a whole, said the auctioneer, formed a 'desirable retreat, favourable for the enjoyment of field sports'.

Nutting Grove appears on early editions of Ordnance Survey maps in the corrupted form 'Nutters Grove', which in view of Dudley's later mental problems is ironic. It continued to be listed as a separate address in *Kelly's Directories of Buckinghamshire* up to the 1890s, but its environment was drastically altered soon after Dudley's death by the erection of a substantial private house, pretentiously named Alderbourne Manor, on the 3-acre home field, with a carriage drive through the adjoining woods to a lodge entrance at the northern end of Fulmer Lane. However, most of the old buildings were apparently retained, to form a service courtyard entrance to the manor. What is believed to be the original 'cottage residence' still stands on the northern side of the courtyard, though the whole complex of buildings, including Alderbourne Manor itself, is now in multiple occupation as flats, and some large modern houses have been built in the grounds. When this area was visited in 1975 Dudley's fields and meadows and their boundary hedges, on either side of Fulmer Lane, could still be identified – they looked as though they hadn't been touched since his time – but a further even more drastic change occurred in the 1980s with the construction of the M25 motorway. This crosses Fulmer Lane exactly where the two fields and two meadows owned by Dudley as 'old enclosures' (that is, pre-1809) were situated, completely obliterating them, though the new road just misses Alderbourne Manor and the original Nutting Grove homestead.

While the Langley Marish enclosure was in progress, in 1810 Dudley decided to enter the bookselling field by reissuing his father's *Treatise on the Globes*, first published forty-four years earlier in 1766 (see figure 10.19). It had been superseded by George junior's *Astronomical and Geographical Essays* in 1789, and after Hannah had sold the copyright of this to W. & S. Jones, they continued to publish new editions up to 1803. Dudley said in a preface[32] that he did not object to this, so long as the book was kept in print, but when it was 'omitted to be published for nearly three years' (which suggests that the 1803 edition was reprinted around 1807) he felt it necessary to take some action. Immediately after his father's death, he said, his mother had given him all of his father's manuscripts, together with the copperplates for the illustrations. He had therefore decided to reissue his father's works, starting with *Treatise on the Globes*. Actually, it was the second (enlarged) edition of 1769 that he used. Inexplicably, he called his reissue 'The Thirtieth Edition', which has resulted in many reference books (including the *Dictionary of National Biography*) stating incorrectly that Adams's *Treatise on the Globes* 'went through 30 editions'. When the Jones brothers issued another edition of George junior's *Astronomical and Geographical Essays* two years later, they added a caustic comment to their preface:[33]

> The first edition [of George senior's book] appeared in 1766, the third in 1772. A reprint of this obsolete work was made in 1810, by the brother (D.Adams) of our late author, which he has denominated, *The Thirtieth Edition*!!!

When Dudley's effects were sold by auction in 1817, one of the sales included 1,000 copies of his 1810 edition of *Treatise on the Globes*. As this would be a typical print run for a book of this nature, it would appear that he managed to sell hardly any copies at all. No doubt that is why he abandoned the idea of reissuing his father's other titles.

In the following year, on 6 June 1811, Dudley's son George was formally apprenticed to him in the Grocers' Company.[34] Had it not been for this entry in the company's books, nothing at all would have been known about this member of the Adams family. A search through subsequent books failed to find any record of his freedom. The fact that young George is not mentioned in any family wills, especially that of his aunt Isabel (as Isabella now called herself) in 1825, strongly suggests that he died while still fairly young. Lack of an heir could well be what caused Dudley to neglect his business, leading ultimately to his bankruptcy.

In May 1814 Dudley issued a *Descriptive Catalogue* of the products sold at his Fleet Street shop. Rather more than just a price list, this ran to forty-four pages and included introductory comments on the various categories. The only known copy is at the University of Virginia Library. Dudley naturally described his own patent spectacles and portable telescopes in some detail, but otherwise there is not a great deal of difference, except in the arrangement, between Dudley's *Descriptive Catalogue* of 1814 and George junior's of 1795 reproduced in appendix II. Items listed in 1814 which were not mentioned in 1795 include Marquois improved parallel rules, Eckhardt's rolling parallel rules,[35] double drawing pens (for example, for drawing roads on maps), Adams's grand compound microscope (30 to 50 guineas), and an improved optical lantern fitted up with a variety of mechanical and plain sliders.

Dudley's telescope draw-tube patent of 1800 would have expired in 1814. In the following year he obtained his third and last patent (no. 3889, dated 7 March 1815), concerning the production of paper and vellum tubes for telescopes. It described a method of making multi-layer or laminated tubes sufficiently strong to be drawn through dies on a draw-bench, like metal ones. The paper tubes were built up layer by layer on a greased mandrel, made hollow to allow a heated iron to be inserted to expedite drying. A final layer of vellum finished off each tube, which was then drawn through a series of dies to a precise size. (Actually, the process may have been more one of scraping to size than drawing, a term which implies some ductility in the material.)

Dudley's last patent coincided with momentous events on the world stage, culminating in the Battle of Waterloo in June 1815. Napoleon's final defeat, marking the end of hostilities between France and Britain which had lasted for more than twenty years, was followed by a long period of high taxation and widespread unemployment. The year 1816 saw riots and mob violence in London as well as in the countryside. In the midst of this social upheaval, the Adams instrument business suddenly collapsed. Dudley's 1814 catalogue contained no hint of impending trouble, and he was evidently able to finance his 1815 patent application without difficulty, but the fact remains: in the spring of 1817, at the age of fifty-four, when he ought to have been enjoying the fruits of a long-established business, he became a bankrupt. Circumstantial evidence suggests that he was both a poor businessman and had been living well beyond his means; also, as indicated above, the loss of his only son may have destroyed all incentive to make the business succeed. The loss of the Ordnance trade may also have been a contributory factor in his downfall, though this happened ten years before the end. Probably he had overextended himself and did not have the financial reserves to cope with a recession, despite having mortgaged 60 Fleet Street for £2,000 (revealed by one of the post-bankruptcy sales of his effects). Alternatively, it is possible that Dudley, brought up in the environment of an eighteenth-century craft workshop, simply failed to move with the times and was overtaken by younger competitors using more modern machinery and methods. In the field of terrestrial and celestial globes, for example (always an important part of the Adams business, especially for Dudley), the new and continually updated products of Cary and Bardin must have seriously reduced Dudley's share of the market.

Whatever the reason, the fact of the matter is plain: the Adams instrument business, begun by George senior in 1734, came to an end in 1817, though Dudley himself lived for another thirteen years – almost to the very end of the Georgian era.

11 Bankrupt: the End of the Adams Instrument Business

Precisely what led to Dudley's downfall cannot be determined from official records, as detailed day-to-day accounts of bankruptcy proceedings of this period have survived in only a small proportion of cases, and his is not amongst them.[1] Nevertheless, a significant amount of information can be derived from the registers, enrolment books, official announcements in the *London Gazette*, and advertisements in other newspapers for auction sales of his various assets.

The bankruptcy of Dudley Adams of Fleet Street, 'optician and mathematical instrument maker, dealer and chapman', was first made public by an announcement in the *London Gazette* on Saturday 24 May 1817, and repeated in abbreviated form in all the other papers a few days later. Meetings of creditors were arranged for 31 May, 3 June, and 5 July, at the Guildhall. This was the standard procedure: if no complications arose, at the first meeting the creditors would prove their debts, at the second choose assignees to take charge of the bankrupt's assets, and at the third decide whether he should be granted his Certificate of Conformity. This was a legal document acknowledging that he had complied with the law in all respects. The bankrupt's assets, when identified, had to be realized by the assignees and distributed to the creditors by way of dividends, for which further meetings would be arranged in due course. This could be a protracted process: in Dudley's case the final dividend was not paid until February 1827 – ten years later.

Because Dudley owned some real-estate in Fleet Street and Langley Marish, and the transfer of property of this nature was rather more complex than the transfer of money or personal effects, the assignee handling his case (the creditors chose only one in this instance) needed to establish his (the assignee's) title to the property in law, so that it could be passed on to the purchaser. To achieve this, the assignee presented a petition to the lord chancellor when the sales of Dudley's assets had been completed, outlining what action had been taken and requesting that it should be enrolled in the records of the court. Though the primary purpose of such enrolment was to establish the assignee's title, for the historian the petition itself provides several interesting pieces of information which, in the absence of detailed files on the case, would otherwise have been lost.

According to the petition,[2] what started the ball rolling was a deposition on oath by Richard Thorne, builder, of Poppins Court, that Dudley Adams owed him £110 for 'goods sold and delivered and for carpentry work done and performed' between 1 January 1814 and 1 May 1817. On the strength of this evidence three commissioners had adjudged Dudley bankrupt and formal notice to that effect had been published on 24 May.

The petition proceeded to relate that at the meeting on 3 June, two people 'being the major part in value of the creditors of the aforesaid Dudley Adams present' had chosen Samuel Legg of Fleet Street (that is, the petitioner), upholsterer,[3] to be the sole assignee. The two people making this decision were named as Isabel Adams and William Taylor.

If, as suggested earlier, Dudley had come to some financial arrangement with his sisters after his mother's death to enable him to be sole owner of the Fleet Street and Langley Marish properties, then naturally Isabel (formerly called Isabella), as the sole survivor by 1817, would have been largely dependent on Dudley for her income and if Dudley subsequently found that he could not continue the payments, she would have been a major creditor. There is no evidence that Dudley and Isabel were estranged; indeed, in her will written in 1825 she left him some property in Dulwich (though in fact she outlived him), so her involvement in the bankruptcy proceedings was not due to malice, but rather necessity, as an elderly spinster dependent on her brother for income.

From the beginning of July to mid-September 1817 a succession of auction advertisements in the newspapers reveals the extent and varied nature of Dudley's assets. (Some of the advertisements do not mention him by name, but are clearly linked to his case by internal evidence.) At least two auctioneers became involved: Mr Scott and Mr Pullen. The first such sale, advertised to take place on 25 July by Mr Scott, auctioneer, 'by Order of the Assignee', was of 'A FREEHOLD ESTATE,

NUTTING GROVE, delightfully situated, agreeably elevated, and healthy, within half a mile of the great road over Gerrard's-cross-common'. The homestead was described as 'a COTTAGE RESIDENCE, with a viranda front in good taste, commanding the lawn, beautiful paddock, pleasure & productive garden and vineyard ...', and the accompanying land as 'six enclosures of rich meadow and arable land, conveniently near, containing in all 25 acres'. Particulars could be had at inns in Gerrard's Cross, Salt-hill, Hounslow, and Uxbridge, or from the auctioneer, of 28 New Bridge Street, the assignee (Mr Legg), of 71 Fleet Street, or the solicitor handling the case (Mr Claborn, of Tokenhouse-yard). The land tax had been redeemed, said the advertisements, and the purchaser could have immediate possession.

The parish records deposited in the Buckinghamshire County Record Office confirm that the land tax had been redeemed, but are inconsistent with regard to ownership.[4] According to the land tax returns, Dudley Adams was both owner and occupier from 1811 (that is, after his mother's death) until 1829; but there are indications that the clerk took little care over estates for which the tax had been redeemed, simply copying the names from year to year. The Church Rate entries are more likely to represent the true position. These have 'B.Way Esq., late Dudley Adams' from 1818 (that is, after the auction sale) to 1829, when Dudley's name is last mentioned. In other words, Benjamin Way Esquire, of Denham Place, Buckinghamshire, was the owner of Nutting Grove from 1818 onwards, presumably having purchased it at the sale. Way or his heirs subsequently built Alderbourne Manor on Dudley's home field, and although the name Nutting Grove continued to be used for the old house for another fifty years or so, the land associated with it in Dudley's time was integrated with Way's other substantial landholdings and the estate lost its separate identity.

Meanwhile another auctioneer, Mr Pullen, was making preparations for a sale of furniture and effects (but not at this stage the stock-in-trade) at 60 Fleet Street. The advertisements for this two-day sale (31 July and 1 August) indicate that Dudley had built up a personal library to replace that sold by Hannah twenty years earlier. As well as household furniture, the sale included a 'valuable LIBRARY, including Repertory of Arts, Rapin's England, Miller's Gardner's Dictionary, Beauties of England and Wales, English Encyclopaedia, and many other scarce books relating to Mathematics, Astronomy, Philosophy, &c.' The furniture included a 'capital office desk with drawers'. Some additional effects, including paintings, were sold later in September.

On 1 August, in parallel with the second day of the above sale, Mr Scott sold by auction the absolute reversion of two investments: £468 Bank of England stock and £130 3 per cent consols. The advertisements refer to the assignee, Mr Legg, and the solicitor, Mr Claborn, so these were almost certainly Dudley's. They were said to be vested under a will, and the purchaser would become entitled to them on the death of a lady aged fifty-nine. Isabel was then in her fifty-ninth year.

On the same day (1 August) Mr Scott also sold the absolute reversion of a mortgage of £2,000 on 60 Fleet Street, described as 'an excellent Freehold House, in a capital trading situation in Fleet Street, modern, erected with superior substantiality'. Like the two investments, this sum would revert to the purchaser 'on the death of a lady aged 59'.

The building itself was sold by Mr Scott three weeks later. Before then, on 11 August, he sold the livestock, furniture, and so on, at Nutting Grove, including 'seven capital or useful blood horses, seven milch cows, 69 sheep and lambs, pigs and poultry, a waggon, carts, ploughs, harrows, a drill-machine'. The contents of the residence included 'some books selected with taste, a few good pictures and prints', and other effects – and some 'valuable dogs'. If the seven blood horses were used by Dudley himself for hunting and other country pursuits, it suggests that he fancied himself in the role of a country gentleman; but 24 acres is unlikely to have been sufficient to support such a lifestyle, so the upkeep of his country retreat must have been a further drain on the resources of the instrument business in London.

On 21 August Mr Scott sold 60 Fleet Street, 'with early possession'. The building was described as being 'very roomy and commodious, four stories above the ground storey, three rooms on a floor, for many years an eminent Optician's'. It was 'judiciously arranged, erected with superior substantiality, completed at an unlimited expense, with convenience considered at every point', and in exceedingly good repair. In 1791 this property had been rebuilt, or at least refurbished, so it was effectively only twenty-six years old at this time. Tallis's London street views (1838) show a typical eighteenth or early-nineteenth century London terrace building, with a central doorway and two shop windows at street level, and three main storeys above, each with two sash windows, and possibly an attic storey in the roof. Counting the latter, this agrees with the auctioneer's description. Despite extensive redevelopment of Fleet Street in the twentieth century for large newspaper offices, numbers 58, 59, 60, and 61 are still individual properties

The End of the Adams Instrument Business 311

61   60   59   58

11.1 60 Fleet Street and its neighbours, photographed in 1991 (number 61 was then shrouded in builders' scaffolding). The brick building constituting number 60 appears to have been rebuilt (or at least refaced) in the mid or late nineteenth century. The large projecting cornice below the top storey suggests that the latter has been added, or possibly created out of a former attic. When this photograph was taken the building was in multiple occupation by several different traders; the ground floor (street level) was a restaurant. (*Photograph by the author*) today. The black-painted brick building which now constitutes 60 Fleet Street is much older and plainer in style than its immediate neighbours (see figure 11.1). Except that it has a large modern display window at first-floor level, a total of five storeys above the shop rather than four, and curved window heads instead of square Georgian ones, it is broadly similar to that depicted in Tallis's view. Perhaps the front was rebuilt in an old style in recent times, or after damage in the Second World War.

After the sale of the building itself, in the following week it was Mr Pullen's turn again, with the longest of this series of sales, the stock-in-trade at 60 Fleet Street. This sale occupied four days, 26–29 August, and was advertised from 12 August right up to the last day of the sale. According to the advertisements, 'The valuable STOCK of Mr.DUDLEY ADAMS, Mathematical Instrument Maker to his Majesty, Optician to his Royal Highness the Prince Regent, &c.' included all the principal types of instruments normally sold by the Adams business. Those picked out for special mention included 'a fine American double barrel air pump', 'Adams's patent and other spectacles', and 'Adams's patent portable telescopes'. Descriptive catalogues, at 1s each, were available, but unfortunately no extant copy has been discovered. Mr Pullen's business address was '80 Fore-street, Finsbury-square'.

That was not quite the end of this series of sales. A further two-day sale by Mr Pullen took place on 16 and 17 September, the advertisements for which reveal that as well as the residential and shop premises at 60 Fleet Street, Dudley had a 'factory' at 16 Gough Square, only a short distance away off the other side of the street. The ratebooks of St Dunstan's show that he had occupied this address since the second quarter of 1806, coinciding with his loss of the Ordnance appointment. This suggests that he had a (perhaps larger) factory somewhere as yet undiscovered for Ordnance work, and when he lost the latter, he found more suitable premises near to his shop for his civilian trade. As the Gough Square property was rented, only the contents, not the building itself, were sold when he became a bankrupt.

This last sale included 'the remaining STOCK of MATHEMATICAL and other INSTRUMENTS; valuable working tools; complete sets of plates for globes, from 3 inches diameter to 28 inches; a 28-inch celestial globe in an elegant frame, supported by figures; a ditto sphere corresponding', and no fewer than fifty 'globes, unfinished'. Dudley's personal property included 100 ounces of plate, gold watches, six paintings by Reinagle, his telescope patent, and 1,000 copies of his book on the globes, as well as several instruments, such as a large reflecting telescope on a carriage, with rackwork. The workshop tools included 'an excellent draw-bench, with racks and triblets; a fly-press; 10 good lathes; benches; and vices'.

The reference to '50 globes, unfinished' suggests a prolific production line brought to a sudden halt; but on the other hand, the fact that so many were in stock could indicate a moribund factory filled with unsaleable products. Some of the globe plates, in particular the 28-inch size, must originally have been Senex's, which had been reworked over and over again in a period of nearly a century, so their sale value in 1817 may not have been very great. Ten lathes, unless some were reserved for special purposes, suggests a workforce of at least that number of men. The pictures by Reinagle, the 100 ounces of plate, and the gold watches, provide a glimpse of Dudley's lifestyle, and when combined with the description of his country estate at Langley Marish suggest that he had been living well beyond his means.

Soon after this sale the assignee (Mr Legg) drew up his petition to the lord chancellor, stating that both the Fleet Street and Langley Marish properties had been sold. It was approved by the lord chancellor on 15 December 1817. A fortnight later, formal notice was given in the *London Gazette* that the commissioners intended to meet at the Guildhall on 22 January 1818 to declare a first dividend.

Not all of the debts had been gathered in by then. In the files of the Royal Institution there is a record of the payment of 17 guineas to the assignee of Dudley Adams on 20 April 1818, for apparatus supplied by him between 1802 and 1805, more than a decade earlier.[5] If this was typical of Dudley's accounting, it is not surprising that he went bankrupt. One wonders how many other uncollected debts the assignee had to chase. Over a year elapsed before a second dividend was declared, on 6 April 1819. Finally, a meeting was held eight years later, on 13 February 1827, to declare a final dividend and close this case.[6]

60 Fleet Street stood empty for a few months, and was then taken by Messrs Howes, Hart & Hall, who opened a 'Shawl and India Warehouse' there (see figure 11.2). The premises remained in their occupation for forty years, until the last survivor, William Hart, announced his retirement in April 1857. A succession of unimportant tenants with various occupations occupied 16 Gough Square. What happened to Dudley's extensive stock-in-trade and workshop tools is uncertain, but part may have been purchased by W. & S. Jones, who had purchased the stocks and copyrights of George junior's books twenty years earlier. Catalogues issued by W. & S. Jones after 1817 included 28-inch globes

> **SUPERB INDIA and BRITISH SHAWLS.—** Every succeeding Season being marked by an increasing demand for elegant SHAWLS, has induced HOWES, HART, and HALL to replenish their Assortment, by continually introducing the most splendid Novelties of Oriental and European Taste. The Nobility and Ladies wishing to select from a variety of superb Shawls, unrivalled in extent, brilliancy and elegance, are invited to inspect the matchless Selection now exhibiting at the India Warehouse, 60, Fleet-street.—N.B. The utmost value given for India Shawls.
>
> **BRADBERRY'S PATENT SPECTACLES,** to be had only at No. 28, Holles-street, the first door out of Oxford-street, on the left hand, facing Hanover-square, where the business is conducted upon the same unerring principles which have given such general satisfaction for upwards of 30 years in London, and all parts of the United Kingdom.— Orders from the country will be immediately attended to.

11.2 Following Dudley's bankruptcy in 1817, 60 Fleet Street, which had been occupied by the Adams business since 1757, passed out of the family and ceased to be associated with scientific instruments. For several decades it was a 'Shawl and India Warehouse'. This advertisement for Howes, Hart & Hall is from *The Morning Chronicle* of 31 January 1821, p. 1 col. 2. The advertisement for Bradberry's patent spectacles beneath it was frequently repeated around then, but no English patent is recorded in that name. On the back page of the same paper is an advertisement for an auction sale of effects 'removed from a mansion in the country' which includes several scientific instruments, notably 'a powerful reflecting telescope by Adams', the only maker mentioned by name. (*Author's collection*)

'with the modern discoveries added in English', but it does not necessarily follow that the Jones brothers owned the plates and tools themselves: a specialist globe-making firm could have been doing the actual manufacture, the Jones brothers being merely retailers.

Though Dudley's bankruptcy marked the demise of the Adams family of Fleet Street as instrument makers, twelve years later the Adams name was revived in rather puzzling circumstances. Francis West, a 'working optician' who had served a seven-year apprenticeship in the Spectaclemakers' Company commencing in 1806,[7] decided at the age of thirty-six to move into the retail sector. He obtained his freedom in the Spectaclemakers' Company on 8 July 1828, and took a shop at 83 Fleet Street, two doors east of Salisbury Court. Though primarily an optician, he began selling all sorts of mathematical and philosophical

TO ALL WHO VALUE THEIR SIGHT.

A
FAMILIAR TREATISE
ON
THE HUMAN EYE:
CONTAINING
PRACTICAL RULES
THAT WILL ENABLE ALL TO JUDGE WHAT
SPECTACLES
ARE BEST
Calculated to Preserve their Eyes
TO EXTREME OLD AGE.

BY FRANCIS WEST,
SUCCESSOR TO MR. ADAMS, OPTICIAN TO HIS MAJESTY
83, FLEET STREET.

ILLUSTRATED WITH THREE CORRECT
DIAGRAMS OF THE HUMAN EYE.

Third Edition.

MAY BE HAD OF ALL BOOKSELLERS AND OPTICIANS.
Price Sixpence.
1829.

11.3 Title-page of a tract dated 1829 by Francis West, optician, of 83 Fleet Street, showing his claim to be 'successor to Mr. Adams', more than a decade after the Adams business had ceased to exist. (*British Library, London, T.1332/5*) instruments, and developed a line of inexpensive products for the educational market.[8]

He also began publishing numerous descriptive and explanatory booklets, some written by himself and some reprinted from the works of long-dead authors such as Benjamin Martin.[9] On the title-pages of these publications West called himself 'successor to Mr. Adams'. He did not specify which Mr Adams, but added initially 'Optician to his Majesty' (see figure 11.3) and later 'Mathematical, Optical, and Philosophical Instrument Maker to his Majesty'. The form of wording used was ambiguous – perhaps deliberately so – and could be taken to mean either that Adams or West himself held the royal appointment.

Both George junior and Dudley held (successively) the appointments of optician to the Prince of Wales and mathematical instrument maker to his majesty during the reign of King George III. Dudley's appointments were correctly given in the advertisements for the sale of his stock-in-trade on 26–29 August 1817, the Prince of Wales having become the Prince Regent by then. On George III's death all appointments to 'his Majesty' lapsed, though in practice many were renewed by George IV early in his reign; for example, about sixty traders were appointed on 5 April 1820, including P. & J. Dollond, opticians, Robert Bancks, mathematical instrument maker, and T. & W. Harris, globe makers and opticians.[10] The latter firm had taken over Dudley's mantle as globe makers to the Prince Regent (in lieu of 'his Majesty') in 1819, though they were not particularly well known in this field in comparison with Bardin, Cary, and Newton. In general these trade appointments, being unsalaried, were not unique: G. & C. Dixey were appointed opticians to his majesty in 1824, and R.B. Bate likewise in 1828, in parallel with Dollond. However, no trace has been found in the lord chamberlain's files of Francis West receiving a royal appointment.

It is possible that West received a royal appointment through some other channel, but even so, why he should have chosen to cite the Adams name remains a mystery. Twelve years elapsed between the closure of the Adams instrument business in Fleet Street and West beginning to call himself 'successor to Mr. Adams' on his publications: the name must have already faded into history by then. A possible explanation is that West had acquired some of Dudley's stock, such as 3-inch globes (a size that West is known to have sold),[11] and adopted the title of Adams's successor to explain why he was selling products bearing the Adams name. His ambiguous citation of the latter's royal appointments was probably designed simply to give his own business a boost, there being little control over such matters until much later in the nineteenth century.[12]

12   Electrician and Political Reformer

Dudley's movements in the year or two following the loss of his business, and his homes in Fleet Street and Langley Marish, are obscure. A peculiar entry in the *Post Office London Directory* for 1819 calls him 'Engineer in Philosophical Instruments and Globes to His Majesty, H.R.H. the Prince Regent, and to all the Royal Family', at 10 Waterloo Place, near Carlton House. However, no evidence that he ever traded from there has been found: it seems to have been just a last desperate attempt to retain some link with his former clientele. In the same year (1819) he wrote several letters from 14 Duke Street, St James's, to the Prime Minister, the Earl of Liverpool,[1] two of which have survived, plus another of uncertain but probably similar date. They reveal that by this time Dudley was beginning to show signs of the (at best) mental confusion which was evident throughout the last ten years of his life. In 1814 the earl had been admitted to the freedom of the Grocers' Company,[2] so perhaps Dudley thought he would listen to the views of a fellow-member; but if the earl had a 'letters from cranks' file, Dudley's would undoubtedly have been deposited therein.

The first dated letter to survive, but not the first that Dudley wrote to the earl, is inscribed 'October 1819' (no day). It is replete with underlining and long dashes:[3]

> Mr. Adams having addressed the Earl of Liverpool, on various subjects; — as, on a <u>Successful Cultivation of the Arts,</u> — including Sculpture, — Engraving, — Perspective, — Architecture, — Drawing in General, — Painting, Light, and Shadow, — grounded on their true Bases, <u>Mathematical</u> and the <u>Geometrical</u> Sciences; — Opening <u>new Sources</u>, for Trade and Commerce; — <u>Improving</u> the Revenue; — On the <u>Effectual support</u> of the Labouring Poor, <u>rise and progress</u> of the Manufacturing Class, which the Poor Laws

can never Effect; — the <u>Modulation and Modification</u> of the Old, — and propositions for new, and Popular Taxes, not affecting the Poor; — a <u>revision</u>, and <u>correction</u> of the present System acted upon, in Agriculture, with a view to great and most beneficial ends, &c.; He (Mr.Adams) solicits the Honour of an Interview with his Lordship thereon.

It is most unlikely that the earl granted the request for an interview. Realizing that he was getting nowhere, on 15 November Dudley went along to the earl's address and handed in a note,[4] saying that as he had received no reply to his several letters, and suspected that none was intended, he would be grateful if the earl would return them. 'Mr.Adams waits on Lord Liverpool, in person, to receive them.'

How long he had to wait on the doorstep can only be conjectured, but it evidently was not overnight, for the next item in the correspondence file is a copy of a note bearing the same date, handed to Dudley with the latter's returned letters; it is signed 'T.C.B.', presumably one of the earl's staff.[5]

Dudley's own note is endorsed 'Letters returned dated Augt 1818 & Oct. 1819', though the particular letter of October 1819 quoted above was evidently missed, or it would not be in the file today. Also in the file is a long rambling three-page letter from Dudley, with no address of origin or date, which is currently placed amongst the correspondence of 1814 but (from its contents) was probably written in 1818 or 1819.[6] In it Dudley refers to the attention he has given in his retirement to the welfare of the state, and says he had predicted the downfall of Napoleon; but the precise nature of his cure for all the nation's ills is not disclosed. He says that he can unfold measures that will 'put down Sedition ... Exhalt the Government and the Nation, placing them in their true Meridian as to Eminence with regard to other Kingdoms and Empires ...', but it was necessary that he should confer with the earl in person to achieve this. It is difficult to imagine any such conference taking place.

At about this time (1819) Dudley, searching perhaps for some means of earning a living, suddenly switched his attention to medical applications of electricity. This may have been prompted by the death of one of the leading practitioners, Francis Lowndes. George Adams junior's *Essay on Electricity* had been published thirty-six years earlier in 1784, the year when Dudley finished his apprenticeship. The first three editions contained a few pages on medical electricity, so Dudley would have had an early introduction to the subject, but (as outlined in chapter 7) it was not until the fourth edition was published in 1792 that Adams's cursory treatment of the subject was enhanced by appending a long letter (fifty-five printed pages) from John Birch, an assistant surgeon at St Thomas's Hospital, giving details of specific cases that had been successfully treated there. When George Adams wrote in the 1780s and 1790s he expected Birch's efforts to result in electricity becoming an accepted tool for the treatment of certain complaints. However, Birch seems to have been an isolated believer in the healing power of electricity: the vast majority of medical men ignored it, and indeed were actively hostile to anything which threatened to challenge their time-honoured reliance on leeches and drugs. *Essay on Electricity* was reprinted only once by the Jones brothers after George junior's death (fifth edition, 1799). Birch himself republished some of the cases described, with additional ones, in 1802,[7] and at his death in 1815 held the post of surgeon-extraordinary to the Prince Regent, but by the 1820s static electricity produced by cylinder or plate machines was virtually obsolete (except for demonstration purposes). The new wonder of the age was 'galvanism', a different form of electrical treatment, which received royal approval in the 1830s.[8]

Early in 1820 Dudley petitioned the court of the Grocers' Company for assistance in 'the Establishment of an Electrical Apparatus'. His petition was considered on 4 February 1820 but was 'ordered to lie upon the table'.[9] Dudley apparently complained about this, for the letter books of the company show that on 29 February the clerk wrote to inform him that his 'Representation' had been laid before the court on the 24th, but had also been ordered to lie upon the table, and that 'it is not probable that the consideration of it will be renewed'.[10]

Despite this setback, somehow or other Dudley managed to acquire the late Francis Lowndes's apparatus, said (by Dudley) to have been the most powerful in Europe, together with the remainder of the lease of his premises at 42 St Paul's Churchyard, to which address he moved forthwith, calling himself a 'Professional Medico-Electrician'. This was to be his principal occupation for the next seven years. In March 1820, only a few weeks after his unsuccessful application to the Grocers' Company, he issued his first publication on the subject, *Electricity is the Fountain* (see figure 12.1). In this fifteen-page tract he explained that he was not trying to compete with the established medical profession: physicians, surgeons, and apothecaries were the proper persons to prescribe and dispense drugs, but in many cases drugs alone could not effect a cure. Because electricity was 'the great vivifying principle of nature', only a combination of orthodox medicine and electrical treatment administered by persons such as himself was the complete answer.

ELECTRICITY

IS THE FOUNTAIN,

THE GREAT VIVIFYING PRINCIPLE OF NATURE;

A SOURCE OF LIFE AND HEALTH:

MEDICINE

THE REQUISITE ACCESSARY TO THAT SCIENCE.

AN ADDRESS

To the Faculty in particular, and to the Public in general, on the subject of *Electricity*, as applicable to the human frame; pointing out the importance and requisite intervention of its power in arresting the progress of Diseases incidental to our nature; and the necessity of its constituting a very essential and leading part of Medical Practice: Designed with a view to the more certain cure and extirpation of Disorders generally, than has hitherto been accomplished: by that means to uphold the art of administering Medicine, as a science susceptible of rendering the most eminently beneficial services to the community, and the rising generation.

BY DUDLEY ADAMS,

PROFESSIONAL MEDICO-ELECTRICIAN,

At No. 42 (the late Mr. Lowndes's), St. Paul's Church Yard, London.

N.B. To be had of the Author, and all Booksellers.

LONDON:

PRINTED BY W. DAVY, 41, JAMES STREET, GROSVENOR SQUARE.

1820.

Entered at Stationers' Hall.

12.1 Title-page of *Electricity is the Fountain*, Dudley Adams's first publication on 'electrical medicine'. At this time (1820) he was occupying Lowndes's former premises at 42 St Paul's Churchyard. (*British Library, London, 1172.k.5/2*)

References in this tract indicate that Dudley had been strongly influenced by Patrick Brydone, FRS (1736–1818), a teacher, traveller, and occasional writer on electricity.[11] Quoting Brydone, Dudley said that electricity was 'a fifth element, distinct from, and superior to, the other four ... a kind of Soul, that pervades, and quickens, every particle of matter'.

Discussing the various conditions for which medical science knew no cure, and for which electricity might provide relief, Dudley particularly mentioned consumption or peripneumonia, 'so peculiar to the climate of England, from the variable state of the weather, the use of pit coal, &c.' He also listed 'the goitre or wen of the Pays du Valais, which Mr.Adams has so often witnessed during a residence of two years at Geneva, and in its environs'. This intriguing comment raises the question, at what period of his life did he spend two years in Switzerland? It surely cannot have been while he was running his globe and instrument business, so there are three possibilities: as a child, before 1777; between the completion of his apprenticeship in 1784 and his marriage in London in 1787; or between his bankruptcy in May 1817 and his residence in London from October 1819 onwards. As the bankruptcy proceedings required his presence in London for most of the year 1817, that hardly leaves time for a two-year residence in Switzerland before he started writing letters to the Prime Minister in 1819. Brydone himself was travelling as a tutor in Italy and Switzerland between 1765 and 1771, but Dudley would have been only nine years old at the latter date. The most likely period would appear to be just after the completion of his apprenticeship: perhaps it was in Geneva that he met the widow with the French-sounding name whom he married in London in 1787.

Neither Lowndes himself nor Dudley described the former's electrical apparatus in detail. Two small publications by Lowndes are known,[12] whose contents consist mainly of a recital of cases successfully treated by electricity rather than details of the technique employed. His literature references show that he was familiar with practically everything written on medical electricity by both British and foreign authors, and published in scientific journals as well as in the form of books or tracts.[13] According to a nineteenth-century editor of William Buchan's popular *New Domestic Medicine*, it was under Buchan's auspices that Lowndes began to practice medical electricity, a form of treatment to which Buchan (1729–1805), unlike the majority of his contemporaries in the medical profession, gave his cautious approval. Posthumous editions

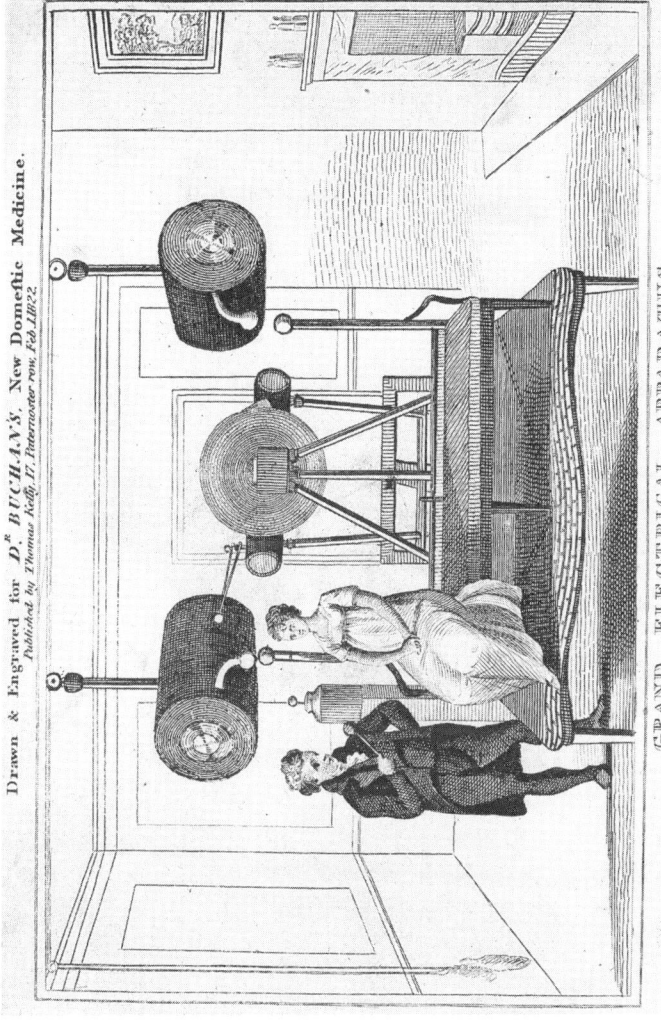

12.2 Lowndes's medical electrical apparatus, as depicted in posthumous editions of Buchan's *New Domestic Medicine*. This plate when first published was dated 25 February 1809. No indication is given of the size of the apparatus, and it seems likely that the human figures and the electrical components are not drawn to the same scale. This reproduction is from the 1822 edition, by which time Lowndes was dead and the apparatus belonged to Dudley Adams, but although the plate was partially reworked to emphasize the background the caption was not altered to acknowledge this change of ownership. (*Vade-Mecum Press, London*)

A

TRIFLE,

USTRATIVE OF THE INSUFFICIENCY

OF

THE MATERIA MEDICA,

AS AT PRESENT CONSTITUTED,

TO PRODUCE EFFICACIOUS RESULTS IN THE

Cure of every Disease;

WITH

The Author's respectful Invitation to the honour of a personal Conference with Gentlemen,

ON THE

MODE WHICH HE PROPOSES TO ADOPT

In Relief, and for the Benefit,

Of that portion of the Community, who are suffering under Complaints, erroneously termed, because hitherto considered as, Incurable.

WITH A LETTER TO THE FACULTY.

By DUDLEY ADAMS,

Medico-Electrician,

No. 22, LUDGATE STREET, LONDON.

London:
Printed by W. Davy, 41, James Street, Grosvenor Square.

1823.

12.3 Title-page of Dudley Adams's second publication on 'electrical medicine', issued in 1823, by which time he was beginning to realize that he would get no co-operation from the medical faculty. (*British Library, London, T.847/1*) of *New Domestic Medicine* published around 1810–25 incorporate the only known illustration of Lowndes's apparatus (figure 12.2).

Lowndes's view – in a nutshell – was that by gently electrifying the whole body, all its vital functions could be accelerated, hastening a cure; specific organs could if necessary be given more concentrated treatment by applying discharges from pointed or rounded electrodes at the appropriate spots. Dudley, in his 1820 tract, referred to patients being placed in an 'electrical bath', but he carefully explained that the bath contained no fluid but electricity, so no article of clothing need be removed. The illustration from Buchan shows a female patient sitting comfortably on an insulated couch (see figure 12.2), but if the globe machine in the background, and the Leyden jars (?) suspended overhead, were drawn to the same scale, Lowndes's apparatus must have been capable of blasting the unsuspecting patient into space. Some artistic licence must surely have been employed by the engraver of this scene.

Dudley and his new occupation at 42 St Paul's Churchyard were listed in *Robson's London Directory* for 1821. Kents, on the other hand, still had him as a mathematical instrument maker at 60 Fleet Street up to 1823, illustrating how out of date some directories could be. By March 1823 he had moved his electrical medicine business to 22 Ludgate Street, near its northern corner with St Paul's Churchyard, approximately where Juxon House now stands. He remained here for at least five years; but his name is not recorded in the ratebooks for this address, which, with a rateable value of £150, was probably a large building in multiple occupation. Dudley presumably rented a small part of it; it is not clear whether he actually lived there, or simply used it for business purposes.

At 22 Ludgate Street, Dudley established his Medico-Electrical Institution, where (he said) he attended from 12 noon to 4 p.m. (except on Sundays) to treat prospective patients. His first publication from there, commenting on the materia medica and its shortcomings (see figure 12.3), was a small booklet of twenty-four pages, dated March 1823. By this time the hopes expressed in *Electricity is the Fountain*, that he and orthodox medical practitioners could work together to advance medical science, had largely evaporated, though he still hoped to arrange a dialogue with members of the faculty. However, with very few exceptions such as John Birch (who had died in 1815 leaving no obvious successor), medical practitioners who had been trained in the orthodox manner were extremely hostile to new-fangled ideas that they didn't understand. Dudley did not help matters by his scathing comments on the inability of doctors relying on the materia medica to cure a wide range of conditions, which he claimed were amenable to treatment by electricity. In footnotes in this tract he cited several specific cases, brought to him as a last resort when treatment by drugs had failed, where he was able to give immediate relief by immersing the patient in his 'medico-electrical bath'. The latter, which Dudley claimed to have invented himself, was apparently a development of Lowndes's insulated couch. He did not describe its precise form, but clues provided by remarks on certain cases indicate that it was situated about a foot above ground level, the patient was seated in it, and its overall dimensions were about 5 feet 4 inches high by 2 feet 6 inches wide. Dudley claimed to have shown by experiment that, contrary to general belief, the electric charge was not confined to the surface but penetrated the entire body of the patient. By breathing the electrified air in the bath, he said, the patient could obtain immediate relief from obstructive complaints such as asthma. This remark suggests that the 'bath' was not a complete Faraday cage, but was more akin to an ionization chamber, which would explain its apparent effectiveness in cases such as asthma and hay fever.

Appended to this booklet were Dudley's terms for administering the 'Electrical Element', starting with three applications for 1 guinea and rising to fifty applications for 8 guineas. The latter was considered to be 'a course for the perfect restoration of health in most cases', but more extended treatment could be spread over a year, either three times per week for 15 guineas or every day except Sundays for 25 guineas. A payment of 100 guineas would buy the right to treatment every day except Sundays for life.

Two letters to the Right Honourable Robert Peel (1788–1850), home secretary in the Earl of Liverpool's administration, indicate that by 1825 Dudley had resumed his practice of addressing members of the government on a combination of electrical and political issues. It is quite likely that this was a continuous activity, his letters to the earl and Robert Peel being just the tip of the iceberg, known simply because their recipients' files have been deposited in the British Library. On 31 March 1825 Dudley sent Peel a copy of his latest publication, with a three-page covering letter explaining that it was the first part of 'a new and splendid work'.[14] As well as dealing with topics such as 'the removal of the existing grievances in the Sister Kingdom' (presumably Ireland), he would, he said, like to draw Peel's attention to his important discovery of the 'Electrico-Ethereal Element', the result of six years' study 'in the face of much interested opposition and discouragement'. To enable Peel to view the effect produced by this 'Wonder of the Age' in comfort, he invited him to make an appointment outside his normal public hours. An endorsement on this letter reveals that the enclosure was the first number of *Religion combined with Science*, a work which is also mentioned in Dudley's last publication in 1827. However, no extant copy of this has been located.[15]

Dudley's letter and enclosures were acknowledged (according to endorsements on the letter) on 2 April, but it is most unlikely that Peel responded to Dudley's invitation to visit 22 Ludgate Street. On 1 December he tried again, with another three-page letter reiterating the urgent necessity to establish a 'new System of Therapeutics' to rescue the medical art from 'the low state of degradation and disrepute' into which it had fallen.[16] He would be happy, he said, to lay his proposals before him, in order that they might be 'submitted to the Executive generally'. Unlike Dudley's previous letter, this one bears no endorsements, and presumably was never answered.

The two tracts issued in 1820 and 1823 respectively (figures 12.1 and 12.3) had a common theme: the utility of electricity as an adjunct to orthodox medicine. Dudley's last known publication, *God Declared*, though still concerned in part with electricity, had a much deeper significance, of a religious nature. A large quarto of forty-eight pages, it was clearly the product of an unsound mind, and as such, defies coherent description. It has a half-title, title, dedication, introduction, and detailed contents list, but none of these bears much relevance to the others, or to the main text. In so far as its theme can be summarized in a single sentence, it is that electricity is the agent by which God provides the vital force in all living things, and motion to all the bodies in the universe; but Dudley touches also on the Irish question, proposals for a universal religion, the need for reform in financial affairs, the deficiences of the medical profession, and the almost miraculous cures achieved at his Therapeutic Institution (hours 12 to 4, Sundays excepted).

He dedicated the work jointly to the king, the government, and the British nation, and explained in the introduction that it was the product of many years' thought; it was not derived from books, as he had long since stopped reading them (apart from the Bible). His mind, he said, was more suited to explore than to be satisfied with what has already been discovered; he was not a reader, but a thinker. But he courted no praise; he would feel a permanent satisfaction in disseminating 'the fruits of my very many lonely hours throughout Britain's vast Empire'.

These introductory remarks (which are datelined 'May 1827') are followed by twelve pages headed 'IMPORTANT PROMULGATION', in which Dudley offers a dissertation in reply to the materialist's question, 'What is God?' The answer, apparently, is:

MIND!
or
GOD'S PRESENCE!
The SUPREME WILL! — ETERNISED ATTRACTION! — ETERNISED RETENTION! — and, ETERNISED REPULSION! — ACTION AND REACTION! or MOTION IN PERPETUITY!

After half a dozen pages expanding this theme, Dudley suddenly switches to Catholic Emancipation (the main political topic at the time), which he was convinced must shortly be granted, despite contrary views expressed by the lord chancellor. (Did Dudley, one wonders, bombard him with letters, too?) He implores the sovereign, together with the Lords and Commons, to call a convocation with the object of combining all religions into one, on a plan which he outlines, and ends this section with a plea for a personal interview with the king and the prime minister 'with a view to the reception of my proposals'.

The main part of this work, however, is concerned with 'ETHEREALISM: the Science of the Elements', which was Dudley's principal contribution to the salvation of mankind (through the application of his Imperial Medico-Ethereal Bath, hours 12 to 4, Sundays excepted). He was no longer attempting to liaise with orthodox medical practitioners, 'a body whose members are rather his oppressors than otherwise', but was addressing this work direct to the public, in his capacity as 'the Discoverer, and first Practical Professor of Real Medical Science'. How could 'the animating quality' be derived from drugs, in deleterious solutions and mercurials? It was obvious that only electricity could reanimate or restore the diminished vital energy of the nerves, which was the key to successful treatment of diseases. Until the medical faculty recognized this, they would remain blindfolded, unenlightened, and uninformed on the very topics they should and ought to investigate, and mankind would be kept permanently in a state of uncertainty of cure.

Diseases of the body were not the only subject covered in Dudley's thesis. At this time (the mid-1820s) the finances of the nation were in a sorry state, so he pledged himself, when called upon by the cabinet or the legislature, to point out the basis on which the financial system should be erected, namely on a foundation possessing the stability of science. After a renewed attack on the medical profession, he concluded by reminding readers of some of the miraculous cures achieved at his Therapeutic Institution, and claimed in a footnote that he could have cured the late Duke of York (George IV's younger brother and heir to the throne) as well if only that personage had applied to him in time.

Shortly after *God Declared* was published Dudley sent a copy to Dr Young, vice-president of the Royal Society, with a long covering letter claiming 'the National reward' for discovering perpetual motion.[17] Apparently he had a confused recollection of the Longitude Prize, first offered in 1714; but although the Board of Longitude still had some funds at its disposal in the 1820s for small awards to inventors, its brief did not cover perpetual motion.

Dudley's Medico-Therapeutic Institution at 22 Ludgate Street was listed in *Robson's London Directory* for 1826/27, but not in any later issues. His publications and correspondence of the 1820s, when read in full, suggest that his mental state was rapidly deteriorating, and it seems quite likely that for the last few years of his life he needed care and attention, perhaps in one of the private mental institutions that abounded in London.[18] The precise date of his death has not been discovered, but on 20 March 1830 he was buried in St Bride's, Fleet Street.[19] The register gives his age at death as sixty-seven, agreeing with a birth date in November 1762. This means, incidentally, that he lived longer than any other male member of the Adams family in his and the previous two generations. Because he was the youngest child of his father's second marriage, the successive lives of his father and himself spanned 121 years, from 1709 to 1830.

The parish of St Bride's, Fleet Street, was where this story of the Adams family began, and where Dudley's father and paternal grandparents were buried, but he never lived in that parish himself. His birth and baptism took place in St Dunstan's in the West, after his father's move to 60 Fleet Street. He married in St Marylebone in 1787. From 1788 to 1796 he lived in St Martin in the Fields, Westminster, before returning to St Dunstan's to reside at 60 Fleet Street until 1817. His Ludgate Street address was in the parish of St Gregory by St Paul's. The burial register, however, gives his address at the time of his death as Friar Street, Blackfriars Road, on the other side of the Thames. This was a small street in the parish of St George's Southwark, just north of

St George's Circus; it was still called Friar Street in directories of the 1920s, but its site today is marked by Webber Street (East). Perhaps its location only a stone's-throw from the new Bedlam Hospital (now the Imperial War Museum) was more than just a coincidence.

Dudley Adams apparently died intestate and lacking any personal possessions worth more than a few pounds, as there is no entry in his name in the indexes to the death duty registers for 1830. However, that is not quite the end of the Adams story. Only a few weeks later Dudley's unmarried sister Isabel Adams (originally called Isabella), born in March 1759, died at the age of seventy-one. From her will signed five years earlier in January 1825 it seems virtually certain that Isabel's and Dudley's sister Hannah (Mrs Matthew Robinson), and also Dudley's son George, must have died before then, as she did not mention them although she was careful to remember numerous other relations. (Hannah, with Dudley, Lucy (died 1805), and Isabel, was named in George Adams junior's will in 1795). Prominent amongst the beneficiaries of Isabel's will were the five daughters of her 'niece' (actually her half-sister's daughter) Mrs Anne Rogers. She also mentioned her brother-in-law Matthew Robinson, whom she thanked for his impartial conduct as executor of her late brother (George junior) and sister (which could mean Lucy but more likely meant Matthew's wife Hannah).

Isabel's will was proved in May 1830.[20] Sometime before then she had added a long rambling codicil which was not dated, witnessed, or signed, so on 10 May 1830 Joseph Blunt the younger, solicitor, of Liverpool Street, London (grandson of Isabel's half-sister Sarah Blunt), and Ann Block of Dulwich swore an affidavit to the effect that the will and codicil were in the deceased's handwriting. Probate was accordingly granted on 17 May to Hannah Rogers (granddaughter of Isabel's half-sister Sarah), who had been appointed executrix in the codicil, it being noted in the grant that Dudley Adams, appointed executor in the will, had died in the lifetime of the deceased. Isabel must therefore have died between 20 March and 10 May 1830.[21]

When she signed her will in 1825 Isabel was living in Dulwich, in a leasehold property belonging to Dulwich College. Apart from her brother Dudley, who was the principal beneficiary, eleven other relatives plus a female servant were mentioned in the will and codicil combined, most of whom were left legacies of a few pounds (none more than £10) and various small items of jewellery, trinkets, or furniture, for remembrance. Amongst the specific bequests to the descendants of her half-sister

Sarah Blunt were several scientific instruments, including an electrical machine and apparatus,[22] a pair of 12-inch globes in the best mounting on tripod stands together with a copy of *Treatise on the Globes*, a 2-foot brass reflecting telescope in a mahogany box,[23] a kaleidoscope,[24] and a zograscope with a collection of prints.[25] Her brother Dudley was left her leasehold property at Dulwich together with another occupied (in 1825) by Adolphus Kent, all her household furniture and effects other than specific bequests, a gold watch with key and seal, and 'the portrait of himself by Ramsay Reinagle'. Dudley was also the residuary legatee. As he died just before Isabel, it is not clear what would have happened to the portrait; it could be still in existence, perhaps unrecognized, somewhere amongst the effects of the Rogers family, who at the time of Isabel's death were living at Clifton, in Bristol.

Another portrait, a miniature in oils set in gold and mother of pearl, apparently of Dudley Adams (the wording is slightly ambiguous) was left to Mrs Anne Rogers. Letters from Dudley (presumably to Isabel), together with family papers relating to business transactions, were left to Hannah Rogers, the executrix. Recent attempts to trace these have not been successful, but the possibility remains that they may turn up one day.[26]

With Isabel's death the surname 'Adams' in the progeny of Morris and Mary Adams of Shoe Lane finally died out. However, George Adams senior had twelve grandchildren through his daughter Sarah and Robert Blunt, and at least one (Anne) through his daughter Hannah and Matthew Robinson. As mentioned earlier, one line of descendants of the Blunt family has been traced to the present day, mainly through the records of the Grocers' Company, each generation in turn having become members by patrimony. Early in the twentieth century a member of the Blunt family investigated some of his collateral relations, particularly the Rogers family of Clifton, Bristol, which has enabled the outline tree shown in table 12.1 to be drawn up. It is quite possible that there exist today numerous families who are unaware that one of their lines of descent stems from George Adams senior, mathematical instrument maker to his majesty in the mid-eighteenth century.

Table 12.1

Principal lines of descent from daughters of George Adams senior, showing family surnames in the fourth generation. Lines known to have terminated are excluded

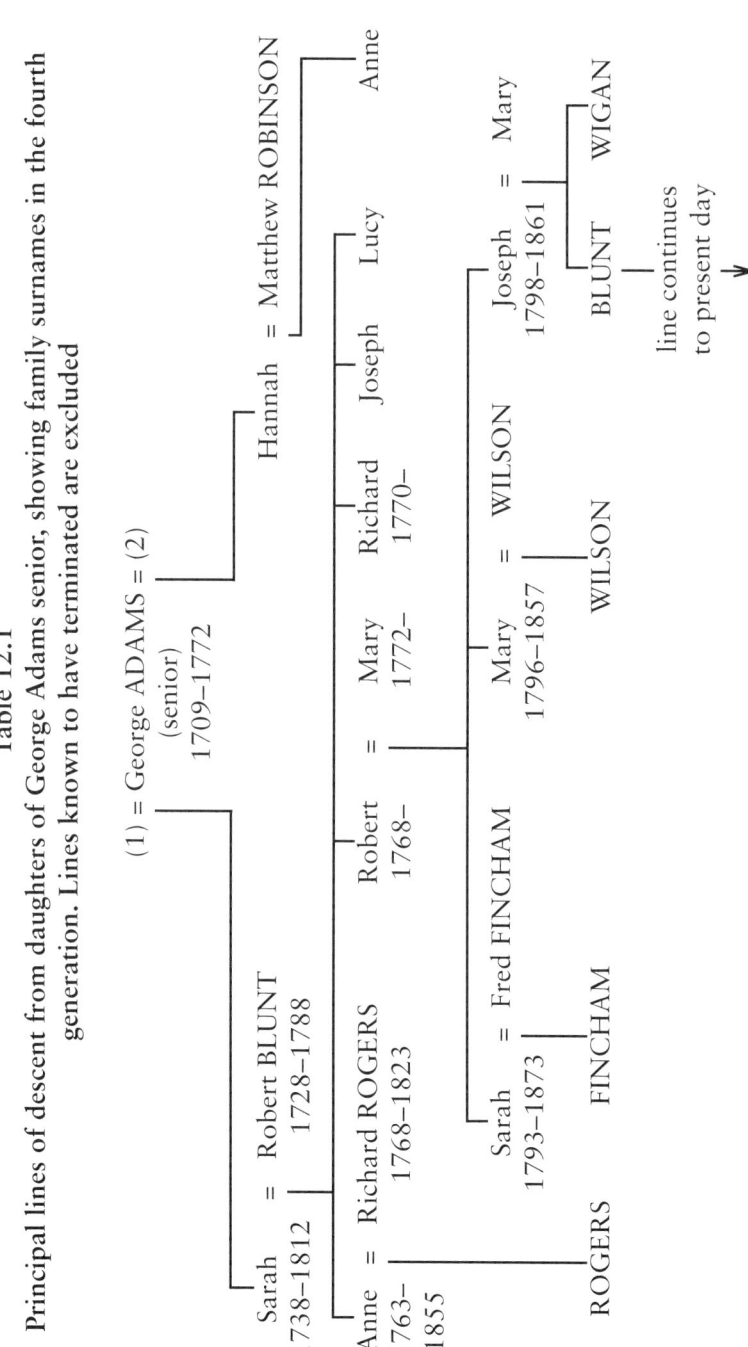

Notes

The following notes are intended mainly for use by students and historians. They consist of references to archival or printed sources, and comments on points which professional readers may consider require justification or amplification. References to printed works are mostly abbreviated: publication details are given in the bibliography.

Notes for chapter 1

1. The registers, ratebooks, and other archives of the parish of St Bridget alias St Bride's, Fleet Street, for the relevant period are deposited in the Department of Manuscripts, Guildhall Library, Aldermanbury, London.
2. Cassini, *Mémoires ... de l'Observatoire royal de Paris* (1810), cited by Joyce Brown, *Mathematical Instrument-Makers in the Grocers' Company* (1979), pp. 81–2. Miss Brown was unable to find any justification for the claim to a French origin for the Adams family.
3. Daumas, *Les instruments scientifiques* (1953), p. 315: [Dollond] 'descendait, comme Adams, d'une famille de protestants français émigrés au siècle précédent.'
4. The principal obstacle is the lack of detail in the burial registers, which in the early eighteenth century generally gave nothing more than the date and name, with no indication of age, address, or occupation. While the birth of a child can generally be assigned to a particular family with reasonable confidence, deaths can often be identified only by cross-checking with other records such as wills and ratebooks, or in the case of a child, by the re-emergence of the same first name in the same family (a common practice).
5. Olivia Brown first drew attention to this in 'The Instrument-making Trade', in R. Porter et al., *Science and Profit in 18th-century London* (1985), p. 22.
6. There were some loopholes in the regulations which have not yet been fully explored. For example, premises owned by the church seem to have been outside the jurisdiction of the City authorities. In eighteenth-century newspaper advertisements for business premises to let, the term 'out of the freedom' occurs quite often.

334   Notes for Chapter 1, pp. 5–10

7. In addition, some uncommon routes to the freedom were occasionally exploited. For example, the lord mayor apparently had the right, if he so wished, to nominate up to three persons for admission to the freedom of the City during his mayoralty; these persons then had to become free of a guild as well if they wished to trade. Benjamin Martin used this method in 1756.
8. Guildhall (London) MS.6561. This is particularly valuable for the early Georgian period, as the vestry minutes do not give occupations until the middle of the century.
9. This term could have several meanings. In this instance it seems to have meant those men who were selected to serve on juries or inquests.
10. Guildhall MS.6561 f. 3; entered as 'Morris Addams'. In the corresponding entry in the vestry minutes (Guildhall MS.6554/4) his name is spelt 'Morres Adams'.
11. Guildhall MS.3435, bundles 1 and 2, Fleet Precinct. The sequence is confused for several years, perhaps due to rebuilding in progress, but Morris Adams is consistently in third place from 1714 onwards.
12. These figures are taken from entries in the court minutes (Guildhall MS.15,835/1) in February and March 1724/25, where the clerk recorded his replies to questions about their membership put to all the guild companies by the lord mayor and the House of Commons. The Loriners' Company had no liverymen until 1712. In 1700 the total membership was 524.
13. Job Adams was probably the man of that name of Aldersgate Street who was buried in St Botolph's Aldersgate on 13 August 1719, leaving a widow Sarah and a son, also Job, then aged fourteen. Letters of administration (Admons) were granted to Sarah Adams, widow relict, on 17 August 1719 (Public Record Office, Chancery Lane, PROB/6/109). Job junior, who would have been George Adams's cousin if their fathers were brothers, was apprenticed in the Upholders' Company on 4 August 1721 and obtained his freedom on 5 March 1728 (Guildhall MS.7142/1); these entries give his late father's occupation as 'Cook', but do not state Job junior's trade. A further grant of Admons. was made in 1733, on Sarah's death, to her daughter Mrs Amy Lilley, which suggests that Job junior had died by then, in his twenties.
14. Joyce Brown, see *Mathematical Instrument-Makers*, p. 36. 'Grocers' Company Register of Freemen and Apprentices 1721–1743', Guildhall MS.11,598/2: 'George Adams son of Morrice Adams'.
15. Loriners' Company, court minutes, 20 September 1722.
16. Annual accounts of quarterage due from the liverymen (only) are inserted loosely in the court minute book, Guildhall MS.15,835/1.
17. For example, when Philip Adams took an apprentice in February 1722/3 he paid 17s 6d for thirty-five quarters due. Similarly Robert Adams (a cutler) paid 14s 0d for twenty-eight quarters due when he took an apprentice in June 1723. At least seven men with the surname Adams were members of the Loriners' Company in the 1720s, but only Morris was a liveryman.
18. Guildhall MS.9168/34. A slender clue is provided by an entry in the records of the Sun Fire Office, which reveals that in 1719 Morris Adams paid a premium of 7s to insure his goods and utensils and so on, equivalent to a capital value for the latter of £700. This was discovered by the late Michael

Crawforth of Oxford in 1987 during a preliminary examination of insurance records for Project SIMON. These records have yet to be fully searched.
19. It is not certain that this is the right man; however, the will refers to his wife Elizabeth as sole executrix, and when one of James Parker's apprentices wished to claim his freedom after his master's death he was presented by Elizabeth Parker, widow.
20. Joyce Brown, see *Mathematical Instrument-Makers*, pp. 35, 36. The turnover date is mentioned in George Adams's freedom entry, Guildhall MS.11,598/2, 10 October 1733.
21. Carwitham, *Description and Use of an Architectonick Sector* (1723), advertisement leaf following text.
22. Ibid., second edition (1733), p. 22.
23. Guildhall MS.9172/136 (original will).
24. Communication from the archivist, Hoare's Bank, October 1985. The bank's ledgers also include an account held by a George Adams in 1727, but this may have been a different man. George, son of Morris, was a minor then.
25. Guildhall MS.11,598/2, 10 October 1733. A complete transcript of this entry is given by Joyce Brown, see *Mathematical Instrument-Makers*, p. 11.
26. This story of George's family background, though undoubtedly correct as far as it goes, is incomplete in some respects which do not affect his instrument-making career but could be important for anyone researching the Adams name. For example, the marriage of Morris Adams and Mary Rogers has not been traced: the only entry in the International Genealogy Index (IGI) that seems relevant refers to Morrice Adams of St Bride's and Elizabeth Lee, married in 1705. The marriage of George's younger brother Henry, thought to have taken place in St Bride's on 14 January 1730/1, is anomalous in that the entry refers to the authority for the marriage being an archbishop's licence (an unusual procedure), yet no record of the issue of such a licence to Henry Adams can be traced. No connection has been found between George Adams and his contemporary Nathaniel Adams, optician, of Charing Cross; he was apprenticed to Edward Scarlett in the Spectaclemakers' Company in 1722, and was recorded in their books as the son of Nathaniel, glover, of St Ann's, Westminster. In view of the different guild and location (Westminster), and the absence of the first name 'Nathaniel' in Morris's family, it seems unlikely that George and Nathaniel junior, though almost the same age, were connected in any way. There was also a contemporary William Adams, 'mathematical instrument maker', mentioned in the Christ's Hospital ledgers in 1735 when he took two boys from there as apprentices (Guildhall MS.12,823/4); in view of his occupation he may possibly have been distantly related to George, but nothing further is heard of him and no instruments signed 'William Adams' are known, so either he died soon after 1735 or he did not trade under his own name.

Notes for chapter 2

1. George's first marriage has not been traced, but it is unlikely to have taken place while he was an apprentice (that is, before 1732), as this was normally prohibited by the terms of engagement.
2. These names were provided by the late Michael Crawforth from his Project SIMON files.
3. See, for example, the Heath & Wing trade card reproduced in Crawforth's paper, 'Evidence from Trade Cards ...', *Annals of Science* (1985) p. 509.
4. Descriptions and illustrations of various instruments proposed around this time are given in Randier, *Marine Navigation Instruments* (1980), pp. 92–103. For Hadley's instrument see his papers in *Philosophical Transactions* in 1731 and 1733, and his tract *A Description of a New Instrument* (1734).
5. Caleb Smith and Ward, *The Description and Use of a New Astronomical Instrument* (1735). The British Library copy, shelfmark 533.h.8(2), is signed by Ward and Smith, and dated 20 January 1734, probably old style. This work was reviewed in *The Present State of the Republick of Letters* Vol. 15 (1735), pp. 75–9.
6. Caleb Smith, *The Description, Use ... Sea Quadrant* (1735).
7. R.S. Whipple, 'Some Scientific Instrument Makers of the 18th Century', *Nature* (1930), p. 245; Clay and Court, *History of the Microscope* (1932), p. 163.
8. *Journal of the Royal Microscopical Society* (1912), p. 125. The microscope itself is described and illustrated on p. 445.
9. Inv.115/1948, Accession no.A.195850. The letter is entered on the modern record cards as dated 1738, but the 1912 interpretation of the date as '1735/6' seems more likely.
10. Clay and Court, 'English Instrument Making in the 18th Century', *Transactions of the Newcomen Society* (1935/6); Whipple made this point in the discussion, p. 52.
11. Adams is not mentioned in the indexes to the letter books or court minutes of this period, but the archives of the East India Company are extensive, so it is possible that his name occurs in other records which were not searched.
12. Joyce Brown, *Mathematical Instrument-Makers*, p. 36; Guildhall MS.11,598/2 (21 April 1736).
13. Rees, *The Worshipful Company of Grocers* (1923), chapter 13.
14. St Bride's vestry minutes, Guildhall MS.6554/5 p. 69. The appointments were confirmed at a further meeting on 20 December, p. 71.
15. Guildhall MS.6561 f. 61 lists the following (to the nearest year): questman 1743, constable 1736, scavenger 1745, collector 1742, sidesman 1743, junior churchwarden 1746, senior churchwarden 1747. More precise dates can be obtained from the vestry minutes.
16. This trade card was reproduced in an article by Broadley in *Knowledge* in 1912, wrongly dated 1775 (possibly a printing error for 1757). A photograph, taken from the article, is cited by Calvert (1971), item 7, with the incorrect 1775 date. The location of the original card used by Broadley in 1912 is not known.

17. For details of Martin's products and publications see Millburn, *Benjamin Martin* (1976), its *Supplement* (1986), and *Retailer of the Sciences* (1986).
18. For the history of orreries in general see H.C. King, *Geared to the Stars* (1978).
19. Science Museum Library, MS.453; signed on the last page 'G.Adams Scripsit Nov.20th 1741.'
20. Joseph Harris, *The Description and Use of the Globes and Orrery*, fifth edition (1740), verso of title. The Duke of Argyll was added to the list of principal customers in the sixth edition, 1745.
21. Deane, *The Description of the Copernican System* (1738), pp. 89–106. In a contemporary handbill (Project SIMON files) Deane said that he would demonstrate the orrery and give lectures on it for 10s 6d per person, which was 1/1,000th of its cost. Subscribers would also receive a raffle ticket, and when 1,000 persons had subscribed the orrery would be disposed of by lottery. It is not known whether this ever took place. It seems hardly likely that Deane would have found 1,000 people willing to pay half a guinea to see the orrery, with only a 1:1000 chance of winning the prize.
22. Calvert, *Scientific Trade Cards* (1971), Pl. 17.
23. Desaguliers, *A Course of Experimental Philosophy* (vol. 2, 1744), pp. 430–38 and plates 31, 32. These models were made to Desaguliers's design by 1739 or earlier, for on 22 November that year he wrote to Cromwell Mortimer confirming that he had shown them to the Royal Society. British Library, Add.MS.4435 f. 107.
24. For locations of extant Adams orreries see King, *Geared to the Stars* (1978), pp. 204–5. Some of these were probably by George Adams junior (that is, after 1772).
25. Olivia Brown, *Whipple Museum Catalogue 4* (1983), item 29; Inv.Wh.1275.
26. I am indebted to the archivist at Mount Stuart, Miss Catherine Armet, for providing me with a copy of the letter and information on Bothwell.
27. *Micrographia Illustrata* (1746), pp. 262–3.
28. E. Dekker, *De Leidsche Sphaera* (1985).
29. Millburn, 'Horology and the Adams Family', *Antiquarian Horology* (1982), 'Addendum' p. 582.
30. Information kindly supplied by Peter de Clerq of Leiden, February 1987.
31. Guildhall MS.11,598/9 pp. 114, 135.
32. Guildhall MS.11,598/2, 8 March 1741/2.
33. Guildhall MS.6561 f. 61.
34. Guildhall MS.6552/4 f. 176.
35. Guildhall MS.11,598/2, 13 July 1734 and 20 September 1739.
36. Like George, Shute Adams had two wives. In 1743 he married a woman with the unusual first name 'Kerrenhappuch'. She died in childbirth in 1744 aged twenty-two, and he quickly married again. Also like George, Shute Adams had numerous children, who were contemporaries of George's.
37. Adams's interest in portable microscopes may also have been prompted by a patent granted to George Lindsay, watchmaker, in 1743 for a microscope 'as portable as a snuffbox'. *London Evening Post* 2/4 June 1743.

38. *Reading Mercury*, November 1742; listed in the December number of *Gentleman's Magazine*.
39. Baker's book was published by R. Dodsley, whose name appears alone in the imprint of the first edition. In the second (1743), third (1744), and fourth (1754) editions, Cuff's name as well is given in the imprint (and another bookseller's, Mary Cooper).
40. G.L.'E. Turner, 'Henry Baker', *Notes and Records of the Royal Society* (1974), pp. 53–79.
41. *London Evening Post* 4/7 June 1743, repeated 14/16 June, 16/18 June, 28/30 June, 12/14 July, 26/28 July.
42. Clay and Court, *History of the Microscope* (1932), pp. 162–3.
43. For an account of Trembley's discoveries see J.R. Baker, *Abraham Trembley* (1952).
44. *London Evening Post* 7/9 January 1746, repeated several times with slight variations.
45. G.L'E. Turner, see 'Henry Baker', p. 63.
46. From January to May 1746 Adams's advertisement appeared at least twelve times in the *London Evening Post*, and occasionally in the *General Advertiser* as well. Baker's book was advertised several times in both papers, mainly in February.
47. The second edition was advertised at least eight times in the *London Evening Post* in July 1747.
48. J.R. Baker, see *Abraham Trembley*, p. 42.
49. F. Watkins, *L'Exercise du Microscope* (1754). According to Watkins, at that time there was nothing available in French corresponding to Baker's *The Microscope made Easy*.
50. *Micrographia Illustrata: or the microscope explained, in several new inventions* (London, 1771). This octavo edition, with 325 text pages and 72 plates, is designated the fourth, but no copy of a third edition is known. See chapter 5 for comments on this publication.
51. In one letter Baker referred to Adams as 'an ignorant but impudent fellow'. At that time (1746) Adams was not a liveryman and held no official appointments.
52. See for example Clay and Court, *History of the Microscope*, chapter 9, and G.L'E. Turner, *Collecting Microscopes* (1981), pp. 57–60.
53. Clay and Court, *History of the Microscope*, p. 136.
54. *Micrographia Illustrata* (1746), p. 244.
55. Daumas, *Les instruments scientifiques* (1953), p. 312.
56. This was probably John Symmonds, 'a watch-case maker of great business, and Colonel of the Lumber Troop', whose death at his house in New Street, Shoe Lane, was reported in the *London Evening Post* for 11/13 November 1746.
57. Guildhall MS.6554/5, 18 November 1746.
58. It is not absolutely certain that this entry refers to George's first wife, but the date is consistent with the births of his children and his second marriage.
59. Guildhall MS.6554/5, 23 April 1747 and 14 April 1748.

60. George Graham, watchmaker and astronomical instrument maker, was one person who opted for payment of the fine when he was chosen in 1735.
61. St Bride's churchwardens' accounts, Guildhall MS.6552/4. The accounts for 1746–47 occupy fifteen pages in this large (16 × 10 inches) volume.
62. Fees for burials and tolling the church bell.
63. 'For removing Poor and attending Hospitals etc.'
64. Mr Jones was the workhouse manager.
65. Sixteen pauper children were placed as apprentices at £2 each.
66. Flowers for the church; salaries of organist and sexton, and so on. Adams also included under this heading 'Paid myself for winding up the Clock ... £2.'
67. Hire of rooms at inns, and refreshment.
68. 'Expenses not reducible under any distinct head.'
69. James Ferguson drew attention to the lack of a lightning conductor for St Bride's in his *Introduction to Electricity* (1770), p. 103.
70. See, for example, *The Portfolio* 98 (11 December 1824), p. 210.
71. During the post-war reconstruction most of the surviving monuments were removed to the crypt, where they are now displayed. Amongst them is a memorial to several members of the Bardin family, globe makers in the late eighteenth century.
72. London marriage licences, Guildhall MS.10,091/88 part II f. 425. Although Ann Dudley, like George Adams, was described as 'of St.Bride's', the bishop's licence was issued for the ceremony to take place at either St Martin in the Fields or St Margaret's Lothbury. Marriage by licence avoided the necessity to publish banns.
73. In Fleet Street, Benjamin Rackstrow, whose shop was next door to John Cuff's, gave electrical demonstrations as a sideline to his main business as a sculptor. As well as the established lecturers in experimental philosophy, other people who advertised public demonstrations of electrical phenomena in the late 1740s were John Bennett (instrument maker), John Neale (watchmaker), Francis Watkins (optician), Richard Oliver (schoolteacher), and Magnus Tyro (surgeon and physician). For the state of electrical science at this time, see Hackmann, *Electricity from Glass* (1978).
74. For an outline of Cole's career see Taylor and Wilson, *At the Sign of the Orrery* (c. 1950), pp. 18–23.
75. Joyce Brown, *Mathematical Instrument-Makers*, p. 39. Taylor and Wilson give his name as 'Ludlow'.
76. *Daily Advertiser*, 25 February 1749; repeated 1 April.
77. *Daily Advertiser*, 14 June 1748. On 2 June, Cole advertised jointly with Liford of the Strand, 'Two new Astronomical Instruments', but these may have been printed volvelles.
78. Thomas Godfrey's paper was communicated to the Royal Society by J. Logan, and printed in *Philosophical Transactions* No.435 (December 1734), vol. 38 pp. 441–50.
79. It was quite common, in cases like this, for one man's advertisement to be printed next to his adversary's, which suggests that the printer of the

newspaper advised advertisers when a rival announcement was to be inserted.
80. *General Advertiser*, 23 December 1748; quoted in full by Taylor and Wilson, see *At the Sign of the Orrery*, p. 23.
81. Crawforth, 'Instrument Makers in the London Guilds' (1987), p. 339.
82. Ordnance bill books, Public Record Office, Kew, WO51/167 p. 165. Payment of this bill was recorded in ledger WO48/89 p. 258.
83. Ordnance bill book WO51/170 p. 64.
84. Ordnance bill book WO51/171 p. 58. It is likely that Deane was still alive at this time, as the corresponding ledger records a payment to him (in settlement of bills nearly a year old) in May 1748, after Adams had submitted his first bill in April.
85. For details of George Adams senior's work for the Office of Ordnance, including references to his 146 bills in the bill books (Class WO51), see Millburn, 'The Office of Ordnance and the Instrument Making Trade', *Annals of Science* (1988). When that paper was written it was thought that the Adams family's connection with Ordnance work ceased on George senior's death in 1772, but it was subsequently discovered that George junior started to receive Ordnance orders around 1780, and both he and his brother Dudley undertook a large amount of Ordnance work during the Napoleonic wars. Some comments on this aspect of the Adams family's business are given here in chapters 7 and 8.
86. These archives were investigated (at the Guildhall Library, London) in 1985 by Mr V.K. Chew, of the Science Museum, London, to whom I am indebted for lending me his notes on the ledgers. Entries relating to Deane appear in ledger 4, and to George Adams senior in ledgers 5 and 6: Guildhall MS.12,823/4–6. Although the entries include cross-references (by serial number) to the bills submitted, the latter apparently were not preserved once the totals had been entered in the ledgers, cash books, and account books.
87. Guildhall MS.12,283/4.
88. *Daily Advertiser*, 29 September 1747. Similar advertisements appeared throughout October, November, and December. From the beginning of 1748, Adams's name was omitted. Sowerby died in October 1749 after only a brief career.
89. *London Evening Post*, 18/20 December 1746; repeated in several later numbers; *Daily Advertiser*, 17 April 1747. The shops of Sisson and Cuff were also mentioned as places where further information could be obtained. The subjects taught included navigation and mechanics as well as pure mathematics.
90. Guildhall MS.11,588/7.
91. *Daily Advertiser*, 27 and 29 June 1749.
92. Patent no. 656, submitted 25 May 1750, signed and sealed by Adams and Jack on 31 August 1750. (Patent serial numbers were allocated retrospectively in the mid-nineteenth century.)
93. *Daily Advertiser*, 25 October 1749; *Public Advertiser*, 3 December 1757.
94. *Daily Advertiser*, 1760. Advertisements for the sale of 'the curious Math-

ematical Library of Mr. RICHARD JACK' appeared almost daily during the first two weeks in February, to take place on 13 February and the following three days. Jack's will was proved in PCC on 8 May 1759; it mentions a wife and young son, Richard junior. No confirmation has been found of the auctioneer's claim that Jack had been an 'Assistant Engineer' (meaning a volunteer – he was not a member of the Corps of Engineers in the Ordnance service) in the expedition against Guadaloupe, which took place in the winter of 1758–59. If true, he probably joined the expedition at his own expense, to gain practical experience of fortification and give added authority to his lectures on this subject. For what is currently known about Jack's career, see W. Johnson, 'Richard Jack', *International Journal of Impact Engineering* (1992).

95. Several titles in the Burney Collection for the year 1751 are represented by only a few scattered numbers. For example, in the second half of the year the only number of the *London Gazetteer* is that for 28 November, which happens to contain a short Adams/Jack advertisement. Similarly, there are only a few numbers of the *General Evening Post* for December, one of which contains a longer Adams/Jack advertisement. Almost continuous runs are available in the Burney Collection of the *General Advertiser* and the *London Evening Post*, and (on microfilm elsewhere in the British Library) the *Daily Advertiser*.
96. See the above comments on availability.
97. Dollond/Watkins advertisements first appeared in the *Daily Advertiser* on 29 April 1758 and were repeated at intervals in this paper for the next year at least. Dollond family archives deposited in the Guildhall Library, London, show that Watkins helped to finance the achromatic lens patent application, but do not reveal whether he participated in the development work. For a discussion of the patent and its effect on the optical trade, see chapter IX of M.M. Robischon's PhD thesis 'Scientific Instrument Making in London during the 17th and 18th Centuries', University of Michigan (1983). A copy is available for consultation in the Science Museum Library, shelved at 681.2:93.
98. Public Record Office (Chancery Lane), IND.22645, 24 November 1750. The principal creditor was Rebecca Deschamps.
99. *Daily Advertiser*, 17 December 1751.
100. This edition and subsequent ones to the eleventh contain advertisements by the Coles for their products.
101. Labaree, *The Papers of Benjamin Franklin* 4 (1961) p. 3.
102. Ibid., citing the charter, laws and catalogue of books of the Library Company of Philadelphia (1757), p. 22.
103. Information supplied by the late Michael Crawforth, from 'Licences to non-freemen to be employed within the City of London', vol. 1; Corporation of London Records Office.
104. Labaree, see *Papers of Benjamin Franklin* 4 p. 515.
105. Joyce Brown, *Mathematical Instrument-Makers*, p. 36. Kettle does not appear to have applied for his freedom.
106. I am indebted to Patricia Fara, a PhD student researching the history of

342 *Notes for Chapters 2–3, pp. 70–83*

        magnetism at Imperial College in 1990, for this information, and for pointing out the reference below.

107. Mountaine and Dodson, *An Account of the Methods ... Variation of the Magnetic Needle* (1758), p. 12.
108. Guildhall MS.11,588/7, p. 303.
109. Ibid., p. 318 (5 May 1756).
110. For details of the Grocers' Company archives deposited in the Guildhall Library see *Guildhall Library Research Guide No. 3: City Livery Companies and related organisations: A guide to their archives in Guildhall Library* (1989), pp. 63–6.
111. For the history of the hall see J.B. Heath, *Some Account of the Worshipful Company of Grocers* (1854), pp. 1–37. In the 1740s the hall was insured with the Hand-in-Hand Fire Office for £6,000; Guildhall MS.11,588/7 p. 49.
112. Ratebooks of St Mildred in the Poultry, Guildhall MS.63. The Poor and Church rates paid by the Grocers' Company together amounted to about £60 per year, about a quarter of the income of this small parish.

Notes for chapter 3

1. According to newspaper reports, the Prince of Wales received an allowance of £35,000 per annum from the age of eighteen, equivalent to around £10 million in today's money.
2. Public Record Office (Kew), Ordnance bill books WO51/198–201.
3. Guildhall MS.6552/4, 4 and 5 February 1757.
4. Guildhall MS.11,316/185, Ward of Farringdon Without, Parish of St Dunstan's in the West. The quarterly ratebooks for this parish for this period are missing.
5. George Graham, who moved in the opposite direction, from St Dunstan's to St Bride's, was excused serving as constable in his new parish as he had already paid a fine for this in St Dunstan's. Guildhall MS.6561 f. 33.
6. Guildhall MS.3016/4, 14 April 1762 and 10 April 1765 respectively. It was usual for a person to reside in a parish for at least four years before being called upon to serve.
7. British Library (formerly Patent Office Library copy); University College, London (Graves Collection); Whipple Museum Library, Cambridge.
8. This explanation may have to be modified if more copies are found, without the later tract attached; but in any case the earlier one must have been printed after Adams became mathematical instrument maker to the Office of Ordnance in 1748, and stocks must have been still available when the later part was printed in 1757 or after, otherwise the catchword 'IN-' would not have been incorporated in the latter.
9. The plate is reproduced in R. Porter et al., *Science and Profit in 18th-century London* (Whipple Museum, Cambridge, 1985), p. 4.
10. Guildhall MS.11,588/7 p. 389. Shute Adams, the druggist, was a warden in 1756.
11. R. Campbell, *The London Tradesman* (1747), p. 305.
12. Guildhall MS.11,588/7: 21, 30 November 1758, 10 May, 14, 18 June, 11

July, 20 September 1759. At a general meeting on 11 July the company took a decision to abolish quarterage, as its collection had become too difficult.
13. Guildhall MS.10,091/101, 21 June 1759. Both parties were 'over 21'.
14. Westminster Public Library, Buckingham Palace Road.
15. In the 1760s Robert Blunt, at the Golden Ball facing the Mews, was a frequent advertiser in the newspapers; he made and sold 'all sorts of Holland and Irish linen shirts', and so on. In London directories the Blunt linen-drapery business continues until about 1820, by which time the proprietress was Mary Blunt, widow of Robert Blunt junior, one of Sarah and Robert Blunt senior's sons (George Adams's grandson). In 1799 Robert Blunt junior was one of about 300 members of the Society for Protection of Trade, formed in 1776 to keep members informed of the activities of 'swindlers and sharpers'.
16. Clay and Court, *History of the Microscope* (1932), pp. 174–6.
17. Chaldecott, *Handbook of the King George III Collection* (1951), item 186.
18. Palmer and Sahiar, *Microscopes* (1971), item 10.
19. Olivia Brown, *Whipple Museum Catalogue 7* (1986), item 114.
20. Ibid., item 113.
21. *Billings Microscope Collection* (1974), figure 349.
22. Donn's proposal is reproduced in facsimile in Ravenhill's introduction to the facsimile reprint of Donn's Map of Devon 1765 (1965), plate II.
23. Ibid., p. 4.
24. Public Record Office (Chancery Lane), LC/3/67 p. 30. 'Dolland' (no first name) was appointed Optician on the same day.
25. The catalogue is now in the Bute archives at Mount Stuart. I am indebted to the archivist, Miss C. Armet, for providing me with copies of pages relating to Adams.
26. G.L'E. Turner, 'The Auction Sales of the Earl of Bute's Instruments', *Annals of Science* (1967).
27. *The Annual Register ... of the Year 1761*, pp. 241–2 of 'Chronicle'.
28. During October 1761 Demainbray frequently advertised his intention to give a course of twenty-four lectures on mechanics, hydrostatics, and so on, in Tanfield Court, Inner Temple, beginning on 3 November. Tickets for the course cost 1½ guineas. Read three times per week, this course would have finished just before Christmas 1761. *Daily Advertiser*, 8, 15, 19, 22, 29 October 1761.
29. In a short account of Demainbray's life published in *The Observatory* (1882), it is stated (p. 282) that he 'undertook the duties of three offices in the Custom House'. Lalande, writing in April 1763, said he then had a good position in the Excise and had abandoned physics: Lalande, *Journal* (1980), pp. 44–5. Hence, Demainbray's switch from lecturing to working for the Office of Excise (and/or Customs) must have occurred in 1762 or early 1763. *Millan's Universal Register* for 1763 gives 'Ste. Demainbray' under 'Customs' as inspector of the East India Warehouse, at £120 per year. The 'Customs' section in Rivington's *Court & City Register* for 1766 gives 'Step. Demainbray' as one of the four examiners of the 'Outport Books', at £100 per year each; 'Dr.Demainbray' as inspector of the 'Unrated East India Goods', at £150 per year; and 'S.Demainbray, Esq.' as surveyor of the East India Warehouses, at

£130 per year. If these all refer to one person, as was assumed by the writer of the account of his life in 1882, he received a total of £380 per year from these sources, a substantial sum. However, out of this he presumably had to pay one or more persons to carry out duties which he did not undertake himself.

30. Science Museum, London, MS.353; Inv.1952–36.
31. Smeaton, 'A Letter … Air-Pump', *Philosophical Transactions* (1753), pp. 417–27.
32. Science Museum, London, MS.354; Chaldecott, *Handbook*, item 319.
33. Chaldecott, *Handbook*, item 318.
34. Ibid., items 320, 321.
35. Morton and Wess, *Public & Private Science* (1993). This volume includes an account of Demainbray's apparatus, and the lecturing scene in London in the mid-eighteenth century, as well as descriptions of the items in the King George III Collection.
36. Ibid., pp. 18–22. A page of the 'fair copy' manuscript is reproduced on p. 18. In the course of compiling this volume, Morton and Wess discovered that a second copy of these 'fair copy' manuscripts, presumably also done by Mr Champion, is at Windsor Castle.
37. Dudley Adams, *Treatise … Globes* (1810), preface.
38. Chaldecott, *Handbook*, p. 82 and item 348, now Science Museum MS.348.
39. Clay and Court, *History of the Microscope*, pp. 176–80.
40. *Micrographia Illustrata* (1771), p. xlix and plate 14 figure 28.
41. Science Museum, London, Inv.1949–116.
42. G.L'E. Turner, see 'Auction Sales', p. 218 (his footnote 18).
43. Lalande, *Journal* (1980). The prior publication was an article 'The Journal of Lalande's visit to England', *The History Teacher's Miscellany*, 4 (1926), pp. 113–18, 140–44, 151–56, 167–71.
44. Lalande, *Journal* (1980), p. 70. Unfortunately the modern editor, Helene Monod-Cassidy, mistook 'Adam' in Lalande's text for Robert Adam the architect, in her footnote 181 and elsewhere. Dr Pringle was physician to the queen's household from 1761. (*Daily Advertiser*, 7 September 1761.)
45. Ibid., p. 72.
46. Chaldecott, *Handbook*, and Morton and Wess, *Public & Private Science*, describe each item individually.
47. Lalande, *Astronomie*, third edition (1792) p. xxx.
48. Most of the observational instruments were transferred to Armagh Observatory, Northern Ireland: for details see Chaldecott, *Handbook* p. 82; Bennett, *Church, State and Astronomy* (1990), pp. 92–8; McFarland, 'Historical Instruments', *Vistas in Astronomy* (1990), pp. 149–210. As this institution already had a transit telescope, the Kew instrument was presumably discarded.
49. King's College, London, MS.K/MUS.18; Chaldecott, *Handbook* item 240.
50. British Library, MS.Kings 279, ff. 7–9.

Notes for chapter 4

1. For Senex's changes of address and his range of products see Tyacke, *London Map-Sellers* (1978), p. 142.

2. The minute books of the Grand Lodge of Freemasons of England for 1723–39 were printed (in transcription) in *Quatuor Coronatorum Antigrapha* 10 (Margate, 1913). At the start of this period John Senex was a Grand Warden, and member of a lodge which met at the Fleece in Fleet Street. Emanuel Bowen, Richard Cooper, and John Pine were amongst other engravers who were Freemasons at this period.
3. John Senex, 'A Contrivance ... Celestial Globe', *Philosophical Transactions* (1741), pp. 203–4 and plate II.
4. Mary Senex, 'A letter ... concerning the Large Globes prepared for her late husband', *Philosophical Transactions* (1752), pp. 290–92.
5. Fenning, *A New and easy Guide to the Use of the Globes* (1754). Adams took six copies, as did Cole, and Gregory; Nairne took four. Mr Thomas Turton, mathematical instrument maker, also took four. A 'Postscript' of ten pages at the end advertised John Neale's patent globes of 3 inches diameter (6 guineas) and 12 inches (16 guineas). These were globes fitted with geared sun and moon attachments to show the apparent geocentric motions of the latter; Neale, a watchmaker by trade, did not make the actual globes. A pair of the smaller size in the Science Museum, London, are by Cushee.
6. The sale was first advertised by the auctioneer, Langford, in the *Daily Advertiser* on 12 September 1755, to take place on 24 September; but on the previous day (the 23rd) he inserted an announcement saying that many complaints had been received that the time allowed for viewing Mrs Senex's extensive stock was too short, so the sale had been postponed. A new date, 15 October, was first advertised on 30 September, and the sale duly took place on this date.
7. For Ferguson's activities at this time, see Millburn, *Wheelwright of the Heavens* (1988), chapters 5 and 6.
8. S. Edell, 'Concave Hemispheres of the Starry Orb', *Bulletin of the Scientific Instrument Society* 7 (1985), pp. 6–8.
9. For Martin's acquisition of the Senex globe business from Ferguson, see Millburn, *Benjamin Martin: Supplement* (1986), pp. 20–22.
10. See, for example, an advertisement by Leonard Cushee in the *Daily Advertiser* on 24 February 1761. Richard Cushee's globes were illustrated in the first edition of Joseph Harris's *Description and Use of the Globes and Orrery* in 1731, and the same copperplate continued to be used in later editions. From 1751 E. Cushee, who has not been positively identified but may have been R. Cushee's widow (Elizabeth?) rather than son, was associated with Benjamin Cole as publishers of this work.
11. *Lloyd's Evening Post*; *London Chronicle* (this transcript is from 29/31 July); *Whitehall Evening Post*; *General Evening Post*; *Daily Advertiser*.
12. The dedication is not, of course, actually signed by Johnson, but Boswell included it in his list of Johnson's writings and its authorship is not doubted.
13. There are some minor differences in spelling and capitalization between the version in the *London Chronicle* and that in Adams's first edition, but these may have been inserted by the newspaper editor.
14. This configuration became common in England from the mid-eighteenth century onwards, whereas on the Continent most globe stands retained the

older form of a horizon platform supported by a number of baluster pillars from a circular or cross-shaped base.
15. These locations are taken from the national lists of extant old globes published from time to time in the pages of *Der Globusfreund*: 21/23 (1973) France; 28/29 (1980) Austria; 31/32 (1983) Holland and so on; 35/37 (1987) Portugal.
16. Bernoulli junior, *Lettres Astronomiques* (Berlin, 1771) pp. 29–30.
17. Wellcome Institute Library, London, MS.67097.
18. Communication from Peter de Clercq, Museum Boerhaave, March 1992.
19. Van der Krogt, *Old Globes in the Netherlands* (1984), refs ADA 8 and 15. Another similar 18-inch pair in the Nederlands Scheepvaart Museum are refs ADA 7 and 13.
20. Public Record Office, Kew, Ordnance bill book WO51/239 p. 154.
21. Ibid., WO51/235 p. 310.
22. The Latin dedications are quoted in full in the British Library Map Catalogue (without capitalization), and also (with English translations) by van der Krogt (1984) pp. 39–41. The pair in the British Library were on public display in the middle of the King's Library in the 1980s. Adams is described in the Latin inscription as 'Artificem Regum'. The latter word is presumably an error for 'Regium', that is, 'appointed by the Crown', and not (as van der Krogt has it) the genitive plural, 'of Kings'.
23. Numbers began to be allocated in Fleet Street in July 1766, in accordance with an Act of Parliament given the royal assent in May. For the progress of the numbering, which began with number 1 on the south side at Temple Bar and proceeded in sequence (not 'odds and evens'), returning to Temple Bar via the north side, see the *London Chronicle* from 24/25 July onwards.
24. Benjamin Martin had his shop bills altered soon after numbers were allocated: see Millburn, *Retailer of the Sciences* (1986) pp. 70–71. In the 1780s George Adams junior, and later Dudley Adams, were still quoting 'Tycho Brahe's Head' on their publications, in addition to 'No.60.'
25. *Der Globusfreund* 21/23 (1973).
26. Van der Krogt, *Old Globes*, refs ADA 5 and 6.
27. Wedgwood archives at the University of Keele, Staffordshire; MS.21171–112 (Adams's bill) and MS.18529–25 (copy of a letter from Wedgwood to Bryerley, April 1774). I am indebted to Dr H.S. Torrens of Keele for locating these.
28. *London Chronicle*; *Lloyd's Evening Post*; *Whitehall Evening Post*; *Gazetteer*; *Daily Advertiser*.
29. Only the second edition, 1759, is listed in R.V. and P.J. Wallis, *Biobibliography* (1986), ref. 748DUN57/59.
30. For Martin's battle with Dunn over planispheres, see Millburn, *Benjamin Martin* (1976) pp. 108–110 and *Supplement* (1986) pp. 29–30.
31. Martin himself had added a network of wires to Senex's celestial globes, to demonstrate the changing positions of the equator and colures with the precession of the equinoxes, described and illustrated in the pages of his *General Magazine* in 1759. It is possible that Adams got the idea for the equatorial wire on his globes from Martin's contrivance; but the latter was intended only for a specific demonstration and did not apply to terrestrial

globes anyway. Martin, *The Young Gentleman and Lady's Philosophy II* (1763) Dialogue IX, pp. 76–88.
32. *Der Globusfreund* (1973); list of old globes in France.
33. Ibid. (1968); list of old globes in Denmark.
34. Bernoulli, *Lettres Astronomiques*, p. 74.
35. Boulton enlisted the aid of James Ferguson in searching for a 6-inch globe, without success; see Millburn, 'James Ferguson's Lecture Tour, 1771', *Annals of Science* (1985), pp. 397–416.
36. Van der Krogt, *Old Globes*, pp. 37–8.
37. Edell, see 'Concave Hemispheres', pp. 7, 14.
38. *Astronomical and Geographical Essays* (1789), p. 244. A footnote on the same page cites Walker's *An easy Introduction to Geography*. In his second edition (1790) Adams gave at this point (pp. 236-7) a long quotation from Walker's book praising the new globes introduced by George Adams senior, particularly his use of an equatorial time scale instead of an external brass hour circle.
39. Gunther, *Early Science in Oxford* vol. 2 (1923), pp. 259–60.
40. A copy of the sale catalogue of Arden's library and apparatus is at the University of Keele, Staffordshire. I am indebted to Dr H.S. Torrens for this reference.
41. For comments on the globes dated 1789 in Teyler's Museum, see chapter 8.
42. One of Pattrick's engraved trade cards, approx. 3 × 5 inches overall, was sold by an ephemera dealer in London in 1993. I am indebted to Peter Delehar for a photocopy of this.
43. For the Cary family see Fordham, *John Cary* (1925). Several advertisements for Cary's globes from 1791 onwards are transcribed in full therein.
44. Details of the range of globes made and sold by the Bardins are given in Millburn and Rössaak, 'The Bardin Family', *Der Globusfreund* 40/41 (1992) pp. 21–57.
45. National Maritime Museum, Greenwich, Inv. G.93.
46. Vince, *A Complete System of Astronomy* (1797), vol. I p. 569.

Notes for chapter 5

1. All except Wright were still alive when Adams died in 1772. Wright died in 1767, Heath in 1773, Ferguson in 1776, and Martin in 1782.
2. See table 12.1 and comments at the end of chapter 12, based on information received from Arthur Graham Blunt, Sarah's great-great-great-grandson. Of her 12 children, only five lived to maturity, but at least one of these (Anne) had a large family, surnamed Rogers. Sarah was ten years younger than her husband Robert Blunt and outlived him by twenty-four years, dying in 1812 at the age of seventy-four.
3. Joyce Brown, *Mathematical Instrument-Makers in the Grocers' Company* (1979), p. 36.
4. Ibid., p. 35.
5. Communication from M.A. Crawforth, based on examination of the Joiners' Company archives, Guildhall MS.8051/5.

348   Notes for Chapter 5, pp. 135–146

6. *Weekly Dispatch* 797, 5 January 1817, p. 7. Charles West 'having been established as a Practical Optician for more than 30 years, & Manufacturer of Glasses &c. to Messrs. Dolland, Adams, Watkins, and others ...'
7. Chenekal, *Watchmakers and Clockmakers in Russia* (1972), p. 16.
8. Communication from Dr A.D.C. Simpson, Edinburgh, August 1985. For Miller's subsequent career see T.N. Clarke et al., *Brass and Glass* (1989), pp. 27–30.
9. Court minutes, Guildhall MS.11,588/7.
10. The three plates in the first edition of *Treatise on the Globes* are not signed. Those added to the second and later editions are signed 'Goodnight sculp', but no artist is mentioned.
11. For the second and later editions this became plate 11, with the figure numbers altered to 26–8. The celestial globe plate, originally plate II, was renumbered plate 10.
12. I am indebted to Dr H.J. Zuidervaart of Middelburg for information on the Dutch editions.
13. Advertised in the *Daily Advertiser* on 12 November 1771 and in the *Gazetteer* on 2 December 1771. No copy of a third edition is known. It may have been lost in the 1757 fire; but perhaps Adams thought the long time that had elapsed since publication of the first and second editions in the 1740s justified an enhancement of the edition number in 1771.
14. For illustrations of Martin's and Watkins's designs see Clay and Court, *History of the Microscope* figures 124, 125, 132 (Martin) and 120, 121 (Watkins).
15. (a) Science Museum, London; (b) Billings Microscope Collection, USA, AFIP 49033-60-4713-206: for details see *Billings* second edition (1974) p. 15 figure 27; (c) Museo di Storia della Scienza, Florence, Inv.1223: *Catalogue of Microscopes* (1991), no. 18; (d) Musée d'Histoire des Sciences, Geneva: for details see A.J. Turner, *Early Scientific Instruments* (1987), p. 118 figure 104.
16. Clay and Court, *History of the Microscope*, p. 190 figure 134.
17. The silver one appeared in the marketplace in the early 1980s, and was subsequently sold by Christie's South Kensington on 19 November 1987, lot 433. It was illustrated in *Bulletin of the Scientific Instrument Society* 11 (1986) p. 27; and 14 (1987) p. 20. Now located at the Royal Museum of Scotland, Edinburgh, this instrument currently bears the name C.W. Dixey of New Bond Street, London, in place of the original Adams signature, which has been erased. Dixey, who was one of the opticians to Queen Victoria, may have carried out repairs or refurbishment in the mid-nineteenth century. Christie's sale catalogue includes a detailed assessment (by Professor G.L'E. Turner) of the microscope's provenance, and its probable connection with the Earl of Bute.
18. G.L'E. Turner, 'The Auction Sales of the Earl of Bute's Instruments, 1793', *Annals of Science* (1967).
19. *Daily Advertiser* Monday 19 October 1772 p. 1. According to this newspaper Adams died on Friday night, that is, the night of 16/17 October. The thrice-weekly *London Chronicle* for 15/17 October reported in the late news column that Adams died 'this morning', that is, in the early hours of 17

October (Saturday). The *Gazetteer* for Monday 19 October reported that he died on Saturday.
20. Table 5.1 derives from an examination of a run of ninety-eight consecutive bill books in Class WO51. For details of the Ordnance books examined and the locations of Adams's bills therein, and an analysis of their contents, see Millburn, 'The Office of Ordnance', *Annals of Science* (1988).

Notes for chapter 6

1. Thomas Wright of Fleet Street, his predecessor as mathematical instrument maker to his majesty (George II), had been retired for twenty years when he died at Hoddeston, Hertfordshire, in 1767.
2. *Daily Advertiser*; *Gazetteer & New Daily Advertiser*.
3. The Duke of Devonshire's signatures as Lord Chamberlain end in January 1761. The Earl of Hertford's follow in the same volume, LC/3/67, commencing in January 1773. It is possible that records for the intervening years were kept in another volume which has since been lost, or is held in the private archives of the relevant lord chamberlain.
4. Surveyor-General's minutes, WO47/80 p. 202.
5. Heath was elected Master of the Grocers' Company on 16 July 1773 but died a few weeks later.
6. Guildhall MS.11,598/4 p. 140.
7. Guildhall MS.11,588/7 p. 630.
8. Ibid., p. 637.
9. Guildhall MS.11,598/4 p. 154.
10. Vestry minutes, Guildhall MS.3016/4, 18 December 1781 and 30 March 1785.
11. Inhabitants list, Guildhall MS.3739.
12. Ibid., Adams.
13. *Daily Advertiser* 10, 12, 14 April 1773.
14. See chapter 4, note 27.
15. WO47/80 p. 282.
16. WO47/81 p. 144.
17. Ordnance bill book WO51/258 p. 20.
18. Part of the delay in payment was due to the fact that, as about two years had elapsed since the warrants for these were issued to George Adams senior, the storekeeper queried whether he should certify that the items had been received strictly in accordance with the orders. The board, having taken into account the circumstances, instructed him to do so.
19. Goodison, *English Barometers* (1977) pp. 124–5.
20. WO47/80 p. 253.
21. WO51/251 p. 211.
22. WO51/253 pp. 310–11.
23. WO51/249 p. 2.
24. WO51/300 p. 38. For the dispute between Dr Pollock and the board see the volumes of Surveyor-General's minutes (WO/47) for the appropriate period, under the heading 'Academy, Royal'.

350 Notes for Chapter 6, pp. 169–176

25. Similar statements appeared in many provincial as well as London newspapers during July 1773, making it difficult to trace a particular comment to its original source.
26. The basic design was patented by J.S. Clais on 30 April 1772; patent no. 1014 in Woodcroft's pre-1852 series. It consisted of a boomerang-shaped beam pivoted near its middle point; a scale pan was suspended from the end of the upper limb, while the lower limb formed a counterweight, the extremity indicating the weight on a calibrated arc. For a woodcut illustration see *Lloyds Evening Post* 9/12 July 1773 p. 34 and adjacent issues. Various different forms of mounting were shown in an advertisement in the same paper on 28/30 July p. 104. An extant example in the King George III Collection has a circular scale instead of a single arc, with three pointers showing the value of gold coins, Troy weight, and Avoirdupois weight simultaneously. Morton and Wess, *Public & Private Science* (1993) p. 388.
27. *Daily Advertiser* 14 July 1773 p. 2 col. 2.
28. For an illustration of two extant examples at the Museum of the History of Science, Oxford, see Millburn, *Retailer of the Sciences* (1986), p. 44.
29. *Lloyds Evening Post* 23/26 July 1773 p. 87 col. 3; 9/11 August p. 143 col. 3.
30. *Daily Advertiser* 4 August 1773 p. 1 col. 1.
31. *Morning Post & Daily Advertiser* 21 August 1773 p. 3 col. 4. Another maker who seems to have advertised once only was John Andrew, of Golden Lion Court, Aldersgate. (*Daily Advertiser* 23 July 1773 p. 1 col. 2).
32. *Daily Advertiser* 26 August 1773 p. 1 col. 3.
33. Ibid., 1 September p. 1 col. 3.
34. Ibid., 29 October p. 1 col. 2.
35. Ibid., 3 November p. 2 col. 1 (a footnote to a lecture advertisement).
36. Ibid., 2 September.
37. Hatton, *Introduction to Clock and Watch Work* (1773).
38. *Daily Advertiser* 9 October.
39. Ibid., 22 October, repeated 10 November. The sellers also included another J. Gilbert, Ludgate Hill, possibly an error.
40. British Library, 1139.k.4(3).
41. *Daily Advertiser*, p. 1 col. 3 on both days.
42. Whipple Museum, *Catalogue 2: Balances and Weights* (1982), no. 17; dated 'c.1750.'
43. Guildhall MS.10,354/2 p. 327. Unusually, the wedding took place in the bridegroom's parish rather than the bride's. Possibly Southwark was only a temporary or lodgings address. The witnesses, apparently named Hinly and Rolfe, have not been identified.
44. See chapter 9, note 5. In 1757, when Hannah (aged seven) was admitted tenant of certain copyhold properties, she was accompanied by guardians rather than parents, suggesting that she was an orphan by then.
45. For Thomas Marsham of the Linnean Society see his obituary in *Gentleman's Magazine* 1819, part II, p. 569.
46. WO47/82 p. 207, 25 November 1773.
47. WO51/265 p. 96. The principal item on this bill was a theodolite costing 25 guineas.

48. WO47/86 p. 190.
49. WO47/87 p. 66.
50. WO47/86 p. 327.
51. Ibid., p. 270.
52. WO47/87 p. 65.
53. Ibid., p. 162.
54. WO47/89 pp. 216, 598.
55. WO47/90 p. 13.
56. WO47/91 p. 69.
57. WO47/96 p. 640, 29 July 1780, when this order was retrospectively authorized.
58. WO47/91 p. 573.
59. WO51/296 p. 42, retrospectively authorized on 15 January 1780.
60. Catalogue appended to the fourth edition of *Treatise on the Globes* (1777), item no. 50.
61. Ibid., item no. 37.
62. WO47/96 p. 989 f. 340. The minute books of this period are indexed by 'page numbers' inserted in the margins, rather than the pages on which they are entered; the numbers evidently refer to some lost volumes of which the surviving ones must be copies, not following the pagination of the originals exactly. Each leaf has a modern folio number for reference, but these numbers, of course, do not correspond with the eighteenth-century index entries.
63. WO47/95 p. 434.
64. WO51/296 p. 41, 12 August 1780, allowed 30 January 1781.
65. WO51/294 p. 294. This bill, for thirty-six quadrants and thirty-six perpendiculars, was allowed on 24 March 1782.
66. WO52/75 p. 123, 28 August 1794; WO52/93 p. 142, 3 November 1795.
67. WO51/295 p. 303, 4 July 1781.
68. WO51/299 p. 311, 4 December 1781.
69. Today, almost all military records of the type discussed here would be 'classified', especially those mentioning destinations overseas. The effect of military security on the instrument trade in the eighteenth century does not appear to have been investigated yet.
70. WO47/97 p. 650 f. 445v.
71. Inv.XXIV/204. Constructed largely of steel bars, about 10 feet long, this device was probably moved around on a wooden carriage resembling that of a manual fire engine. In 1784 Samuel Phillips, the Ordnance 'enginemaker', made 'a New Carriage to the Instrument for trying the calibres of 18 Po. Guns', for 5 guineas: WO47/103, 10 May 1784, under index heading 'Artificers'.
72. WO51/295 p. 303, 4 July 1781.
73. WO52/20 p. 42.
74. Fowler Bean was freed on 1 March 1781: Guildhall MS.11,598/4 p. 257. Perhaps he was turned over to another master before then, but no record of this has been found.
75. Guildhall MS.11,598/4 p. 208. (Extra duty was payable on premiums of £50 upwards.) Possibly Ann was charging George and Hannah rent for the

use of 60 Fleet Street, and George retaliated by charging his mother for his brother's training.
76. Guildhall MS.11,598/4 p. 219.
77. Trade Card, Science Museum, Inv.1934–104 (and in author's collection).
78. Joyce Brown, *Mathematical Instrument-Makers in the Grocers' Company* (1979); table of highest and lowest premiums pp. 87–8.
79. Communication from the late M.A. Crawforth, 14 February 1986, citing the licences for foreign journeymen, Corporation of London Records Office.
80. Joyce Brown, see *Mathematical Instrument-Makers*, cites a Fowler Bean, surgeon, apprenticing his two sons to himself in 1791; possibly the same man, as Adams's apprentice was a son of a surgeon (deceased).
81. Guildhall MS.11,598/4 p. 278, 5 September 1782.
82. Guildhall MS.11,598/5, 3 December 1789. His address then was given as 64 Charing Cross, his late father's shop.
83. Guildhall MS.11,588/8 p. 267, 29 May 1795.
84. A.G. Blunt became free of the Grocers' Company by patrimony in 1931, a liveryman in 1947, member of the court in 1956, and Master in 1962 (communication from the beadle, 19 June 1990). When the first draft of this chapter was being written in 1990–91 he was living in retirement in New South Wales, and kindly provided me with much useful genealogical information on the Blunt family and their Rogers relations.
85. *Transactions ... Society of Arts* vol. III (1785) pp. 184–6.
86. Ibid., vol. II (1784) p. 255.
87. Smeaton, 'Description of an improvement in the Application of the Quadrant of Altitude to a celestial Globe, for the resolution of Problems dependant on Azimuth and Altitude', *Philosophical Transactions* vol. 79 (1788) pp. 1–6 and plate I.
88. *Analytical Index to the first twenty-five volumes of the Society instituted at London for the Encouragement of the Arts, Manufactures, and Commerce* (London, 1808).
89. Museum of the History of Science, Oxford, MS Gunther 4. Octavo, approx. 170 leaves (not numbered). This manuscript has been microfilmed.
90. For further remarks on Magellan and his activities as an agent for overseas buyers of English instruments, see chapter 8 and the references thereto.
91. At the time of writing this chapter, the manuscript had not been paginated or foliated. Entries are identified by the date of the meeting.
92. Printing on glass from copperplates was not applied successfully to the production of lantern slides until the 1820s. An impression from the plate was taken on a thin film of glue which was transferred to the glass and the design then burnt-in. See John Barnes, 'The projected image', *The New Magic Lantern Journal* (October 1985).
93. Communication from Dr Anita McConnell, June 1993.
94. One non-standard Adams instrument at Coimbra University is a brass Copernican armillary sphere mounted in an all-brass stand; for an illustration see *Bulletin of the Scientific Instrument Society* 32 (March 1992) p. 26.
95. Toksvig, *Emanuel Swedenborg* (1949). The 'official' biography published by the Swedenborg Society is by Trobridge.

96. First published in Latin in 1771, this work was given wide publicity in modern times by its inclusion in Dent's Everyman's Library (no. 893) in 1933, with an introduction by Helen Keller. All of Swedenborg's works are maintained in print by the Swedenborg Society, instituted in London in 1810 for that purpose and now located at 20/21 Bloomsbury Way, WC1A 2TH.
97. Hindmarsh, *Rise and Progress of the New Jerusalem Church* (1861).
98. Hindmarsh gives the text of the editors' note in Latin with an English translation, p. 31. Adams was one of the signatories.
99. Tafel, *Documents ... Swedenborg*, 3 (London, 1890). On p. 995 Tafel states: 'The Doctrine of the Lord was translated into English by Mr. P. Provo, and published in 1784 by the London Printing Society. A second edition, corrected by Mr. George Adams, Optician, was printed at his expense in 1786.' This statement was repeated in the American *Annals* (see note 101 below), I p. 130. Hannah Provo, god-daughter, was a beneficiary under the will of Hannah Adams in 1810.
100. [Hindmarsh] *New Magazine of Knowledge* I (London, 1790), pp. 138–40.
101. Odner, *Annals of the New Church*, vol. I: *1688–1850* (1904), p. 149. This comment was presumably prompted by Adams's theme, that the principal purpose of human vision was to enable us to see the works of the Creator.
102. 'Thomas Bugge's Travel Diaries'; typescript translation at the Museum of the History of Science, Oxford.
103. British Library, Add.MS.8097 ff. 28–9. The writer was proposing to use the variation for finding the longitude, with the aid of globes (to be published by himself) showing lines of constant variation. Adams had told him that the variation could not be measured with sufficient accuracy at sea for this method to be practicable.
104. British Library, Add.MS.33,272 ff. 19–22, especially f. 20r and f. 22r.
105. Ibid., ff. 101–2.
106. Royal archives, Windsor Castle, ref. RA.29,084 and 29,086. These were for items supplied to the Prince of Wales, not the king.
107. The precise date of this appointment has not been ascertained, but from the title-pages of his publications described in chapter 7 it must have been sometime in 1787.

Notes for chapter 7

1. See contemporary newspapers such as the *London Chronicle* from the end of March onwards for reports of parliamentary debates.
2. *London Chronicle* 23/25 December 1783 p. 612 col. 2.
3. Sources cited by Adams include *Philosophical Transactions*, *Memoires de l'Academie de Berlin*, *Journal de Physique*, the *Swedish Transactions*, and the *Monthly Review*; and writings (or work carried out) by Archard, Atwood, Beccaria, Beckett, Birch, Brooke, Brydone, Canton, Cavallo, Cavendish, Cullen, De Luc, Eeles, Fell, Franklin, Gamble, Haas, Hauksbee, Henley, Higgins, Hunter, Kinnersley, Lyons, Marat, Marsham, Milner, Morgan,

Nicholson, Nollet, Partington, Priestley, Pringle, Ronayne, Saussure, Symmer, Townsend, Villette, Volta, Walsh, Watson, and Wilson.
4. Ingenhouz comments on this point in a letter to van Marum dated 12 February 1790, printed (in translation) in *van Marum* vol. VI pp. 177–8.
5. This plate is not numbered; it is labelled 'To face the end of the Supplement'.
6. This John Hill (who should not be confused with the author of *The Construction of Timber*) was born in 1756 and died in 1819; an outline of his place in the Hill pedigree is given in *Burke's Landed Gentry*, 1894 edition, under Hill of Gressenhall Hall. When he was in contact with George Adams around 1790 his father and grandfather of the same name were still alive, hence 'the third'. His will (PCC PRO PROB/11/1617 Q.278–9) is a lengthy document which indicates that he was then a wealthy landowner, well able to afford the latest scientific instruments and apparatus.
7. Author's collection. Unfortunately the appendix had been rather roughly removed (probably many years ago) before this volume came on to the market in 1968, but its contents can be inferred from Hill's numerous references to it.
8. *Morning Herald* 20 February 1788 p. 1. Walker also gave regular courses at his house in George Street, Hanover Square, but the one at Founders' Hall seems to have been an annual event aimed at a larger audience.
9. Gage, *History of the Linnean Society* (1938) pp. 6–26.
10. 'Mr.Marsham' is also mentioned in *Essay on Electricity* in connection with an experiment on melting wires with the Leyden jar (expt. CCXXXV, second edition, 1785, pp. 377–9).
11. Bradbury, 'The Quality of the Image ...', in *Historical Aspects of Microscopy* (1967), pp. 151–73. See also Frison, 'Adams Microscopes ...' *The Microscope* (1951) pp. 199–206, 221–4. Frison tested an instrument similar to that examined by Dr Bradbury, and found that with powers up to ×92 the image was 'hazy' but the highest power (×152) was 'rather good'.
12. For Prince see Upham, 'Memoir of John Prince' (1837) and Schechner, 'John Prince' (1982). Upham had access to Prince's notebooks and correspondence, which have since been lost. On pp. 216–17 he quoted a long letter to 'Mrs Adams' written on 25 January 1796 when Prince had just heard of George junior's death, in which he said 'I ranked myself among his friends, and was gratified by the tokens he gave me of his friendship, which I endeavoured to return, by promoting his interest and reputation here among my friends in the line of his profession.' This letter was presumably addressed to George junior's widow Hannah, who was then still living at 60 Fleet Street, though it could equally well have been sent to his mother.
13. *Gentleman's Magazine* 1796 II (November) pp. 897–9 and plate II. Hill quoted long extracts from Adams's letter to him about the use of the instrument.
14. Clay and Court, *History of the Microscope* p. 223. It is described there as 'the original form of this instrument as made by Adams', but in fact the instrument shown has Prince's suggested mounting arrangement and Hill's lamp: it is not Adams's own form with a tripod foot as shown in his *Essays on the Microscope*. Clay and Court did not say in which of their collections (if either) this particular instrument was then located.

15. For example, two at the Whipple Museum, Cambridge, Inv.Wh.201 and Wh.1796.
16. *British Critic* 12 (July–December 1798) pp. 49–59.
17. Originally introduced by the National Maritime Museum for its catalogue of instruments in the Department of Navigation & Astronomy (1970), this term conveniently defines a specific type of model with separate tellurian and lunarium attachments fitting on to a basic planetarium.
18. Wheatland, *The Apparatus of Science* (1968), pp. 48–51.
19. James Ferguson, *Lectures on Select Subjects* (1760), plate XX.
20. Adams did not mention Martin by name, but the latter's *Essay on Visual Glasses* (1756) was one of the fifteen sources listed.
21. Goodison, discussing this claim, states that the floating gauge may have been invented by Ramsden.
22. Stone, *The Construction and Uses ...* (1723). To each of Bion's chapters, Stone added a few pages describing specifically English instruments. An augmented edition appeared in 1758, repeating the original text with additional information on English products. This augmented edition was reprinted in facsimile in 1972 by Holland Press.
23. H.C. King, *Geared to the Stars* (1978) pp. 204–5.
24. This is a reduced copy of the frontispiece in *Essays on the Microscope* (1787), with a slightly amended title to suit the more general subject of *Lectures*. (The latter contains nothing about microscopes except a ray diagram showing the optical principle.)
25. Author's collection. An advertisement in this volume mentions also a third edition of *An Essay on Vision*, but no copy so designated has been found.

Notes for chapter 8

1. Available as a set or individual fiches from Microform Academic Publishers, East Ardsley, Wakefield, Yorkshire, England. The arrangement is alphabetical by writer, of whom there were about 300; the total number of letters is about 3,000. Only a few of van Marum's replies are included, in the form of copies made by his wife.
2. *Van Marum* vol. VI pp. 266–72 (English translation), 'Notes on a voyage to London in 1790'.
3. *Van Marum* vol. VI pp. 224–5 (English translation of part of the letter, which in the original French runs to four pages of manuscript). Brooks' electrometer is also mentioned in an earlier letter dated 29 July 1785.
4. Ibid., pp. 225–7 (English translation).
5. Fiche 141, partly illegible.
6. Fiche 1.
7. *Van Marum* vol. VI p. 1; also reproduced in facsimile in plate facing p. 4.
8. Ibid., pp. 175–7 (English translation).
9. Ibid., pp. 296–8. The equatorial mounting by Ramsden is catalogue no. 266.
10. Fiche 1.
11. *Van Marum* vol. VI pp. 378–81 (English translation).
12. Fiche 1.

13. *Van Marum* vol. VI p. 161.
14. John R. Millburn, 'Horology and the Adams Family', *Antiquarian Horology* 13 (1982), pp. 368–76, 582.
15. *Astronomical and Geographical Essays* (1789) p. 539. A copy of van Marum's letter (in French) is on fiche 1, following Adams's letter dated 17 December 1790.
16. Fiche 1.
17. *Van Marum* vol. VI pp. 2–3.
18. *Lectures*, Plate 'Astronomy XIII', dated 4 December 1793.
19. Fiche 1.
20. *Van Marum* vol. VI pp. 4–5.
21. This is a reference to Blair's fluid achromatic lenses; see H.C. King *History of the Telescope* (1955) pp. 189–90. Van Marum also consulted Peter and John Dollond about Blair's work, but they replied on 29 September 1791 to the effect that they did not expect it to result in any improvement in telescope performance.

Notes for chapter 9

1. He does not appear to have been buried in either of his 'home' parishes (St Bride's and St Dunstan's), or in Southampton town.
2. Public Record Office, PROB/11/1265 Q.531.
3. The *Gentleman's Magazine* for 1809, part II p. 1179, records the death of Mrs Ann Adams on 15 November, at Dulwich, in her eighty-seventh year. Ann Adams was a very common name, but the year corresponds with a change in ownership of the Langley Marish estate (see chapter 10), and the age of the deceased is consistent with that given by Ann Dudley on her marriage to George Adams senior in 1748. Dr Joseph Allen of Dulwich is called 'my worthy friend' in George senior's will; Lucy Adams was living in Dulwich when she signed her will in 1802; and Isabel, the youngest daughter of Ann and George senior, was living in Dulwich, where she owned two leasehold houses, in the 1820s. No will or Admons for Ann Adams in 1809 has been traced in the London or Surrey records, but as George senior's widow had only a life interest in her late husband's estate this is not significant.
4. Nothing has been discovered about this freehold property. After Hannah's death it was to go to her brother Thomas Marsham.
5. This was copyhold of the manor of Great Baddow, Essex. The manorial records show that on 2 June 1757 Hannah Marsham, aged seven, was admitted as 'tenant' of six pieces of land (with cottages and so on) inherited from her grandmother Ann Marsham under a will dated 27 June 1752. In July 1774 George Adams (junior) and Hannah Marsham made an agreement in contemplation of marriage which apparently passed the property to George Adams, Dr Joseph Allen being named as trustee. Six years later, on 20 November 1780, one-half of the property (that is, a half-share in each piece) was surrendered by George and Hannah to Dr Allen, but no mention of this division is made in George's will. Dr Allen was admitted 'tenant' of this half

on 7 June 1781. Allen's death in 1796 was not noticed by the manorial court until 1810 (probably as a result of Hannah's death that year), when his heir was advertised for, as a result of which his nephew Joseph Allen of the East India House, London, claimed the property and was admitted. The records at this point are not easily interpreted, but it would appear that the property was then sold for £1,200 to Richard Dixon. Essex Record Office, D/DWv M63 and M64.

6. For example, Mary Senex continued her late husband's printselling and globe-making business for fifteen years after his death in 1740. When Benjamin Martin bought the Senex globe plates from James Ferguson in 1757 he advertised that he had engaged the same workmen who had previously made globes for Senex himself, and presumably for his widow also.
7. WO45/38. Volumes WO45/36–39 cover the four quarters of the year 1795. Each double-page opening has columns for the letter writer's name, date of letter, date of receipt, date of minute arising (if any), and abstract of subject (in a few words).
8. Dudley Adams, *Treatise on the Globes* (1810), pp. vii–viii.
9. Volume S-C.S.28 item 6.
10. John R. Millburn, *The Library of George Adams* (Aylesbury, 1988); 16pp booklet + two 98-frame fiches. Copies are in the British Library and the principal science libraries in the UK.
11. William Jones explained the circumstances of this purchase in a note dated 1 August 1812, added to the sixth edition (1812) of George junior's *Astronomical and Geographical Essays*, probably prompted by Dudley's remarks in his 1810 edition of *Treatise on the Globes*. Jones said that George junior's widow Hannah not having received an adequate offer for the books, 'on her application to the Editor and his Brother, by an advanced tender, they became, impartially, the purchasers'.
12. Public Record Office, PROB/11/1515 Q.486; proved 20 October 1810. Details of the market values of each legacy are given in the death duty register IR26/164 pp. 12–13. Death duties at that time did not apply to land, so the Essex and Surrey estates are not mentioned. The rate of duty on other bequests depended on the relationship of the beneficiary to the testator.

Notes for chapter 10

1. Guildhall Library, MS.10,351. The baptism registers for this parish at this date give the date of birth as well.
2. Guildhall MS.11,598/4 p. 208.
3. The witnesses were Thomas Bird, probably a clerk or verger, and another person thought to be 'P. Martin'.
4. Guildhall MS.11,598/5.
5. Public Record Office, PROB/11/1164 Q.178. George Adams was one of the executors.
6. Westminster Public Library, Archives Department, F.592. The premises were empty for the first two quarters of 1788. Although house numbers were allocated in Westminster in 1762, the rate collectors followed their estab-

lished 'walks' and in general did not take any notice of house numbers until the early nineteenth century. Dudley Adams is the eighty-second entry under the heading 'Spring Gardens' in Charing Cross Ward, in the parish of St Martin in the Fields. Drummond's Bank is the eighty-seventh entry. One house or shop, at the northern side of the entrance to Spring Gardens, next to number 53, was apparently called 1 Spring Gardens so does not figure in the numbering of Charing Cross.

7. See also Hugh Phillips, *Mid-Georgian London* (1964) pp. 96–8 for a description of this region and illustrations of its appearance in the mid-eighteenth century.
8. Guildhall MS.11,598/5.
9. For example, Christie's 29 March 1990, lot 17. Several Adams globes of this date located in Holland are listed by van der Krogt, *Old Globes*, pp. 40–41.
10. For details of this mission see Sir George Staunton, *An Authentic Account of an Embassy* (1797).
11. British Library, Add.MS.33,979 f. 169.
12. J. Seeley, *Stowe* (1797), p. 46.
13. *Catalogue of the Contents of Stowe House, near Buckingham; which will be sold by auction, by Messrs. Christie and Manson, on the premises, on Tuesday, August 15th, 1848, and extending over thirty-seven days, commencing at twelve o'clock precisely each day.*
14. Public Record Office, LC/3/68 p. 13.
15. Guildhall MS.11,598/5.
16. Guildhall MS.11,588/8 p. 390.
17. This gunnery instrument, with a plumb line rather than a spirit level, is amongst the collection of miscellanea at Snowshill Manor, Broadway, Worcestershire.
18. WO52/92 p. 271; a gunner's quadrant for Alderney, £2 12s 6d; WO52/93, p. 259, two pairs of brass calipers for Dominica, £3 10s 0d.
19. WO52/82 p. 93. The steel compasses were for measuring the lengths of fuses.
20. This is the earliest that has been found, but he may have advertised elsewhere in July.
21. The ratebooks state that Robert Blunt was to pay from midsummer. Directories indicate that the Blunt linen-drapery business was at 53 (instead of 64) Charing Cross from 1797. Robert's mother Sarah died in 1812.
22. This was 'Printed for and sold by the author' in Bristol, and sold in London by booksellers and by the mathematical instrument makers Adams and Troughton in Fleet Street, and Jones in Holborn. As the dedication is dated 12 September 1796, 'Mr.Adams' must have been Dudley.
23. For details of the patenting system before 1852, see C. MacLeod, *Inventing the Industrial Revolution* (1988).
24. British Museum, Department of Prints and Drawings, Heal 105.1.
25. At the time of writing the Longleat instruments had not been examined by an instrument historian. Information given here is derived from a BBC-TV programme 'Arthur Negus Enjoys', first transmitted on 26 August 1984 and repeated several times since then.
26. Royal Archives, Windsor Castle, RA.29119. This item was found by Mr V.K. Chew, formerly of the Science Museum.

27. British Library, Add.MS.35,648.
28. Pages 118–21 of the 1977 edition.
29. WO47/2588.
30. WO49/248.
31. Buckinghamshire County Record Office, Aylesbury, parish of Langley Marish. Several changes have taken place in parish boundaries since the eighteenth century: most of the relevant area is now part of the civil parish of Fulmer.
32. *Treatise on the Globes* (1810) pp. vii–ix. In the preamble to this, Dudley said that his father died in 1773, an error which was perpetuated in reference books until Nicholas Goodison (private communication, 1968) gave the correct date from the 1772 *Gentleman's Magazine*. This error of only one year delayed the discovery of George Adams's will for decades, as nobody had thought of looking before 1773.
33. *Astronomical and Geographical Essays* sixth edition (1812), preface p. iii.
34. Guildhall MS.11,598/6. In this entry Dudley is called a mathematical instrument maker rather than an optician.
35. Eckhardt's rolling parallel ruler was patented by him in 1772 and the patent rights transferred to Dollond, but by this time the patent had expired.

Notes for chapter 11

1. Dudley Adams is not listed in the index to Class B3 at the Public Record Office, containing the surviving detailed files of bankruptcy cases.
2. Public Record Office, B5/13, pp. 266–70. This entry is signed by the lord chancellor on p. 267 'Be it as prayed' and dated 15 December 1817.
3. In London directories he is also called 'Upholder', a term that could mean undertaker. Later directories give Samuel Legg & Son, carpenters, at this address.
4. Buckinghamshire County Record Office, PR.123/5/2 and Q/RP1/8/38–51.
5. Quoted by Dr A. McConnell in a letter to the *Bulletin of the Scientific Instrument Society* 20 (1989) p. 26.
6. No other references to Dudley Adams have been found in the indexes to the *London Gazette* up to the mid-1830s.
7. Guildhall MS.5213/4.
8. This aspect of West's activities has been investigated by Brian Gee in an unpublished PhD thesis.
9. For example, in 1830 he reprinted Martin's tract on the 'opake' solar microscope, which had originally been published about sixty years earlier.
10. Public Record Office, LC/3/69 pp. 8–9.
11. They are listed in his published catalogues, but globes of this size in the 1830s could have been Newton's rather than Adams's.
12. Unauthorized use of royal appointments or the royal arms was made a specific offence in the Patents, Designs and Trade Mark Act of 1883 (46 & 47 Vic. Cap. 57). From this date onwards royal warrants have always been strictly personal to the named recipient; warrants are not granted in the name of a firm or company, though the individual may trade as such. The Royal Warrant Holders' Association, which today acts as a watchdog to

ensure that the regulations are strictly enforced, was originally founded in 1840 as a dining club. For the early history of royal warrants, see Tim Heald, *By Appointment* (1989).

Notes for chapter 12

1. Robert Banks Jenkinson (1770–1828), Prime Minister 1812–27.
2. J.B. Heath, *Some Account of the Company of Grocers* (1854), pp. 344–6.
3. British Library, Add.MS.38,280 f. 268.
4. British Library, Add.MS.38,281 f. 27.
5. Ibid., f. 29.
6. British Library, Add.MS.38,257 ff. 72–3.
7. J. Birch, *Essay on the Medical Applications of Electricity* (1802).
8. In 1831 Michael La Beaume, a London practitioner, was appointed 'Medical Galvanist and Electrician' by William IV. (LC/3/70 p. 37) This appointment was renewed by Queen Victoria in 1839. (LC/3/71 p. 99)
9. Guildhall MS.11,588/12 p. 199.
10. Guildhall MS.11,605/2 p. 169.
11. Brydone had three contributions on medical electricity published in the *Philosophical Transactions*: 50 pp. 392, 695; and 63 p. 163.
12. See Lowndes, *Observations on Medical Electricity* (1787) and *The Utility of Medical Electricity* (*c*. 1791).
13. The authors cited include Becket, Birch, Bohadtch, Cavallo, De Haen, Deshaies, Diembrock, Duncan, Ferguson, Fothergill, Hall, Home, Hunter, Jallabert, Knox, Love, Mauduyt, Percival, Priestley, Spry, Symes, Wardrope, Watson, Wesley, Wilkinson, and Zetzell.
14. British Library, Add.MS.40,375 ff. 275–6.
15. It is possible that the copy sent with this letter is filed somewhere else amongst the Peel papers.
16. British Library, Add.MS.40,383 ff. 207–8.
17. Board of Longitude Papers, RGO.14/54 Sect.2 f. 13; now at Cambridge University Library (a microfilm is at the Public Record Office, Kew). I am indebted to Dr J.A. Chaldecott for a transcript of this letter.
18. Parry Jones, *The Trade in Lunacy* (1972).
19. Guildhall MS.6543/3.
20. Public Record Office (Chancery Lane), PROB/11/1770 Q.292–3.
21. The death duty registers do not help here, as the date of death therein is left blank.
22. To Lucy Blunt, youngest child of Robert and Sarah, born 1778.
23. The globes, textbook, and telescope were left to Hannah Rogers (the executrix), third daughter of Anne and Richard.
24. To Lucy Blunt Rogers, fourth daughter of Anne and Richard.
25. To Mary Isabel Rogers, fifth daughter of Anne and Richard. For a modern account of the Zograscope, see Allan Mills, 'The Optical Diagonal Machine', *Bulletin of the Scientific Instrument Society* 28 (1991).
26. Hannah Rogers died unmarried on 8 December 1854, aged sixty-one, before her mother Anne, who lived to ninety-two and died on 9 November 1855.

Anne's husband, Richard Rogers, originally a 'Meal Factor' in Hackney, had died on 9 February 1823 aged fifty-nine. In the second quarter of the nineteenth century Mrs Anne Rogers and her daughters ran a boarding school for young ladies at 3–4 Royal York Crescent, Clifton. The 1851 census shows fifteen scholars and six servants resident there, in addition to Mrs Rogers ('retired'), four of her daughters ('School Mistresses'), and another unrelated female teacher. The residents also included Sarah A. Rogers, unmarried, aged twenty-five, born in Essex, whose relationship to Mrs Rogers is given as 'grand-daughter'. Her position in the family tree has not been ascertained, but her presence at Clifton raises the possibility of Hannah's effects (including the Adams papers and artifacts) being passed down to at least one more generation in the Rogers family. I am indebted to Mr T.J. Bryant of Bristol for supplying me with the above details, and for locating the Rogers family grave in Clifton parish churchyard. The stone on the latter records the deaths of Richard 1823, Anne (wife) 1855, Anne (daughter) 1846, and Hannah 1854.

Appendix I   George Adams Senior's Catalogue 1766

This catalogue of eight octavo pages is normally found with George Adams senior's *Treatise Describing and Explaining the Construction and Use of New Celestial and Terrestrial Globes* (1766), though the leaves are separately signed '*' and there is no catchword specifically linking it to the main text. It is generally bound at the end, but in the British Library's copy (shelfmark 717.f. 19) it is bound at the front of the book after the contents list.

A CATALOGUE of Mathematical, Philosophical, and Optical INSTRUMENTS. Made and sold by GEORGE ADAMS, Mathematical Instrument-Maker to the King, at his Shop the Sign of Tycho Brahe's Head, in Fleet-street, London.

Where Gentlemen and Ladies may be supplied with such Instruments as are either Invented or Improved by himself; and constructed according to the most perfect Theory.

The Study of the Mathematicks being now become a necessary Part of every Gentleman's Education, and as it contains an inexhaustible Fund of useful Knowledge, wherein the greatest Geniuses may exert their utmost Faculties, and the meanest find something within his Reach, the Theory of it displays an ample Field, and the Practice of it has always been productive of many great Advantages to Men of Action and Business.

Therefore Mathematical Instruments are the Means by which the Sciences of Geometry, Astronomy, Philosophy, and Opticks, have been rendered useful in all the common and necessary Occurrences of human Life. They not only enable us to connect Theory with Practice, but also instruct us how to turn bare Contemplation into the most substantial Use, by making one of the most servicable Branches of Learning the natural Way of rendering this Knowledge general and diffusive.

Mathematical Instruments

	£	s	d
Cases of Drawing Instruments, the Scales and Sector either Wood, Ivory, or Brass, from 10s.6d. to	5	5	0
Magazine Cases of Instruments, from £7 7s. to	26	0	0
Silver Cases of Instruments, from £2 12s.6d. to	21	0	0
Proportionable Compasses	1	11	6
Ditto	3	3	0
Sectors of a new Construction	2	2	0
Ditto	4	4	0
Ditto	6	6	0
Ecliptical [sic] Compasses of a new Construction	5	5	0
Gunners Calippers	1	15	0
Gunners Quadrants	2	12	6
Perpendiculars for gunnery	1	11	6
Parallel Rules, two and three Feet Pocket Rules, Gauging and other Sliding Rules, Gunter Scales, Surveying and Navigation Scales, with all other Kinds of Scales and Rules, at the usual Prices.			
Plane Tables for Surveying Land	3	13	6
Ditto of a new Construction, with a Brass-Headed Staff	15	0	0
Theodolites with plain Sights	5	5	0
Ditto with Telescopic Sights, and Vertical Arch	12	12	0
Ditto, of the best Sort	21	0	0
Gunter's Chains	0	7	6
Levels, of the latest Construction	7	7	0
Ditto, larger	12	12	0
Protractors from 2s.6d. to	1	10	0
A new Protractor for setting off Angles to a Minute	4	4	0
A new Plotting Scale with a Nonius Division	4	4	0
Pantographers for reducing and enlarging Plans and Drawings	4	14	6
Measuring Wheels	6	6	0
Pedometers	5	5	0
Way Wisers for Wheel Carriages	5	5	0

	£	s	d
Astronomical Quadrants	12	12	0
Ditto	31	10	0
Ditto	52	10	0
Transit Instruments, from £10 10s. to	31	10	0
Equal Altitude Instruments, from £21 to	50	0	0
Astronomical Sectors, from £50 to	100	0	0
Horizontal Sun-Dials, from 10s.6d. to	10	10	0
Universal Equinoctial Dials, from 10s.6d. to	3	13	6
Ditto, on a Pedestal, &c.	12	12	0
Eliptical Double Dials, from £5 5s. to	10	10	0
Pocket Horizontal Dials, for several Latitudes	2	2	0
Hadley's Quadrants	2	2	0
Ditto	2	12	6
Ditto	3	3	0
Ditto	4	4	0
Dr. Knight's new Steering Compass	2	12	6
——— his new Azimuth Compass	5	15	6
Orreries, from £25 to	150	0	0
A Planetarium, or Manual Orrery for the Copernican and Ptolemaic System	3	3	0
Ditto	8	8	0
A new Planetarium	18	18	0
A new Tellarium	36	0	0
Adams's New Globes, 18 Inches Diameter	9	9	0
Ditto, in Mahogany	11	11	0
Ditto, in Mahogany and Brass Mounting	26	0	0
New Globes, 12 Inches Diameter	5	5	0
Ditto, in Mahogany	6	16	6

Optical Instruments

	£	s	d
Reflecting telescopes, 1 Foot	5	5	0
Ditto, 18 Inches	8	8	0
Ditto, 2 Feet	12	12	0
Ditto, on a Rack Stand	20	0	0
Small Reflecting ditto, 6 Inches	3	13	6
Ditto, 4 Inches	2	2	0
The new Achromatic Refracting Telescope with seven Glasses, two Feet	2	2	0

	£	s	d
Ditto, three Feet	3	3	0
Ditto, four Feet	4	4	0
and so on in Proportion for any other Length, all other Sorts of Refracting Telescopes, at the usual Prices			
The large Double Constructed Microscope	6	6	0
Ditto, in a flat Mahogany Case	7	7	0
Ditto, of an inferior Sort	3	13	6
Wilson's Pocket Microscope	2	2	0
Ditto	2	12	6
Ellis's Microscope	2	2	0
A solar Microscope	4	14	6
Ditto	5	15	6
A Microscopical Apparatus in a Mahogany Box	12	12	0
all other Sorts of Microscopes at the usual Prices			
A Pocket Camera Obscura	0	10	6
Ditto	1	1	0
Ditto, large	1	5	0
Ditto, larger	2	2	0
A Book Camera, of a new Construction	5	5	0
A new Instrument for taking Perspective Views	6	6	0
Concave and Convex Mirrors, from 7s.6d. to	26	0	0
Prisms, from 7s.6d. to	2	2	0
Zograscopes, for viewing Prints, from 18s. to	2	12	6
Opera Glasses, from 8s. to	1	1	0
Reading Glasses, from 2s.6d. to	3	3	0
Spectacles for the Nose	0	1	0
Ditto for the Temples	0	3	0
Ditto	0	5	0
Ditto	0	7	6
Ditto, in Silver	0	14	0
Ditto, Double Joints which neither press the Nose, or Temples	0	9	0
Ditto	0	12	0
Ditto, in Silver	1	1	0
Spectacles of Brazil Pebbles, in Steel or Silver, at the usual Prices.			

Philosophical Instruments

	£	s	d
Air Pumps, exclusive of any Apparatus [no price]			
A single Barrel Ditto	2	2	0
A small double Barrell'd ditto	4	4	0
A ditto larger	6	6	0
A large Table Air Pump	10	10	0
A large Standing Air Pump	21	0	0
and Apparatus to either of these according to the Desire of the Purchaser.			
Barometers	2	2	0
Ditto, with Thermometers	2	12	6
Ditto	5	5	0
Ditto, with Hygrometers	5	15	6
A new invented Barometer of a peculiar Make	7	7	0
Diagonal Barometers	5	5	0
Wheel Barometers	8	8	0
Triple Diagonal Barometers	16	16	0
Botanick Thermometers	0	10	6
Ditto	0	18	0
Farenheit's Pocket Thermometers	1	1	0
———— Thermometers for Brewing Distilling, &c.	1	11	6
Electrical Machines	6	6	0
Ditto	10	10	0
Hydrostatic Ballances	2	12	6

A new and curious Hydrometer is now in Hand, of which the Price cannot be yet ascertained.

These, with all the Variety of Instruments required in the Practice of the several Branches of the Mathematicks, Philosophy, and Optics, are Made and Sold as above, at the most Reasonable Prices.

Appendix II   George Adams Junior's Last Catalogue 1795

The catalogue reproduced in facsimile in the following pages formed gathering Pp, pages 565–79 of the third edition (1795) of George Adams junior's *Astronomical and Geographical Essays*, published in the year of his death. The copy reproduced here (from the author's collection) is on paper watermarked 1794. Though many of the descriptions are too brief for an extant instrument to be positively linked with an entry in this catalogue, it illustrates the wide-ranging nature of the Adams instrument business at the peak of its activities. Globes from 28 to 12 inches diameter are included, although at this date they would have been signed by Dudley Adams at Charing Cross. The omission of smaller sizes may be a mistake.

Some of the prices have a wide spread, as certain items, such as cases of drawing instruments, would have been assembled from a range of optional components to suit the customer's requirements. Some others, such as convex and concave mirrors, would have been made to almost any size that the customer specified. The fixed prices quoted in the catalogue would have been for fairly standard items either available from stock or made to a predetermined pattern.

Military instruments, such as gunner's quadrants in three varieties, are included but not priced, presumably to emphasize to prospective customers the wide range of the firm's capabilities. It is not clear who, apart from the Board of Ordnance (and the East India Company), would have bought (say) 'General Williamson's instruments for howitzers', though some artillery officers may have purchased their own calipers and levels, as well as drawing instruments.

(565)

A CATALOGUE

OF

Mathematical and Philofophical Inftruments,

MADE AND SOLD BY

GEORGE ADAMS,

Mathematical Inftrument Maker to His Majefty, and Optician to the Prince of Wales,

No. 60, Fleet-Street, London.

Optical Inftruments.

	£.	s.	d.
THE beft double-jointed filver fpectacles, with glaffes	1	1	0
The beft ditto, with Brazil pebbles	1	16	0
Single joint filver fpectacles, with glaffes	0	15	0
Ditto, with Brazil pebbles	1	11	6
Double-joint fteel ditto, with glaffes	0	7	6
Beft fingle-joint fpectacles	0	5	0
Ditto, inferior frames, from 2s. 6d. to	0	3	6
Beft Tortoifefhell fpectacles	0	10	6
Nofe fpectacles, mounted in filver	0	7	0
Ditto in Tortoifefhell and filver	0	4	0
Ditto in horn and fteel	0	1	0
Spectacles for couched eyes			

P p   Spectacles

566   A CATALOGUE OF INSTRUMENTS.

	£.	s.	d.
Spectacles with shades			
Concave glasses in horn boxes, for short-sighted eyes			
Ditto, mounted in tortoiseshell and silver, pearl and silver, in various manners, and at different prices			
Reading glasses from 2s. 6d. to	2	2	0
Opera glasses from 10s. 6d. to	4	4	0
Ditto to be used at sea by night	1	11	6
Telescopes of various lengths, sizes, and prices			
Acromatic telescopes, portable and convenient for the pocket; the sliding tubes are of brass, and therefore not subject to the inconveniences of those that are made with vellum drawers, from 1l. 11s. 6d. to	10	10	0
Telescopes to be used at sea by night			
Acromatic perspective glasses for the pocket, from 10s. 6d. to	2	12	6
A thirty-inch acromatic telescope, with different eye-pieces for terrestrial and celestial objects. This is one of the most pleasant telescopes that is made for general purposes, from 8l. 8s. to	11	11	0
An acromatic telescope, about three feet and an half long, with different eye-pieces	18	18	0
A three-feet reflecting telescope, with four magnifying powers, with rack-work	42	0	0
A ditto, two feet long, with ditto	21	0	0
A two-feet reflecting telescope, with two powers	12	12	0
An eighteen-inch ditto	8	8	0
A twelve-inch ditto	5	5	0
Adams's LUCERNAL MICROSCOPE, for opake and transparent objects: it does not fatigue the eye, is in all cases a proper substitute for the solar microscope, and on many occasions superior to it, from 21l. to	25	0	0
A small double-reflecting microscope	2	12	6
A larger ditto	3	13	6
An improved universal double microscope	6	6	0
Ditto, fitted up in a different form, from 8l. 8s. to	14	14	0
Ellis's aquatic microscope	2	2	0
Ditto, with an adjusting screw	2	12	6
Withering's and other botanical microscopes, from 10s. 6d. to	3	3	0
Small pocket microscopes, from 6s. to	3	13	6
Solar microscopes	5	5	0

Ditto

A CATALOGUE OF INSTRUMENTS.

	£.	s.	d.
Ditto	6	6	0
Solar microscopes for opake and transparent objects, from 16l. 16s. to	21	0	0
Curious collections of objects for the microscope, either opake or transparent			
Collections of salts, properly prepared for the microscope			
Magnifying glasses for botanical, anatomical, and other purposes, from 2s. to	1	11	6
Small magic lanthorns, with twelve small glass sliders, from 10s. 6d. to	1	1	0
Small magic lanthorns, with the sliders better painted			
Large ditto, from 1l. 5s. to	1	11	6
Optical machines for viewing perspective prints, from 18s. to	1	16	0
Scioptric balls	0	10	6
Small camera obscuras, from 10s. 6d. to	3	3	0
Book and pyramidical camera obscuras, from 3l. 3s. to	7	7	0
An artificial eye for illustrating the principles of vision	5	5	0
Prisms, mounted in various manners			
Concave and convex mirrors, from 7s. 6d. to	18	18	0

For articles to illustrate the principles of optics, see my " Lectures on Natural Philosophy."

Geographical and Astronomical Instruments.

PRICES OF GLOBES.

28 inches diameter, mounted in mahogany frames	52	10	0
28 ditto frames carved and ornamented			
18 ditto mounted in the common manner	6	6	0
18 ditto mounted in the common manner in mahogany frames	8	8	0
18 ditto mounted in the best manner, in stained frames	10	10	0
18 ditto mounted in the best manner, in mahogany frames	12	12	0

18 inches

A CATALOGUE OF INSTRUMENTS.

	£.	s.	d.
18 inches diameter, mounted in the best manner, in carved frames	16	16	0
16 ditto mounted in the common manner	6	6	0
12 ditto mounted in the common manner	3	3	0
12 ditto mounted in the common manner, in mahogany frames	4	4	0
12 ditto mounted in the best manner, in stained frames	5	15	6
12 ditto mounted in the best manner, in mahogany frames	7	7	0
9 ditto mounted in the common manner	2	2	0
An armillary sphere, shewing at one view the real apparent motion of the heavens	31	10	0
A manual planetarium	2	2	0
A planetarium with wheel-work, by which the order and motion of the planets, their situation with respect to the earth at different times, and the reason of their appearing to be sometimes stationary, and to move at other times in contrary directions, are rendered obvious to the eye	7	17	6
A tellurian for illustrating the phenomena of the earth, and forming a proper companion for the former	10	10	0
A correct and elegant planetarium, lunarium, and tellurian	18	18	0
Ditto with the diurnal motion to the earth	24	0	0
The most complete planetarium, tellurian, and lunarium	36	15	0
Orreries, from 18l. 18s. to	1000	0	0
A small quadrant, with an horizontal and vertical motion, for the instruction and amusement of young people	0	18	0
A ditto framed in brass, more accurately made and graduated	3	13	6
A small equatorial, being at once an accurate universal dial, and a portable observatory; with it a great number of curious and interesting problems may be solved, and any person soon rendered master of the elements of practical astronomy	10	10	0
Ditto on a larger scale, with a telescope, and more adjustments	26	5	0

Transit instruments, astronomical quadrants, circular instruments, according to their sizes, the variety and accuracy of their adjustments.

Mathematical Inſtruments, for Geometry, Drawing, &c.

	£.	s.	d.
Plain compaſſes for meaſuring lines, from 1s. to	0	5	0
Drawing compaſſes with moveable points, from 5s. 6d. to	3	3	0
Drawing pens, from 1s. to	0	5	0
Bow compaſſes for deſcribing ſmall circles, from 3s. to	0	5	0
Hair compaſſes for taking extents with accuracy	0	7	6
Beam compaſſes for deſcribing large circles, laying down diviſions, &c.	3	13	6
Triangular compaſſes for transferring three points at once from any plan or drawing, from 18s. to	1	11	6
Proportionable compaſſes for diminiſhing plans, or drawing in any aſſigned proportions	1	11	6
Elliptical compaſſes with friction rollers, for drawing ellipſes	4	4	0
Spiral compaſſes for deſcribing ſpirals, anſwering alſo as beam and elliptical compaſſes, from 6l. 6s. to	11	11	0
Parallel rules of various conſtructions			
Plain ſcales and ſectors of different ſizes			
Square protractors, ſemicircular and circular			
Pocket caſes of drawing inſtruments, from 7s. 6d. to	5	15	6
Magazine, or complete collection of drawing inſtruments, at 11l. 11s. at 17l. 17s. and from thence to	50	0	0
Perſpective compaſſes for aſcertaining the relative proportion of objects	0	18	0
A new-invented inſtrument for drawing in perſpective, which will facilitate the operations of the artiſts, and greatly aſſiſt beginners	7	7	0
Pantographers, for copying, reducing, and enlarging drawings, from 2l. 12s. 6d. to	6	6	0

Surveying Instruments.

	£.	s.	d.
Cases of drawing instruments, from 12s. to	21	0	0
Parallel rules			
Plotting scales			
Sets of plotting scales			
A parallel rule and protractor			
Protractors for laying down angles, according to the size			
Ditto with a nonius and moveable limb	2	12	6
Ditto rendered more accurate, as the limb is made to move with a tooth and pinion	4	14	6
Measuring wheels	6	16	6
Measuring chains of 50 and 100 feet			
Gunter's chain	0	8	6
Tape boxes, according to the length			
Surveying cross or square	1	11	6
Plain tables, with an index and sights; by these the plan is taken on the spot, and does not require a future protraction	3	13	6
Beighton's improved plain table, with an index of a peculiar make; in this instrument the line of sight is always over the center of the table, the station lines are also drawn parallel to those measured on the ground	14	14	0
Theodolets, or instruments for measuring angles, distances, &c. are made in various ways, some being more simple and portable, others more accurate, and with a greater number of adjustments			
Theodolets, with four plain sights, and a compass box, from 5l. 5s. to	5	15	6
A small theodolet, with telescopic sights and vertical arch, from 10l. 10s. to	21	0	0
Larger ditto, with the vertical arch affixed to a long axis	14	14	0
A larger ditto, like the preceding, only moving with rack-work	21	0	0
The latest improved theodolet, with double telescope, and every requisite adjustment	36	16	0
Circumferenters, the principal instrument used in America, from 2l. 2s. to	4	14	6

A CATALOGUE OF INSTRUMENTS.

	£.	s.	d.
An improved circumferenter, so contrived, that the operator need not rest the truth of his work entirely on the needle; it may be also used to take altitudes	4	4	0
Air or spirit levels with telescopic sights, from 5l. 5s. to	12	12	0
Air or spirit levels with plain sights	1	11	6
Station staves with sliding vanes for levelling	2	12	6
A small surveying compass, with sights, a nonius division, and three-legged staff	4	14	6
A small surveying compass and single stick	2	2	0
(The two foregoing instruments are portable and light, and may be put with ease in the pocket)			
A small theodolet, with telescopic sights and vertical arch, from 10l. 12s. to	21	0	0
Larger ditto, with the vertical arch affixed to a long axis	14	14	0
A larger ditto, like the preceding, only moving with rack-work	21	0	0
Miners compasses, used for carrying on works under-ground, from 10s. 6d. to	2	2	0
Ditto with a small telescope fixed to one side	1	11	6
Optical square, a small instrument for surveying by right angles; it requires no staff, and may be easily corrected or adjusted	1	16	0
An optical instrument for determining with accuracy when objects are in a strait line	1	7	0

For a knowledge of the various instruments used in geometry, levelling, surveying, &c. see my " Graphical Essays."

572   A CATALOGUE OF INSTRUMENTS.

Military Inftruments.

£.   s.   d.

Gunner's levels or perpendiculars
Gunner's callipers
Beam callipers
Shot guages
Shell ditto
Gunner's quadrants with a plummet
Ditto with a level
Ditto with an adjufting fcrew
Ditto and perpendicular combined together
General Williamfon's inftruments for howitzers, mortars, &c.
Surveying compaffes, furveying croffes, cafes of inftruments, telefcopes, plain tables, theodolets, &c.

Inftruments for Navigation.

Cafes of inftruments, and telefcopes of different kinds, fizes, and prices
Night telefcopes, from 1l. 11s. 6d. to —   2 12   6
Opera glafs for the fame purpofe   ——   1 11   6
A telefcope for an eye-glafs micrometer, for determining the diftance of a fhip at fea
Hadley's quadrants in black ebony frames
Hadley's fextant in wood   ——   6 16   6
Ditto in brafs, on the moft improved plan, from 1l. 11s. to   ————   ————   15 15   0
Knight's fteering compafs, with improvements   2 12   6
Knight's azimuth ditto   ————   5 15   6
Ditto on friction wheels   ————   10 10   0
Marine barometers; by thefe ftorms have been foretold at fea fome hours before they happened 9 9 0
Dipping needles from 12l. 12s. to   ——   31 10 0
Walker's azimuth compafs for afcertaining the variation without calculation, &c. &c.

Inftru-

A CATALOGUE OF INSTRUMENTS. 573

Inftruments for Electricity.

	£.	s.	d.
A fmall electrical machine, with a felect apparatus 8		8	0
An electrical machine and medical apparatus in a box; the machine is mounted in a plain but ftrong manner, and fo as to act with power; the apparatus is the moft convenient and fimple hitherto contrived for medico-electrical purpofes 6		16	6
Improved electrical machines, from 3l. 13s. 6d. to 40		0	0
Electrical batteries, from 2l. 12s. 6d. to			
Electrical jars of different fizes			
Electrical jars with an electrometer affixed to them	0	18	0
Medical bottles, with a tube for qualifying the fhock	0	7	6
Ditto larger	1	1	0
Directors with glafs handles for medical purpofes	0	10	6
Jointed difchargers with glafs handles	0	12	6
Plain difcharging rods	0	3	6
An univerfal difcharger and prefs	1	16	0
Kinnerfly's electrical thermometer	1	1	0
Quadrant electrometer	0	10	6
Cavallo's atmofpheric electrometer, from 15s. to	1	1	0
Ditto with additions by De Sauffure	1	11	6
Bennet's gold leaf electrometer	0	18	0
An apparatus for making Canton's and Wilfon's experiments on electric attractions, &c.			
Compound apparatus, by which a great number of neat and fatisfactory experiments may be performed	3	13	6
Ditto without the exhaufted flafk and conductor	1	11	6
Leyden vacuum	0	10	6
Luminous conductors	1	1	0
Spiral tubes, from 4s. 6d. to	0	10	6
Coloured fpirals	0	9	0
Sets of fpirals	1	11	6
Luminous words, from 10s. 6d. to	1	11	6
Spotted bottle	0	9	0
Belted bottle	1	1	0
Double bottle, a pleafing and ufeful part of an electrical apparatus, to gain a clear idea of the Franklinian theory	0	18	0

Plates

574   A CATALOGUE OF INSTRUMENTS.

	£.	s.	d.
Plates and ſtands for dancing images, pith images, and pith balls	0	9	0
An artificial ſpider	0	2	6
A ſmall head, with hair	0	4	6
An electrical piſtol for inflammable air	0	7	6
An electrical cannon for gunpowder	0	6	0
A thunder houſe	0	7	0
Ditto with a drawer	0	8	6
A powder houſe	0	18	0
A pyramid	0	15	0
A bone ball, and a ball of box wood fitted on braſs wires	0	5	0
Electric flyer and points			
A plain ſet of bells	0	9	0
Five bells mounted on a ſtand	0	18	0
A ſet of muſical bells	1	7	0
Magic picture	0	10	6
Electrical ſtools, from 10s. 6d. to			
An electrophorus, from 10s. 6d. to			
A variety of other articles, too numerous to be inſerted here.			

Apparatus for Experiments on Magnetiſm.

An apparatus for explaining the principal phenomena of magnetiſm, from 3l. 3s. to	15	15	0
Magnets, from 10s. 6d. to	3	3	0
Small compound magnets	0	15	0
Horſeſhoe magnets, from 2s. to	21	0	0
Compound ditto, from 15s. to	21	0	0
Dipping needles, from 12l. 12s. to	31	10	0
Variation compaſſes, from 2l. 12s. 6d. to	21	0	0

Inſtru-

A CATALOGUE OF INSTRUMENTS.

Inftruments for Experiments on Pneumatics.

	£.	s.	d.
A fmall fingle-barrel air-pump	2	12	6
A fmall double-barrel ditto	5	5	0
A larger ditto	7	7	0
A table air-pump	11	11	0

The American double-barrelled air-pump, the lateft improvement on this inftrument, in which the air receives no impediment from the action of valves or cocks, exceeding Smeaton's in accuracy and fimplicity, and far fuperior in both refpects to feveral later contrivances

A condenfing engine. This may be, if defired, combined with the former, but the rational and practical experimentalifts will find many advantages in having them detached from one another.

Apparatus for an Air-pump.

	£.	s.	d.
The Magdeburgh hemifpheres, from 12s. to	1	16	0
A flat plate and collar of leathers for placing on open receivers	0	17	6
Guinea and feather apparatus, for experiments on the refiftance of the air, from 18s. to	1	11	6
A fet of mills for ditto	1	11	6
A ditto on a better conftruction	4	4	0
Bell apparatus, for fhewing that a vacuum does not communicate found	0	7	6
Ditto on a better conftruction			
Ditto with wheel work, by which the bell may be put in motion or ftopped at pleafure	3	13	6
A new apparatus for ftriking flint and fteel in vacuo			
An apparatus for firing gunpowder in vacuo			
A copper bottle, beam and ftand, for weighing of air	2	16	0
A box bladder and lead weights, to fhew the elaftic power of the air	0	15	6

Ditto

576 A CATALOGUE OF INSTRUMENTS.

	£.	s.	d.
Ditto on an improved plan	1	1	0
A model of a pump, illustrating at the same time the nature of pumps, and proving that there is no such thing as suction	1	5	0
A small receiver and plate, which clearly evinces that receivers are kept on the pump by pressure, not suction	0	14	0
A filtering cup	0	6	6
A plate and piece of wood	0	6	6
(The two last articles are for shewing the porosity of vegetables)			
The torricellian experiment	0	16	0
Fountain in vacuo	0	6	6
Ditto on a different construction	0	18	0
Lungs glass	0	6	6
A single transferer plate and pipe for a fountain	0	16	6
A double transferer, for communicating a vacuum from one receiver to another	3	3	0
A burnt air-pipe, for experiments on infected air	0	18	0
Breaking square and cage			
A small bladder and lead weight			
A small balance-beam and stand	0	7	6
Ditto on an improved construction			
Receivers of different sizes, with various other articles.			

A CATALOGUE OF INSTRUMENTS.

Meteorological Inſtruments.

	£.	s.	d.
A plain portable barometer	2	2	0
Ditto with a thermometer	3	3	0
A plain barometer, covered frame, and glaſs door	2	12	6
Ditto with a thermometer	3	13	6
A barometer with a long cylindric thermometer	4	4	0
A ditto with ditto, and De Luc's hygrometer	7	7	0
A barometer and thermometer, with a guage, the indexes moving by rack-work	5	15	6
A barometer for meaſuring the altitude of mountains, &c.	9	9	0
Marine barometers	9	9	0
Diagonal, wheel, and ſtatical barometers			
Fahrenheit's thermometers, from 1l. 1s. to	2	12	6
Ditto for botanic purpoſes	0	18	0
Ditto for the brewery	1	1	0
De Luc's hygrometer. Theſe are the only inſtruments by which comparative obſervations can be made on the dryneſs and moiſture of the air, from 3l. 3s. to	7	7	0
Rain guages			
Hygrometers with the beard of the wild oat	0	10	6
Fontana's eudiometer for aſcertaining the purity of the air	2	5	0

Inſtruments for illuſtrating the Mechanic Powers, the Laws of Motion, &c.

	£.	s.	d.
A conciſe apparatus for illuſtrating the nature of the ballance, the pulley, the different kinds of levers, the inclined plane, the wheel and axle, the ſcrew, a compound engine, and a compound lever; alſo, a double cone to move up an inclined plane, and other pieces to ſhew the properties of the center of gravity, from 21l. to	26	5	0
A ditto on a more enlarged ſcale			

Atwood's

578   A CATALOGUE OF INSTRUMENTS.

	£.	s.	d.
Atwood's apparatus for demonstrating with accuracy the laws of accelerated and retarded motion. It is one of the most pleasing and scientific instruments in mechanics, as well from the variety of experiments that may be made with it, as the accuracy with which they are performed	26	5	0
A machine for illustrating the theory of central forces; in this machine the times are marked by found, the spaces are shewn by an index, the errors arising from friction are so far lessened as to be scarcely sensible ——— —	31	10	0

Pullies of various combinations and constructions
A small carriage, inclined plane, wheels of different sizes, &c. for experiments on wheel carriages
Compound steelyard
An apparatus for experiments on collision
Ditto for illustrating the composition and resolution of motion
Pyrometers on various constructions
   With many other articles and models for experiments on friction, pendulums, &c. too numerous to be comprized in a small catalogue

Instruments for Experiments in Hydrostatics and Hydraulics.

Hydrostatic ballances, from 2l. 2s. to —	10	10	0
A concise apparatus for experiments on hydrostatics	21	10	0

An apparatus for shewing that fluids have weight
Ditto for shewing that the particles of fluids exercise their pressure independently one of the other
Ditto to shew that fluids press in every direction
Ditto to demonstrate the lateral pressure of fluids
Ditto to shew that, *cæteris paribus*, the pressure of fluids is as their perpendicular height
The hydrostatic paradox
The hydrostatic bellows
Apparatus for illustrating the laws of pressure and equilibrium between heterogeneous fluids

Ditto

A CATALOGUE OF INSTRUMENTS. 579

	£.	s.	d.
Ditto for illuftrating the actions of fluids upon bodies immerfed in them			
An apparatus for experiments on fpouting fluids			
Hydrometers for proving fpirits, from 1l. 11s. 6d. to	4	14	6
An apparatus for making experiments on capillary tubes			
The model of the diving bell			
A glafs model of the lifting pump			
A ditto of the lifting and forcing pump, De la Hire's pump, &c.			
A japanned copper fountain to act by condenfed air with a variety of jets			
Apparatus for experiments on fyphons.			

For other articles neceffary to illuftrate mechanics and hydroftatics, fee my "Lectures on Natural Philofophy."

Appendix III Aids to Dating Adams Instruments and Publications

When the signature on an instrument, or the imprint of a publication, includes an address and/or an appointment, a definite range of possible dates can be assigned to it by reference to the table below. The absence of such means of identification, however, does not necessarily mean that the item lies outside these dates: the presence or absence of an address or appointment depends on factors such as the space available, the price, and the intended market or customer. Many extant instruments are signed simply 'G.Adams, London', on which evidence they could be dated anywhere between 1734 and 1795. Other factors such as style or manufacturing technique, or documentary evidence, must be considered in such cases. When the signature is abbreviated still further to 'Adams London' the range of dates could theoretically be extended to 1734–1817. However, it is generally considered that instruments so signed were most probably sold by Dudley Adams after 1795, when the goodwill in the surname Adams was more important than Dudley's own first name. It is possible that some instruments signed thus were supplied by subcontractors in the period 1788–95, when George junior's and Dudley's businesses overlapped.

George Adams senior owned (with Richard Jack) only one patent, George junior none at all, and Dudley three. No telescopes or quadrants bearing the inscription 'Patent' and George Adams's name or address have been reported, but if any do turn up they can be dated 1750–64 on this evidence. Dudley's three patents were in force from 1797–1811 (spectacles), 1800–1814 (multiple-tube telescopes) and 1815–29 (drawn vellum tubes). The last was sold as part of his effects in 1817. Instruments may have continued to be made in accordance with a patent after it had expired.

George Adams junior did not use the qualification 'junior' on any of his instruments or publications, as his father was already dead when he became a freeman himself: he simply continued to use the name George Adams, and to describe late editions of his father's works as 'Printed for the Author'.

The shop sign 'Tycho Brahe's Head' was used throughout the firm's existence in Fleet Street, but when used by itself it generally indicates a date prior to 1766.

Significant Dates

George Adams (senior or junior)

1734   Instrument business in Fleet Street founded by George senior 'near the Castle Tavern'.
1738   Address changed to 'Corner of Racquet Court', until 1757.
       Both of the above addresses are sometimes given as 'over-against [opposite] St. Bride's'.
1748   Mathematical Instrument Maker to his Majesty's Office of Ordnance (or 'the Ordnance'), until 1772.
1748   Mathematical Instrument Maker to the Royal Mathematical School (Christ's Hospital), until 1795.
1757   Address changed to 'near Water Lane, Fleet Street', until numbers were allocated in July 1766. This address is sometimes given as 'near the Bolt-in-Tun, Fleet Street'.
1757   Mathematical Instrument Maker to his Royal Highness George, Prince of Wales, until the latter's accession in October 1760.
1760   Mathematical Instrument Maker to his Majesty (George III), until 1795.
1766   Address became 60 Fleet Street when numbers were allocated in July.
1772   George senior died 17 October. For a few years shop bills and advertisements may be headed 'Ann & George Adams'.
       Note: The appointment to the Office of Ordnance lapsed on George senior's death, but those to 'his Majesty' and the Royal Mathematical School continued.
1780   (No specific appointment date.) George junior became one of the regular suppliers to the Office of Ordnance, in parallel with Ramsden and others, until 1795.

1787   George junior became Optician to his Royal Highness George, Prince of Wales (in addition to Mathematical Instrument Maker to his Majesty), until 1795 (died 14 August).

Hannah Adams

1795   (August) Mathematical Instrument Maker to his Majesty, and to the Royal Mathematical School, until November 1796.

Dudley Adams

1788   Established separate business at 53 Charing Cross, until 1796.
1794   Globe maker to his Majesty (George III) until 1817.
1795   Mathematical Instrument Maker to the Office of Ordnance, until 1806 (in parallel with others).
1796   (July) Address changed to 60 Fleet Street, until 1817.
1796   (November) Mathematical Instrument Maker to his Majesty (George III), Optician to their Royal Highnesses the Prince of Wales and the Duke of York, until 1817.
Note: the optician appointments may have commenced on the death of George junior in August 1795. The Prince of Wales became the Prince Regent in 1811.
1817   (May) Bankrupt. Adams instrument business closed.

Note: Royal Appointments

No member of the Adams family was:

a)   Optician or Mathematical Instrument Maker to Frederick, Prince of Wales (died 1751), or King George II (died 1760);
b)   Optician to the George, Prince of Wales, who became King George III in 1760;
c)   Mathematical Instrument Maker to the George, Prince of Wales, who became King George IV in 1830.

Hence, 'Mathematical Instrument Maker to the Prince of Wales' indicates a date in the range 1757–60, and 'Optician to the Prince of Wales' a date in the range 1787–1817 (Prince Regent from 1811). As the Adams business closed in 1817, before the accession of King George IV, all their appointments to 'his Majesty' refer to King George III.

Appendix IV Short-Title List of Publications by the Adams Family

Unless specified otherwise, all publications in this list are octavo, a description which includes works printed on half-sheets in eight-page sections. All English editions were published at London; overseas editions have the place of publication indicated. Works with large numbers of plates are sometimes found today with the plates folded and bound with the text, and sometimes with the plates in a separate volume. Prices, where quoted, are taken from contemporary advertisements if they are not printed on the books themselves. Posthumous editions by W. & S. Jones usually had a non-integral catalogue of Jones instruments inserted when the volume was bound, which may differ in date by many years from the book itself: these catalogues are not noted in the collations, but catalogues or price lists which are integral with the text are included.

The arrangement of this list is chronological by first edition, indicated by a date in **bold** in the left-hand margin. Later editions are tabulated below, starting with an indented date. Significant changes in title between editions are noted. Titles in general are abbreviated, but they are not paraphrased.

George Adams senior

1746   *Micrographia Illustrata, or, the knowledge of the microscope explain'd, together with an account of a new invented universal single or double microscope* Printed for and sold by the author. Small quarto (approx. 225 × 170mm). 16, 263 pp. Frontis. + 64 plates. Price 16s bound.

   Pages 243–63 are a 'catalogue' (that is, list) of 335 instruments, not priced but numbered for reference.

1747 *Second edition*
Printed for and sold by the author, and sold by S. Birt.
Size and contents as for first edition. Apart from the title-page, the two editions appear to be identical, and are possibly made up from the same sheets.

? [*Third edition*] No copy known.

1771 *Micrographia Illustrata: or, the microscope explained, in several new inventions, particularly of a new variable microscope*
Fourth edition
Printed for and sold by the author. (16), lix, 325, 14 pp. Frontis. + 71 plates. Price 10s 6d half-bound.
The last fourteen pages are a catalogue of instruments, partially priced, integral with the text.

1748 *The Description and Use of a New Sea Quadrant, for Taking the Altitude of the Sun from the visible Horizon*
Printed by John Hart. 40 pp., 1 plate. Price 1s.

1753 *The description and use of the Universal Trigonometrical Octant, invented and applied to Hadley's Quadrant by George Adams*
Printed for the author 'and given only with the instrument'. (2), 135, (3) pp. 2 plates.
The last three pages are a catalogue of instruments, not priced, integral with the text.

c. 1757 *Instructions for the use of Hadley's Quadrant ... with a description and use of the nonius divisions*
Published by the author [in 2 sections]. [1] 8 pp. 1 plate. [2] 12 pp.
See chapter 3 for comments on publication dates; the twelve-page section was issued first.

1766 *A Treatise Describing and Explaining the Construction and Use of new Celestial and Terrestrial Globes*
Printed for and sold by the author. xii, 242, 8 pp. 3 plates (pl. I as frontis.). Price 5s bound.
The last eight pages are a priced catalogue of instruments, separately signed, but evidently intended to be part of the book. In the British Library copy they are bound amongst the prelims. (See appendix I for transcript.)

1769 *A Treatise describing the construction, and explaining the use ...*
Second edition

Printed for and sold by the author. xxviii, 345, (7) pp. 14 plates.

> In this and all later editions the title is altered as above, and extended (under the edition number) to include a description of the solar system, and the use of the globes for solving spherical triangles. The last seven pages are a priced catalogue of instruments, integral with the text.

1772 *Third edition*

Printed for and sold by the author. xxviii, 345, 17 pp. 14 plates.

> The last seventeen pages are a priced catalogue of instruments, integral with the text.

1777 *Fourth edition*

Printed for and sold by the author. xxviii, 345, 14 pp. 14 plates.

> The last fourteen pages are a partially priced catalogue of instruments, integral with the text. Although the imprint refers to the book being sold by 'the author', this edition was produced and sold by George Adams junior, with little, if any, alteration.

1782 *Fifth edition*

Printed for and sold by the author. xvi, viii, 345, (14) pp. 14 plates.

> See notes to the fourth edition. The last fourteen pages are a catalogue of instruments (unpriced), numbered [3]–16 but complete. [See also first item in part II]

1810 *Thirtieth edition* [sic]

See under part III: Dudley Adams, 1810.

This title was superseded in 1789 by George Adams junior's *Astronomical and Geographical Essays*, q.v.

Translations

1770 (Dutch) *Gronden der Sterrenkunde van George Adams*

Amsterdam, Cornelius van Tongerlo. 84, 472 pp. 14 plates.

> Not examined; information from a Dutch correspondent.

1771 (Dutch) *Gronden der Sterrenkunde, gelegd in het Zonnestelzel bevatlijk*

… [translated by] *Jacob Ploos van Amstel, M.D.*

Amsterdam, David Klippink & Hendrik Gartman.

> Not examined; information from a copy of the title-page, Utrecht University Library.

n.d. *The Geometrical Walking Cane, made and sold by George Adams*
[undated pamphlet]

This was seen in a London bookseller's *c.* 1970; its present location is unknown. From the typography it was almost certainly published by George Adams senior rather than junior.

George Adams junior

n.d. *A Catalogue of Optical, Philosophical, and Mathematical Instruments, made and sold by George Adams, Mathematical Instrument Maker to his Majesty, at Tycho Brahe's Head, (No.60.) Fleet-Street, London*
16 pp. (British Library, Cup.504.c.22.)

This is the same as the catalogue in the fifth edition of *Treatise on the Globes* (1782), plus a title-page as above. The entries are not priced, and differ in arrangement from earlier issues. It was probably the first catalogue issued by George junior in his own right, rather than as a reprint of his father's.

1784 *An Essay on Electricity; in which the theory and practice of that useful science are illustrated ... to which is added, an Essay on Magnetism*
Printed for and sold by the author. xvi, 367, iv pp. 6 plates (pl. I as frontis.). Price 5s in boards.

The last four pages are a priced catalogue of instruments, signed '*' but apparently printed with the text pages and intended to form part of the book.

    1785 *... explaining the theory and practice of that useful science ...*
    Second edition, corrected and considerably enlarged
Printed at the Logographic Press for and sold by the author. x, 476, (18) pp. Frontis. + 7 plates; vignette on engraved title.

The last eighteen pages are integral with the text, and consist of 13pp. index and 5pp. priced catalogue of instruments.

    1787 *Third edition, corrected and considerably enlarged*
Printed by R. Hindmarsh for and sold by the author. lxxxvi, 473 pp. Frontis + 8 plates; vignette on engraved title. Price 6s in boards.

Pp. 469–73 are a priced catalogue of instruments.

    1792 *... explaining the principles of that useful science ... to which is now added, a Letter to the Author, from*

> Mr. John Birch, surgeon, on the subject of Medical Electricity
>
> *Fourth edition*
>
> Printed by R. Hindmarsh for and sold by the author. xii, 588 pp. Frontis. + 5 plates. Price 6s in boards.
>
>> In this edition the essay on magnetism is omitted. Pp. 575–88 are a priced catalogue of instruments. P. xi is a proposal advertisement for *Lectures* (1794). See also 1792 (below) for Birch's letter.
>
> 1799 *... by the late George Adams ...*
>
> *Fifth edition, with corrections and additions, by William Jones*
>
> Printed by J. Dillon & Co. for and sold by W. & S. Jones. xiv, 594 pp. Frontis. + 5 plates. Price 8s in boards.
>
>> The text of this edition follows the arrangement of the fourth (with additions by Jones), omitting magnetism and including the letter from Birch. Pp. xiii and xiv are advertisements for Jones's editions of Adams's works.

Translation

> 1785 (German) *Versuch über die Elekricität, worinn Theorie und Ausübung ...*
>
> Leipzig, im Schwickertschen verlag.
>
>> Not examined; information from the *National Union Catalogue*.

1787 *Essays on the Microscope; containing a practical description of the most improved microscopes: a general history of insects ... a description of three hundred and seventy-nine animalcula ... a view of the organization of timber ...*

> Printed by R. Hindmarsh for and sold by the author. Quarto (approx. 275 × 210 mm) xxiii, 724 pp. Frontis.
>
> *Plates for the Essays on the Microscope*
>
> Oblong folio (approx. 255 × 370 mm) Title + 32 plates (1–31, 26A). Price £1 6s in boards (text + plates).
>
>> The last four pages of the text volume are a priced catalogue of instruments.
>
> 1798 *... three hundred and eighty-three animalcula ... illustrated with thirty-two folio plates. By the late George Adams*
>
> *Second edition, with considerable additions and improvements, by Frederick Kanmacher, F.L.S.*

Printed by Dillon and Keating, for the editor; and for W. & S. Jones.

Quarto (approx. 270 × 210 mm; or large paper, 295 × 230 mm) xviii, (6), 724, [16] pp. Frontis. + 32 plates. Price £1 8s in boards.

P. [714] is a list of prices of instruments by Jones matching those shown in the plates. Although this edition has the same number of pages overall as the first, this is just a coincidence, as the contents differ appreciably. Kanmacher claimed that his added material was equivalent to 100 new pages. Contemporary advertisements indicate that the plates (several of which are new, replacing Adams's), though the same size as those in the first edition, were intended to be folded and bound with the text; but Kanmacher said in his preface that the purchaser could choose whether to have them thus or in a separate volume. No title-page appears to have been provided for the latter, though.

The [sixteen] pages at the end of the above collation represent a catalogue of instruments and so on by W. & S. Jones, printed on the same paper as the text (watermarked 1796) and evidently intended to be bound as part of the book. The letterpress of the catalogue is laid out for an octavo-size page, and was probably set for use in a separate octavo catalogue as usually issued by W. & S. Jones, but the individual pages were evidently re-formed specifically for use in this quarto volume. The catalogue may therefore be considered to be an integral part of the book. The last leaf of the catalogue is (as usual) a list of books by George Adams published by W. & S. Jones; in the author's copy it is dated 2 October 1797, but other (later) dates are found, indicating that the Jones 'quarto' catalogue was updated from time to time.

1787 *A Catalogue of Mathematical and Philosophical Instruments, made and sold by George Adams*

Printed by R. Hindmarsh. 8 pp. (approx. 205 × 135 mm). (Museum of the History of Science, Oxford).

This is a separate printing, not part of one of the above volumes. The instrument contents almost match those in the third edition (1787) of *Electricity*, except for some small price differences. Adams's appointments include Optician to the Prince of Wales, so the catalogue may have been issued to draw attention to this new appointment.

1789 *Astronomical and Geographical Essays: containing, I. A comprehensive view of the general principles of Astronomy. ...*

IV. *An Introduction to Practical Astronomy; or the use of the quadrant and equatorial*
Printed by R. Hindmarsh for and sold by the author. xix, 665, 15 pp. Frontis. + 21 plates.

The last fifteen pages are a priced catalogue of instruments, integral with the text. Subsequent editions have slightly different wording in the subtitles, not noted below, for example '... *a full and comprehensive view, on a new plan ...* '

1790 *Second edition*

Printed by R. Hindmarsh for and sold by the author. xix, 599, 15 pp. Frontis. + 21 plates.

The last fifteen pages are a priced catalogue of instruments, integral with the text.

1795 *Third edition*

Printed by R. Hindmarsh for and sold by the author. xx, 579 pp. Frontis. + 16 plates. Price 10s. 6d. in boards.

Pp. 565–79 are a priced catalogue of instruments. Between publication of the second and this edition, most of the plates had been used in *Lectures* and still bear the plate numbers for that work. Plate XV is a new one, not in previous editions.

Part IV, *An Introduction to Practical Astronomy*, was also issued separately in 1795, q.v.

1799 *Fourth edition, corrected*

Printed by J. Dillon & Co. for and sold by W. & S. Jones. xvi, 532 pp. Frontis. + 16 plates. Price 10s 6d in boards.

P. 532 is a list of prices of instruments by Jones matching those shown in the plates.

1803 *Fifth edition, corrected and enlarged, by William Jones*

Printed by W. Glendinning for and sold by W. & S. Jones. xvi, 518 pp. Frontis. + 16 plates. Price 10s 6d in boards.

P. 518 is a list of prices of instruments by Jones matching those shown in the plates.

1812 *Sixth edition, corrected and enlarged, by William Jones*

Printed by W. Glendinning for and sold by W. & S. Jones. xvi, 518 pp. Frontis. + 16 plates. Price 12s 0d.

P. 518 is a list of prices of instruments by Jones matching those shown in the plates.

American editions

    1800   *... by the late George Adams*
            Fourth edition, with the author's last improvements, illustrated with elegant copper-plates
            Whitehall: Printed for William Young, Philadelphia. xvi, 194, (ii), viii, 195–564 pp. Frontis. + 15 plates.

            This is based on the fourth London edition (Jones, 1799), and retains 'fourth' in the title, but is apparently the first edition to be published in America. The pages are numbered in two parallel sequences, one consecutive at the foot and the other section by section at the head. A note on page vii states that the sections could be purchased separately if desired, or bound into a single volume. The second section (only), *An Essay on the use of the Celestial and Terrestrial Globes*, has its own title-page and prelims, resulting in the break in the pagination shown in the collation.

    1808   *Fifth edition, with the author's last improvements, illustrated with copper plates*
            Philadelphia: Published by William W. Woodward. Dickinson, printer.

            The above is from the title-page of essay II. It is not clear whether the entire volume was republished in 1808, or whether '1808' volumes have only essay II replaced by a fifth edition. The *National Union Catalogue* entry implies that the latter is the case.

1789 *An Essay on Vision, briefly explaining the fabric of the eye, and the nature of vision*
            Printed by R. Hindmarsh for and sold by the author. viii, 153, 14 pp. 1 plate.

            The last fourteen pages are a priced catalogue of instruments, integral with the text.

    1792   *Second edition*
            Printed by R. Hindmarsh for and sold by the author. xi, 172 pp. 1 plate. Price 3s. in boards.

            Pages 159–72 are a priced catalogue of instruments.

Translations

    1792   (Dutch) *Verhandeling over het zien*
            Amsterdam: H. Gartman. viii, 174 pp. 1 plate.
            Not examined; information from a correspondent.

1794 *(German) Anweisung sur Erhaltung des Gessihts und zur Kenntnis der Natur des Sehens*
Leipzig.
Not examined; information from a correspondent.

1800 (Dutch, another edition) Gotha: C.W. Ettinger.
Not examined; listed in the *National Union Catalogue*.

1789 *Description, use and method of adjusting Hadley's Quadrant and Sextant*
Printed by R. Hindmarsh for and sold by the author. 70, (2) pp. 1 plate.
The last two pages are a priced catalogue of instruments, integral with the text.

1790 *A short dissertation on the Barometer, Thermometer, and other Meteorological Instruments*
Printed by R. Hindmarsh for and sold by the author. viii, 60, (1) pp.
The last page is a priced catalogue of meteorological instruments, integral with the text.

1791 *Geometrical and Graphical Essays, containing a description of the mathematical instruments used in Geometry, Civil, and Military Surveying, Levelling, and Perspective*
Printed by R. Hindmarsh for and sold by the author. xvi, 500 pp. Frontis. + 32 plates. Price 13s in boards.
Pp. 486–500 are a priced catalogue of instruments.

1791 *An Appendix ... by John Gale*
Not examined; the British Library copy (530.e.16) has been lost. According to comments in the text volume, the *Appendix* consists of tables for use in John Gale's method of surveying, printed separately for the benefit of readers who wished to practise this.

1797 *... by the late George Adams*
Second edition, corrected and enlarged by William Jones
Printed by J. Dillon & Co. for and sold by W. & S. Jones. (iv), xii, 518 pp. Frontis. + 34 plates. Price 14s in boards.
Pp. 516–18 are a list of prices of instruments by Jones matching those shown in the plates. The plates are usually bound in a separate octavo-size volume with a title-page *Plates to the Geometrical and Graphical Essays*, but are sometimes included in the text volume. The frontispiece in the text volume in this and all Jones editions is a representation of Ramsden's 'Great Theodolite', replac-

ing Adams's 1791 frontispiece, which depicts a different theodolite with a complex mounting by Adams.

1803 *Third edition, corrected and enlarged by William Jones*
Printed by W. Glendinning for and sold by W. & S. Jones. xvi, 518 pp. Frontis. + 34 plates.
> Pp. 516–18 are a list of prices of instruments by Jones matching those shown in the plates. The plates are usually bound separately as above.

1813 *Fourth edition, corrected and enlarged by William Jones*
Printed by C. Baldwin for and sold by W. & S. Jones. xii, 534 pp. Frontis. + 34 plates.
> Pp. 532–4 are a list of prices of instruments by Jones matching those shown in the plates. The plates are usually bound separately as above.

Translations

1795 (German) *Geometrische und graphische Versuche, oder Beschreibung der mathematischen instrumente ...* [translated by] *J.G. Geissler*
Leipzig, S.L. Crusius. 39 plates.
> Not examined; listed in the catalogue of the Bibliothèque Nationale, Paris.

1985 (German) *George Adams. Geometrische und graphische Versuche*
Darmstadt: Wissenschaftliche Buchgesellschaft. 439 pp. 126 illus.
> This is a rearranged reprint of the text of the German edition, with facsimile reproductions of the individual figures on the plates, plus an introduction, glossary and so on.

1792 *A Letter to Mr. George Adams, on the subject of Medical Electricity; from Mr. John Birch, surgeon*
57 pp. (Museum of the History of Science, Oxford)
This is a separate printing of the Letter added to the fourth edition of Adams's *Essay on Electricity*.

1794 *Lectures on Natural and Experimental Philosophy, considered in its present state of improvement ... displaying the goodness, wisdom, and power of God.*
In five volumes. The fifth volume consisting of the plates and Index
Printed by R. Hindmarsh for and sold by the author (and by advance subscriptions).
Vol. I: xlvii, 548 pp. Frontis.

Vol. II: vii, 561, (2) pp.
Vol. III: vii, 579 pp.
Vol. IV: viii, 576 pp.
Vol. V: *Index and Plates to Lectures on Natural and Experimental Philosophy* 44 pp. 39 plates (air 6, optics 8, mechanics 5, hydrostatics 3, astronomy 15, electricity and magnetism 2).

Pre-publication price to subscribers £1 4s, after publication £1 10s.

Pp. xxv–xlvii in volume I are a list of subscribers (approx. 1,000). The two unnumbered pages at the end of volume II are references to the plates, and errata; these pages are an integral part of the book. Several of the plates had previously been published in other works by Adams. The frontispiece in volume I is the same design as in *Essays on the Microscope*, re-engraved in octavo size, with a different caption to suit the contents of this work.

1799 ... *by the late George Adams*
Second edition, with considerable corrections and additions, by William Jones
Printed by J. Dillon & Co. for and sold by W. & S. Jones.
Vol. I: xxxii, 592 pp. Frontis.
Vol. II: viii, 576 pp.
Vol. III: vii, 583 pp.
Vol. IV: viii, 576 pp.
Vol. V: *Plates and Index to Lectures ...* 48 pp. 43 plates.

For this edition Jones left Adams's text substantially unchanged but inserted new material in footnotes and appendices, clearly identified.

American edition

1806/1807 *Carefully revised and corrected by Robert Patterson*
Whitehall: Printed for William W. Woodward, Philadelphia.
4 vols. 43 plates.

This version is based on the second London edition (Jones, 1799). Volume IV contains a subscribers list with approx. 190 names (247 copies), and an appendix by Patterson.
Not examined, except subscribers list; information from a correspondent and the *National Union Catalogue*.

Translation

1798/89   (German) *Vorlesungen über die Experimental-Physik nach ihrem gegenwärtigen Zustande in unterhaltenden un fasslichen Erklärungen der vornehmsten Erscheinungen in der Natur ...* [translated by] *J.G. Geissler*
Leipzig: S.L. Crusius.
Not examined; information from the *National Union Catalogue*.

1795   *An Introduction to Practical Astronomy*
94, (2) pp. Price 2s. 6d. sewed.
This is part IV of *Astronomical and Geographical Essays*, 1795, sold as a separate item.

Dudley Adams

c. 1790   *A Catalogue of Optical, Philosophical, and Mathematical Instruments, made and sold by Dudley Adams, globe manufacturer and mathematical instrument maker, West side of Charing-Cross, London*
Octavo, 15 pp. (Museum of the History of Science, Oxford)
This has a caption title as above occupying the upper half of p. 1. P. [16] is blank. Most of the entries, though not all, are priced.

1810   *A Treatise, describing the construction and explaining the use of new Celestial and Terrestrial Globes ...*
By George Adams, Sen. Long deceased. Father to the late George Adams.
The Thirtieth edition ... Now Published by Dudley Adams ... brother to the late George Adams
Printed (by C. Baldwin) for and sold by the publisher. xvi, 242 pp. 14 plates (pl. I as frontis.)
This edition follows George Adams senior's text of the second edition, 1769 (not the first edition), with illustrations printed from his plates. The plates of the celestial and terrestrial globes have 'George' erased and replaced by 'Dudley'.

1814   *A descriptive Catalogue of Mathematical, Philosophical, and Optical Instruments, manufactured and sold by Dudley Adams ... (son of George Adams senior, long deceased ...)*
Printed by C. Baldwin. 44 pp. Price 1s (University of Virginia Library)

More than just a price list, this includes extended descriptions of some items, especially those in which Dudley Adams had a particular interest, such as globes and his patent telescopes.

1820 *Electricity is the Fountain, the great vivifying principle of nature; a source of life and health*
Printed by W. Davy. 15pp. (British Library)
This was Dudley's first publication after he had purchased Francis Lowndes's medical electrical apparatus and business, following his own bankruptcy as an instrument maker.

1823 *A Trifle, illustrative of the insufficiency of the Materia Medica*
Printed by W. Davy. 12mo? (130 × 80 mm) 24pp. (British Library)

1825 *Religion combined with Science. No.1*
Not examined, not located. This is mentioned in a letter to Robert Peel dated 31 March 1825, and in *God Declared* (below). It was evidently intended to be the first of a series, but no copies have been located of this or any later numbers.

1827 *God Declared. Our Saviour proved to have been God. ... Etherealism, the science of the elements, exemplified*
[Printed by] T.C. Hansard, Pater-noster-row Press. Price 10s 6d. Large quarto (approx. 305 × 245 mm) xiv, (2), 30pp. (British Library)
From the evidence of this publication, Dudley's mental state had sadly deteriorated by this time.

Bibliography

(excluding works by members of the Adams family, for which see Appendix IV)

Archinard, Margarida, 'The Scientific Instruments of Horace-Bénédict de Saussure', in *Studies in the History of Scientific Instruments*, edited by C. Blondel et al. (London, 1989), pp. 83–95.
Baker, Henry, *The Microscope made Easy* (London, 1742), and later editions to 1759.
Baker, John R., *Abraham Trembley of Geneva, scientist and philosopher* (London, 1952).
Bennett, Abraham, *New Experiments in Electricity* (Derby, 1789).
Bennett, J.A., *The Divided Circle: a History of Instruments for Astronomy, Navigation and Surveying* (Oxford, 1987).
Bennett, J.A., *Church, State and Astronomy in Ireland: 200 Years of Armagh Observatory* (Armagh and Belfast, 1990).
Bernoulli, Jean, *Lettres astronomiques où l'on donne une idée de l'état actuel de l'astronomie pratique dans plusieurs villes de l'Europe* (Berlin, 1771).
The Billings Microscope Collection of the Medical Museum Armed Forces Institute of Pathology, second edition (Washington, DC, 1974).
Birch, John, *Considerations of the Efficacy of Electricity in removing Female Obstructions* (London, 1779), and second edition, 1780.
Birch, John, *An Essay on the Medical Applications of Electricity* (London, 1802).
Bradbury, S., 'The Quality of the Image produced by the Compound Microscope, 1700–1840', in *Historical Aspects of Microscopy*, edited by S. Bradbury and G.L'E. Turner (Cambridge, 1967), pp. 151–73.
Bransby, John, *The Use of the Globes: containing an Introduction to Astronomy; a Description of Globes and Maps* (Ipswich, 1791), and second edition, 1808.
Broadley, A.M., 'The Knowledge of Makers of Scientific Instruments in the Seventeenth and Eighteenth Centuries: their Trade-cards and other Rariora', *Knowledge* 35 (new series 9) (1912), pp. 306–11.
Brown, Joyce, *Mathematical Instrument-Makers in the Grocers' Company, 1686–1800* (London, 1979).

Brown, Olivia: *see* Whipple Museum catalogues.
Buchan, William, *The New Domestic Medicine ... To which is now first added, Memoirs of the Life of Dr. Buchan ... By William Nisbet*, new edition (London, 1814).
Calvert, H.R., *Scientific Trade Cards in the Science Museum Collection* (London, 1971).
Campbell, R., *The London Tradesman. Being a compendious View of all the Trades, Professions ... now practised in the Cities of London and Westminster* (London, 1747), facsimile reprint 1969.
Carwitham, T., *The Description and Use of an Architectonick Sector, and also of the Architectonick Sliding Plates* (London, 1723), and second edition, 1733.
Cassini, Jean-Dominique de, *Mémoires pour servir à l'histoire des sciences et à celle de l'Observatoire royal de Paris* (Paris, 1810).
Chaldecott, J.A., *Handbook of the King George III Collection of Scientific Instruments* (London, 1951).
Chenakal, Valentin L. (translated by W.F. Ryan), *Watchmakers and Clockmakers in Russia, 1400 to 1850* (Antiquarian Horological Society, 1972).
Clarke, T.N., A.D. Morrison-Low and A.D.C. Simpson, *Brass and Glass: Scientific Instrument Making Workshops in Scotland as illustrated by instruments from the Arthur Frank Collection at the Royal Museum of Scotland* (Edinburgh, 1989).
Clay, Reginald S. and Thomas H. Court, *The History of the Microscope, compiled from original instruments and documents, up to the introduction of the Achromatic Microscope* (London, 1932).
Clay, Reginald S. and Thomas H. Court, 'English Instrument Making in the 18th Century', *Transactions of the Newcomen Society* 16 (1935/36), pp. 45–54.
Cole, Benjamin, *The Description of a New Quadrant, for finding the Latitude at Sea* (London, 1748), and second edition, with an appendix, 1749.
Court, Thomas H. and M. von Rohr, 'Contributions to the History of the Worshipful Company of Spectaclemakers', *Transactions of the Optical Society* 31, no. 2 (1929/30), pp. 53–90.
Crawforth, M.A., 'Evidence from Trade Cards for the Scientific Instrument Industry', *Annals of Science* 42 (1985), pp. 453–554.
Crawforth, M.A., 'Instrument Makers in the London Guilds', *Annals of Science* 44 (1987), pp. 319–77.
Daumas, Maurice, *Les instruments scientifiques aux XVIIe et XVIIIe siècles* (Paris, 1953), and English edition translated and edited by Mary Holbrook, 1972.
Deane, William, *The Description of the Copernican System ... being an Introduction to the Description and Use of the Grand Orrery* (London, 1738).
Dekker, E., *De Leidsche Sphaera: Ein uitzonderlijk planetarium uit de zevetiende euw* (Leiden, 1985).
Desaguliers, J.T., *A Course of Experimental Philosophy* (London, vol. 1, 1734, vol. 2, 1744), and second edition, 1745, third edition 1763.
Donn(e), Benjamin: *see* Ravenhill.
Edell, Stephen, 'Concave Hemispheres of the Starry Orb', *Bulletin of the Scientific Instrument Society* 7 (1985), pp. 6–8.

Fenning, Daniel, *A new and easy Guide to the Use of the Globes: and the Rudiments of Geography* (London, 1754).

Ferguson, James, *Lectures on Select Subjects in Mechanics, Hydrostatics, Pneumatics, and Optics: with the Use of the Globes* (London, 1760), and later editions to the tenth, 1803.

Ferguson, James, *An Introduction to Electricity, in six sections* (London, 1770).

Fordham, Sir Herbert George, *John Cary: Engraver, Map, Chart and Print-Seller and Globe-Maker, 1754 to 1835* (Cambridge, 1925), facsimile reprint 1976.

Franklin, Benjamin: *see* Labaree.

Frison, E., 'Adams Microscopes and Microtomes: with notes on instruments by Magny', *The Microscope* 8 (July/August 1951), pp. 199–206 and (September/October) pp. 221–4.

Gage, A.T., *A History of the Linnean Society of London* (London, 1938).

[Thomas Godfrey] 'An Account of Mr.Thomas Godfrey's Improvement of Davis's Quadrant, transferred to the Mariner's Bow, communicated to the Royal Society, by Mr.J.Logan', *Philosophical Transactions* No.435 (December 1734), in Vol.38 for the years 1733–34 (London, 1735), pp. 441–50.

Goodison, Nicholas, *English Barometers, 1680–1860*, second edition (Antique Collectors' Club, 1977).

Gunther, R.T., *Early Science in Oxford* vol. 2 (Oxford, 1923), facsimile reprint 1967.

Hackmann, W.D., *Electricity from Glass: the History of the Frictional Electricity Machine 1600–1850*, Science in History Series, 4 (Alphen aan den Rijn, 1978).

Hadley, John, 'A Description of a New Instrument for taking Angles', *Philosophical Transactions* No.420 (August/September 1731), in Vol.37 for the years 1731, 1732 (London, 1733), pp. 147–57.

Hadley, John, *A Description of a New Instrument for taking the Latitude or other Altitudes at Sea. With Directions for its Use* (London, 1734).

Hadley, John, 'A Spirit Level to be fixed to a Quadrant for taking a Meridional Altitude at Sea, when the Horizon is not visible', *Philosophical Transactions* No.430 (November/December 1733), in Vol.38 for the years 1733, 1734 (London, 1735), pp. 167–72.

Hambly, Maya, *Drawing Instruments* (Sotheby's Publications, London, 1988).

Harris, Joseph, *The Description and Use of the Globes and the Orrery* (London, 1731), and later editions to the twelfth, 1783.

Hatton, Thomas, *An Essay on Gold Coin ... with a Description and Use of the most improved Weighing Instruments*, second edition (London, 1774).

Hatton, Thomas, *An Introduction to the Mechanical Part of Clock and Watch Work* (London, 1773), facsimile reprint 1978.

Heald, Tim, *By Appointment: 150 years of the Royal Warrant and its Holders* (London, 1989).

Heath, John Benjamin, *Some Account of the Worshipful Company of Grocers of the City of London*, second edition (London, 1854).

Hill, John, *The Construction of Timber, from its Early Growth, explained by the Microscope, and proved from Experiments, in a great Variety of Kinds* (London, 1770).

Hindmarsh, Robert, *Rise and Progress of the New Jerusalem Church, in England, America, and other parts* (London, 1861).

Johnson, W., 'Richard Jack, Minor Mid-18th Century Mathematician: Writings and Background', *International Journal of Impact Engineering* 12 (1992), pp. 123–40.

Keith, Thomas, *A new Treatise on the Use of the Globes, or a philosophical View of the Earth and Heavens* (London, 1805), and later editions.

King, Henry C., *The History of the Telescope* (London, 1955), reprinted 1976.

King, Henry C., *Geared to the Stars: the Evolution of Planetariums, Orreries, and Astronomical Clocks* (Toronto and Bristol, 1978).

Kirby, Joshua, *The Description & Use of a New Instrument called an Architectonic Sector* (London, 1761).

Labaree, Leonard W. et al. (editors), *The Papers of Benjamin Franklin* (New Haven and London, 1959–?).

[Lalande, Jerome] *Journal d'un voyage en Angleterre, 1763, with an introduction by Helene Monod-Cassidy*, Studies on Voltaire and the Eighteenth Century, 184 (Oxford, 1980).

(Lalande), Jérôme le Français, *Astronomie*, third edition (Paris, 1792).

Lefebvre, E. and J.G. De Bruijn (editors), *Martinus Van Marum, Life and Work* (Leiden, 1969–76), 6 vols.

Lowndes, Francis, *Observations on Medical Electricity* (London, 1787).

Lowndes, Francis, *The Utility of Medical Electricity Illustrated, in a series of cases, and practical observations* (London, c. 1791).

MacLeod, Christine, *Inventing the Industrial Revolution: the English Patent System, 1660–1800* (Cambridge, 1988).

Maddison, F.R., *A supplement to a catalogue of scientific instruments in the collection of J.A. Billmeir, exhibited by the Museum of the History of Science, Oxford* (Oxford and London, 1957).

Malie, Thomas, *A New and Accurate Method of Delineating all the Parts of the different Orders of Architecture, by means of a well contriv'd and easily manag'd Instrument* (London, 1737).

Marriner, Sheila, 'English Bankruptcy Records and Statistics before 1850', *Economic History Review* 33 (1980), pp. 351–66.

Martin, Benjamin, *Micrographia Nova: or a new Treatise on the Microscope, and Microscopic Objects* (Reading, 1742).

Martin, Benjamin, *An Essay on the Nature and Superior Use of Globes, in conveying the First Principles of Geography and Astronomy to the Minds of Youth* (London, 1758).

Martin, Benjamin, *The Young Gentleman and Lady's Philosophy, by way of Dialogue Vol. I* (London, 1759), *Vol. II* (London, 1763) [Part I of *The General Magazine of Arts and Sciences*, 1755–64].

Martin, Benjamin, *The Monied Man's Vade-Mecum. Being an Explanation of the Nature, Structure, and Use of a New Portable Steelyard for weighing Gold Coin* (London, 1773).

Martin, Benjamin, *Appendix to the Description and Use of the Globes* (London, 1766).

Martin, Benjamin, *The Description and Use of an Opake Solar Microscope* (London, 1774).
McFarland, J., 'The Historical Instruments of Armagh Observatory', *Vistas in Astronomy* 33 (1990), pp. 149–210.
Millburn, John R., *Benjamin Martin: Author, Instrument-Maker, and 'Country Showman'*, Science in History Series 2 (Leiden, 1976).
Millburn, John R., 'Horology and the Adams Family', *Antiquarian Horology* 13 (1982), pp. 368–76, and addendum p. 582.
Millburn, John R., 'James Ferguson's Lecture Tour of the English Midlands in 1771', *Annals of Science* 42 (1985), pp. 397–416.
Millburn, John R., *Benjamin Martin: Supplement* (London, 1986).
Millburn, John R., *Retailer of the Sciences: Benjamin Martin's Scientific Instrument Catalogues, 1756–82* (London, 1986).
Millburn, John R., 'The Office of Ordnance and the Instrument Making Trade in the mid-eighteenth Century', *Annals of Science* 45 (1988), pp. 221–93.
Millburn, John R., *The Library of George Adams, Mathematical Instrument Maker to King George III* (Aylesbury, 1988), booklet + 2 microfiches.
Millburn, John R. with Henry C. King, *Wheelwright of the Heavens: the Life and Work of James Ferguson, FRS* (London, 1988).
Millburn, John R. and Tor E. Rössaak, 'The Bardin Family, Globe-makers in London, and their associate, Gabriel Wright', *Der Globusfreund* 40/41 (1992) pp. 21–66 and figures 1–10.
Morton, Alan Q. and Jane A. Wess, *Public & Private Science: the King George III Collection* (Oxford, 1993).
Mountaine, William and James Dodson, *An Account of the Methods used to describe Lines, on Dr.Halley's Chart of the Terraqueous Globe; Shewing the Variation of the Magnetic Needle about the Year 1756, in all the known Seas; their Application and Use in correcting the Longitude at Sea; with some occasional Observations relating thereto* (London, 1758).
Multhauf, Robert P., *A Catalogue of Instruments and Models in the possession of the American Philosophical Society* (Philadelphia, 1961).
Odner, Carl Theophilus, *Annals of the New Church: vol. I. 1688–1850* (Bry Athyn, Pennsylvania, 1904).
Palmer, F.W. and A.B. Sahiar, *Microscopes, to the end of the nineteenth century* (London, 1971).
Parry Jones, W.L., *The Trade in Lunacy* (London, 1972).
Phillips, Hugh, *Mid-Georgian London* (London, 1964).
Porter, Roy, Simon Schaffer, Jim Bennett, and Olivia Brown, *Science and Profit in 18th-century London* (Cambridge, Whipple Museum, 1985).
Randier, Jean, *Marine Navigation Instruments*, translated by John F. Powell (London, 1980).
Ravenhill, W.L.D., *Benjamin Donn: A Map of the County of Devon, 1765* (London and Bradford, 1965)
Rees, J. Aubrey, *The Worshipful Company of Grocers, an historical retrospect, 1325–1923* (London, 1923).
Schechner, Sara J., 'John Prince and Early American Scientific Instrument Making', *Sibley's Heir* 59 (1982), pp. 431–501.

[Seeley, J.] *Stowe. A Description of the House and Gardens of the Most Noble & Puissant Prince, George-Grenville-Nugent-Temple Marquis of Buckingham* (Buckingham, 1797).

Senex, John, 'A Contrivance to make the Poles of the Diurnal Motion in a Celestial Globe, pass round the Poles of the Ecliptic', *Philosophical Transactions* No.447 (1738), in Vol.40 for the years 1737, 1738 (London, 1741), pp. 203–4 and plate II.

Senex, Mary, 'A letter from the widow of the late Mr.John Senex, FRS ... concerning the Large Globes prepared for her late husband, and now sold by herself', *Philosophical Transactions* No. 493 (1748/9), in Vol.44 for the years 1749, 1750 (London, 1752), pp. 290–92.

Skempton, A.W. (editor), *John Smeaton, FRS* (London, 1981).

Smeaton, J., 'A Letter from Mr.J.Smeaton to Mr.John Ellicott, FRS, concerning some improvements made by himself in the Air-Pump', *Philosophical Transactions* Vol.47 for the years 1751 and 1752 (London, 1753), pp. 415–27 and plate XVIII.

Smith, Caleb, *The Description, Use, and excellency of a New Instrument, or Sea Quadrant, invented by Caleb Smith, for taking Altitudes of the Sun, Moon, and Stars* (London, [1735]).

Smith, Caleb and William Ward, *The Description and Use of a New Astronomical Instrument, for taking Altitudes of the Sun and Stars at Sea* (London, 1735).

Staunton, Sir George, Bt, *An Authentic Account of an Embassy from the King of Great Britain to the Emperor of China ... taken chiefly from the papers of his Excellency the Earl of Macartney* (London, 1797).

Stone, Edmund, *The Construction and Principal Uses of Mathematical Instruments. Translated from the French of M. Bion* (London, 1723).

[Swedenborg] *The True Christian Religion, containing the Universal Theology of the New Church, from the Latin of Emanuel Swedenborg*, Dent's Everyman's Library 893 (London and Toronto, 1933).

Tafel, R.L., *Documents concerning the Life and Character of Emanuel Swedenborg*, vol. 3, Swedenborg Society (London, 1890).

Taylor, E. Wilfred and J. Simms Wilson, *At the Sign of the Orrery, the origins of the firm of Cooke, Troughton & Simms, Ltd.* (York, c. 1950).

Toksvig, Signe, *Emanuel Swedenborg, scientist and mystic* (London, 1949).

Trobridge, G., *Swedenborg, Life and Teaching*, fourth edition (1935), reprinted by the Swedenborg Society (London, 1974).

Turner, Anthony, *Early Scientific Instruments: Europe 1400–1800* (London, 1987).

Turner, G.L'E., 'The Auction Sales of the Earl of Bute's Instruments, 1793', *Annals of Science* 23: 3 (September 1967), pp. 213–42.

Turner, G.L'E., 'Henry Baker, FRS, Founder of the Bakerian Lecture', *Notes and Records of the Royal Society of London* 29: 1 (October 1974), pp. 53–79.

Turner, G.L'E., 'The Portugese Agent, J.H. de Magellan', *Antiquarian Horology* 9 (1974), pp. 74–6.

Turner, G.L'E., *Collecting Microscopes* (London, 1981).

Turner, G.L'E., *The Great Age of the Microscope: the Collection of the Royal Microscopical Society through 150 years* (Bristol and New York, 1989).

Turner, G.L'E., *Catalogue of Microscopes* (Florence, Museo di Storia della Scienza, 1991).

Turner, G.L'E. and T.H. Levere, *Van Marum's Scientific Instruments in Teyler's Museum* (Leiden, 1973), vol. 4 of *Martinus Van Marum: Life and Work*.

Tyacke, Sarah, *London Map-Sellers, 1660–1720* (Tring, 1978).

Upham, Charles W., 'Memoir of Rev.John Prince, LL.D., late Senior Pastor of the First Church in Salem, Mass.', *The American Journal of Science and Arts* 31 (1837), pp. 201–222.

van der Krogt, Peter, *Old Globes in the Netherlands*, translated from the Dutch by Willis ten Haken (Utrecht, 1984).

van Marum: see Lefebvre; G.L'E. Turner.

Vince, S., *A Complete System of Astronomy* (Cambridge, 1797).

Wallis, R.V. and P.J. Wallis, *Biobibliography of British Mathematics and its Applications*, Part II: *1701–1760* (Newcastle upon Tyne, 1986).

Watkins, Francis, *L'Exercise du Microscope, contenant un abregé de tout ce qui a été ecrit par les meilleurs Autheurs* (London, 1754).

Wheatland, David P., *The Apparatus of Science at Harvard, 1765–1800* (Harvard, 1968).

Whipple Museum (Cambridge) catalogues:
1. *Surveying*, by Olivia Brown (1982).
2. *Balances and Weights*, by Olivia Brown (1982).
3. *Astronomy and Navigation*, by J.A. Bennett (1983).
4. *Spheres, Globes & Orreries*, by Olivia Brown (1983).
6. *Sundials and Related Instruments*, by David J. Bryden (1988).
7. *Microscopes*, by Olivia Brown (1986).

Whipple, Robert S., 'Some Scientific Instrument Makers of the 18th Century', *Nature* 126 (August 1930), pp. 244–6 and 283–6.

Index

Adams, Ann (daughter of George Adams snr), 21, 24, 30, 42, 83, 133, 135
Adams, Ann (wife of George Adams snr), 14, 40, 78, 159, 162, 167, 174, 184, 273
  death, 264, 293, 356
Adams, Charlotte (daughter of George Adams snr), 78, 83
Adams, Dudley (son of George Adams snr), 102, 110, 125, 126, 129, 133, 159, 174, 184, 264, 265
  apprenticed, 273
  apprentices, 276, 277
  bankruptcy, 307, 359, 385; cause, 305–6; library, sale of, 309; petition, 307–8; property, sale of, 309–10, 312; stock, sale of, 312–13
  barometers, 279; stick, illustration, 296
  bill, illustration, 299
  business acumen, lack of, 313
  A Catalogue of ... Instruments, 276–7, 305, 397
  compasses, proportional, illustration, 290
  death, 329
  A descriptive Catalogue of ... Instruments, 397–8
  dial, universal inclining, illustration, 297
  Duke of York, optician to, 280, 385
  Earl of Liverpool, letter to, 318–19
  early life, 273, 275
  Electricity is the Fountain, 320, 322, 325–6, 398; title-page, 321
  Fleet Street, move to, 280, 385
  George III, instrument maker to 384, 385
  globes, 184, 186, 276, 277; 12-inch, illustrations, 286, 287, 301, 302; 18-inch, illustrations, 87, 254–8; 18-inch, manufacture, 260; 28-inch, 277; George III, supplier to, 277
  God Declared, 398
  Grocers' Company, liveryman, 277
  hygrometers, description, 292
  illness, mental, 318–19, 327
  instruments: purchasers of, 286, 289; surveying, 295
  marries Margaret Sophia de Langlade, 275
  medical electricity, practitioner, 319–20, 322, 325–7
  microscopes, compound, 286; illustration, 298
  opera glasses, 286
  Ordnance Office: loss of business with, 292–3; supplier to, 179, 265–266, 279, 384
  perpendicular (level), gunner's, illustration, 288, 289
  Prince of Wales, optician to, 280, 316, 385
  protractor, brass, illustration, 292
  Religion combined with Science, 327, 398
  sets up business, 275–6
  spectacles: illustration, 281; patent, 284, 383
  surveyor's level, illustration, 294
  Switzerland, possible visit to, 322

Adams, Dudley (cont'd)
  telescopes: portable, illustrations, 282–3; portable, patent, 284, 305, 383; portable, prices, 285; reflecting, illustration, 278; tubes, patent, 305, 383
  thermometers: description, 292; illustration, 274
  trade cards: design, 280–81, 284–5; illustrations, *frontispiece*, 275
  A Trifle ... Materia Medica, 398; contents, 328–9; title page, 324
Adams family
  connections, 335
  country estate, Nutting Grove, 293, 299, 302–4; bankruptcy sale, 309
  dates, significant 384–5
  genealogy xx–xxi, 332
  information sources on xiii–xvi
  instruments, dating of 383–5
  library: contents, 268; sale, 266–8
  property transfers, 356–7
  publications: dating of 383–5; list, 386–98
  royal appointments 385
Adams, George, innkeeper, 57
Adams, George, jnr, 24, 27, 56, 78, 83, 91, 95, 105, 127, 129
  accommodation problems, 259–60
  air-pumps, 181, 189
  apparatus, price list, 378–9
  apprenticed, 133, 135
  apprentices, 162, 181, 183–4, 273
  Astronomical and Geographical Essays, 110, 126, 129, 217–23, 255, 259, 260, 261, 300, 304, 347, 356, 357, 359, 367, 391–2; 6th edition, 249
  barometers, 231; stick, illustration, 185; wheel, 164
  Benjamin Martin, rivalry with, 171–2, 225, 229
  bills: illustration, 196; location, 196–7
  A Catalogue of ... Instruments, 184, 276, 305, 367–82, 389, 391; 2nd–6th editions, 392; American editions, 393
  Chapter House Philosophical Society, member, 186, 188
  coin balances, 171–2
  death, 263, 385
  Description, use ... of ... Hadley's Quadrant, 82, 229–30, 394
  An Essay on Electricity, 196, 199–201, 204–7, 252, 354, 389; 2nd edition, 204, 206, 389; 3rd edition, 204–5, 389; 4th edition, 204, 389–90; 4th edition, frontispiece, illustration, 205; 5th edition, 207, 390; illustrations, 202–3; medical electricity, 319–20; sources, 353–4; translations, German, 390
  An Essay on Vision, 193, 224, 229, 355, 393; 2nd edition, 393; sources, 225; translations, Dutch, 393, 394; translations, German, 394
  Essays on the Microscope, 190, 207–9, 211–13, 216–17, 252, 286, 354, 355, 390; 2nd edition, 216–17, 249, 390–91; advertisement, 208; engravings, 211; illustration, 210
  eudiometers, 255–6
  Geometrical and Graphical Essays, 70, 194, 235–6, 244, 394; 2nd edition, 241, 245, 295, 394–5; 3rd edition, 395; 4th edition, 395; sources, 235; translations, German, 395
  globes, 173
  Grocers' Company, freeman, 162
  hygrometers, 231
  instruments: astronomical, price list, 370–71; drawing, illustrations, 194–5; for electricity, price list, 376–7; geographical, price list, 370–71; for hydraulics experiments, price list, 381–2; for hydrostatics experiments, price list, 381–2; for magnetism experiments, price list, 377; mathematical, price list, 372; meteorological, price list, 380; military, price list, 375; navigation, price list, 375; optical, price list, 368–70; for physics experiments, price list, 380–81; for pneumatics experiments, price list, 378; surveying, price list, 373–4
  An Introduction to Practical Astronomy, 397

Adams, George, jnr (cont'd)
  Lectures on Natural and Experimental Philosophy, 189, 199, 218, 231, 245–9, 260, 395–6; 2nd edition, 248, 396; American edition, 248, 396; sales, 245; sources, 247; subscribers, 247–8, 262; translation, German, 397
  microscopes: compound, description, 209, 211, 261; compound, illustrations, 190, 215; lucernal, description, 213, 216, 252; lucernal, illustrations, 212, 214
  octants, illustrations, 232
  Ordnance Office, supplier to, 167, 176, 177, 179, 340, 384
  orreries, illustrations, 221
  paints, illustrations, 194–5
  perambulator wheel, dial, illustration, 244
  Powder Magazine, Purfleet, supplier to, 163
  Prince of Wales, optician to, 316, 196, 385
  publications, 198–9; list, 200
  quadrants, illustrations, 232–3
  A short dissertation on the Barometer, 230–31, 394
  spheres, armillary, illustrations, 225
  and Swedenborgian movement, 192–3
  telescopes, 177; equatorially-mounted, illustrations, 226–8; refracting, illustration, 178
  theodolites, 177; illustrations, 187, 241
  Thomas Hatton, alliance with, 172
  van Marum, Martinus, supplier to, 250–52, 255–6, 258–62
  will, 263–5
Adams, George (of Bristol), 31
Adams, George, snr, 8
  advertisements, business, 15, 18–19, 32, 79
  air-pumps, 100, 104–5; illustration, 99
  apprenticed, 8, 11
  apprentices, 20–21, 30
  barometers, stick, illustrations, 154–5, 160

Benjamin Cole, rivalry with, 48–50, 52, 60, 66, 115
Benjamin Martin, rivalry with, 121, 123–4, 138, 199
birth, 3
business career: early, 24; summary, 146–7, 151, 153
calipers, gunner's, illustrations, 148, 149
churchwarden, 40–42
coin balances, 169; illustrations, 170
collector of Poor Rates, 30–31
compasses: beam, 105; magnetic, illustrations, 156; mariners', 70; variation, 193
cosmotheorion, 27, 30, 38
death, 146, 348–9, 385
Description and use of a New Sea Quadrant, 45, 387; illustrations, 50, 51
Description and use of the Universal Trigonometrical Octant, 30, 69, 387
dial, ring, illustration, 55
draughtman's tool, illustration, 73
and East India Company, 19–20
employees, details, 68
executor of mother's will, 12–13
fire at premises, 76–7, 112
geometrical solids, illustrations, 58–9
The Geometrical Walking Cane, 389
George III, supplier to, 94, 146, 385
globes: 3-inch, 108; 3-inch, cost, 127; 3-inch, description, 127; 3-inch, illustration, 128; 6-inch, cost, 127; 9-inch, 129; 12-inch, 109, 120, 127; 12-inch, illustrations, 119, 125; 18-inch, 120, 124, 127; 18-inch, cost, 118; 18-inch, dating of, 126; 18-inch, description, 113–14; 18-inch, illustrations, 116–17; advertisements for, 109, 112; catalogue, 112–13, 114; celestial, illustration, 87; template, illustration, 122
Grocers' Company: Court of Assistants, elected, 137; full member, 13; liveryman, elected, 60; steward, 72; warden, 137; warden, junior, 82–3

410 *Index*

Adams, George, snr (cont'd)
  house in Buckinghamshire, 159
  Instructions for the use of Hadley's Quadrant: description, 79, 82, 387; frontispiece, illustration, 77
  instruments: catalogue, 362–6; drawing, illustration, 165; mathematical, price list, 363–4; optical, price list, 364–5; philosophical, price list, 366
  marriage to Anne Dudley, 42
  Micrographia Illustrata, 22, 27, 33, 34, 36, 37–8, 44, 68, 102, 107, 115, 167, 173, 337, 344, 386; 2nd edition, 387; 4th edition, 139, 140, 141, 146, 207–8, 209, 338, 387; 4th edition, title page, illustration, 136; engravings, 211–12; frontispiece, illustration, 22
  microscopes: compound, illustrations, 23, 28, 86; Cuff-type, illustration, 88; double, illustration, 22; features, 139, 144; lucernal, description, 212; pocket, description, 144; pocket, illustration, 140; silver, description, 102, 104; silver, illustration, 103; simple, illustrations, 29; solar, 32–3; solar, illustration, 35; trunnion-mounting, description, 89, 91; trunnion-mounting, illustrations, 84, 86; variable, description, 144; variable, illustrations, 141, 145
  Napier's Bones, illustration, 71
  Ordnance Office: supplier to, 52–4, 56, 68, 70, 75, 146, 147, 340, 384; value of business with, 147, 151, 152, 349
  orreries: construction, 25–7, 30, 246; illustration, 17
  pantograph, 105
  parish constable, 21
  perambulator, illustration, 93
  perpendicular (level), gunner's, illustrations, 150, 175
  plane tables, illustrations, 80–81
  Prince of Wales, supplier to, 75, 83, 85, 91–2, 384
  patent application, 60–61, 383
  quadrants, 45, 48, 69; illustrations, 50, 175
  questman, 31
  Racquet Court, move to, 21, 24, 384
  Royal Mathematical School, supplier to, 56, 68, 124, 146, 384
  Royal Military Academy, Woolwich, value of business with, 166–7
  St Dunstan's parish, move to, 78–9, 112
  sectors, illustrations, 39, 90
  sidesman, 31
  site of first shop, 7, 14
  subcontractors, use of, 135, 137, 151
  sundials, 72; illustration, 67
  telescopes, 105
  Tower of London, supplier to, 76
  trade card, illustration, 134
  Treatise on the Globes, 118, 127, 199, 218, 331, 357, 359, 387–8, 397; 2nd edition, 138, 387; 3rd edition, 167, 173, 174, 198, 388; 4th edition, 351, 388; 5th edition, 217, 388; criticism, 124, 304; description, 113, 124; frontispiece, illustration, 111; title-page, illustrations, 110, 300; translations, Dutch, 388;
  will, 159, 161, 359
Adams, George (son of Dudley Adams), 285–6, 304–5
Adams, Golburne (brother of George Adams snr), 3
Adams, Hannah (wife of George Adams jnr), 174, 263, 264, 292, 350, 353, 354, 357
  abandons instruments business, 268–9, 279–80
  barometers, 265
  death, 269, 330
  Royal Mathematical School, supplier to, 266, 385
Adams, Henry (brother of George Adams snr), 3, 11, 13, 24
Adams, Isabella (Isabel) (daughter of George Adams snr), 83, 264, 285, 304, 308, 356
  death, 330
  will, 330
Adams, Job (brother of George Adams snr), 3, 8, 334
Adams, Lucy (daughter of George Adams snr), 78, 83, 264, 285, 356
  death, 293

Adams, Margaret (wife of Dudley Adams), dies, 285
Adams, Mary (daughter of George Adams snr), 14, 21
Adams, Mary (mother of George Adams snr), 3, 13
　death, 12
Adams, Morris (Maurice) (father of George Adams snr), 3, 36, 334
　cook, 5, 8
　death, 10–11
　freeman of London, 4, 8
　and the Loriners' Company, 9, 10
Adams, Philip, 334
Adams, Robert, cutler, 334
Adams, Sarah (daughter of George Adams snr), 24, 42
　marriage to Robert Blunt, 83, 133
Adams, Shute, druggist, 31, 337, 342
Adams, Sophia (daughter of George Adams snr), 83
Adams, William (son of George Adams snr), 78, 83, 91, 133
air-pumps, 100, 105, 181, 189, 344
　description, 100, 104–5
　illustration, 99
Aix-la-Chapelle, Treaty of, 53
Allamand, J.N.S., 115
Allen, Dr Joseph, 356, 357
American War of Independence, 180, 181
　cost, 198
Andrew, John, coin balance maker, 350
Andrews, William, 68, 69
Anscheutz & Schlaff, Denmark Street, 169
Arden, John, 129, 347
Argyll, Duke of, 36, 94, 146, 337
Armet, Catherine, archivist, 337, 343
Ashburner, Leonard, 21
Atwood's machine
　description, 259
　illustration, 253
Austrian Succession, War of, 53
Ayscough, James, optician, 66

backstaff (type of quadrant), 18
Baker, Henry
　The Microscope Made Easy, 32, 33, 34, 36, 49, 139, 338
　The Natural History of the Polype, 33

Baker, J.R., *Abraham Trembley*, 338
Bancks, Robert, instrument maker, 316
Bank of England, 74
Banks, Sir Joseph, 131, 193, 277
Bardin family, 339, 347
Bardin, William, globe maker, 130–31
Barnes, John, 352
barometers, 231
　stick, illustrations, 154–5, 160, 185, 296
　wheel, 164
Bass, George, 15
Bate, R.B., optician, 316
Baynham, Charles, 30, 38
Bean, Fowler, 162, 183, 351
Bennett, John, instrument maker, 339
Bere, Revd M.A., 291
Berge, Matthew, supplier to Ordnance Office, 292–3
Bernoulli, Johann, 114–15, 127
　Lettres Astronomiques, 346, 347
Billings Microscope Collection, 343, 348
Birch, John, surgeon, 196, 206, 320, 325
　Essay on the Medical Applications of Electricity, 360
　A Letter to Mr. George Adams on ... Medical Electricity, 395
Bird, John, 4
Birt, S., bookseller, London, 34
Blagden, Sir Charles, 193
Blenkinsop, John, 277, 279
Block, Ann, 330
Blunt, Arthur Graham, 184, 347, 352
Blunt family
　descendants of Adams family, xx–xxi
　linen-drapery business, 276, 347
Blunt, Joseph, 330
Blunt, Lucy, 360
Blunt, Mary, 343
Blunt, Robert, jnr, 277, 280, 343, 358
　apprenticed to George Adams jnr, 183–4, 275
Blunt, Robert, snr, 24, 343
　marriage to Sarah Adams, 83, 133, 275
Blunt, Sarah, 330, 331, 343
　death, 358
Blunt, William, 83
Bontein, Archibald, 54

Boscawen, Admiral Edward, 52
Boswell, James, 118, 345
Bothwell, Archibald, 27, 337
Boulton, Matthew, 127, 347
Bowen, Emanuel, engraver, 345
Bowles, T., engraver, 36, 211
Bradbury, Dr S., 209
  Historical Aspects of Microscopy, 354
Brahe, Tycho, 15
Bransby, John *Use of the Globes*, 131
Broadley, A. M., 336
Brown, Joyce, *Mathematical Instrument-Makers in the Grocers' Company*, 333, 334, 335, 336, 339, 341, 347, 352
Brown, Olivia, 333, 343
Bruti, Revisio, 91
Brydone, Patrick, 322, 360
Buchan, William, *Domestic Medicine*, 322, 323, 325
Buckingham House, 104
Bugge, Thomas, astronomer, 193, 353
Burr, Aaron, 69
Bute, Earl of, 27, 94, 146
Byron, Commodore John, 127

calipers, gunner's, illustrations, 148, 149
Calvert, H.R., 336
  Scientific Trade Cards, 337
Campbell, R., *The London Tradesman*, 342
Carwitham, T., 91
  Description and Use of an Architectonic Sector, 335
Cary family, 347
Cary, John
  globes: 12-inch, cost, 130; 21-inch, cost, 130
catadioptrical quadrant, 18–19
Cavallo, Tiberius, 188
Cavendish, Henry, 205
Chaldecott, J.A., 101, 360
  Handbook of the King George III Collection, 343, 344
Chapter House Philosophical Society, 186, 188–9, 191, 252
Charles II, King, 72
Charlotte, Queen, 104
Chenekal, Valentin L., *Watchmakers and Clockmakers in Russia*, 348

Chew, V.K., 266, 340, 358
Chizov, Nikolay Galaktionovich, 137
Christ's Hospital, *see* Royal Mathematical School
chromatic aberration, 65
Churchman, John, 193
circumferentors, 92, 235–6
  illustrations, 238
Clais, J.S., 350
Clarke, T.N., *Brass and Glass*, 348
Clay, Reginald S., *History of the Microscope*, 33, 84, 85, 89, 102, 104, 144, 216, 298, 336, 338, 343, 344, 348, 354
Clockmakers' Company, 5
Coggs, John, 15
Coimbra University, 191
coin balances, 169, 172–3
  illustration, 170
  makers of, 169, 171
  reason for, 168
  sales, 171
coinage, legislation, 168, 173
Cole, Benjamin, 26, 56, 69, 339, 345
  quadrants, 44–5; illustrations, 46, 47
  rivalry with George Adams snr, 48–50, 52, 60, 66, 115
Collier, William, 30
Collinson, Peter, 68
Commissary Court of London, 10, 11, 12
compasses
  beam, 105
  magnetic, illustrations, 156
  mariners, 70
  military, 70, 72
  proportional, illustration, 290
Cook, Captain James, 126
Cooper, Mary, bookseller, 338
Cooper, Richard, engraver, 345
Copernican system, 25
cosmotheorion, 27, 30, 38
Court, Thomas H., 33, 84, 85, 89, 102, 104, 144, 216, 298, 336, 338, 343, 344, 348, 354
Courtauld & Cowle, toyshop, 171
Crane Court, 14
Crawforth, Michael, 334–5, 336, 340, 347, 352
Crisp, Sir Francis, 102
Cuff, John, 14, 32, 37, 65–6, 139, 339

Cunn, Samuel G., *Use of the Sector*, 17
Cushee family, 108, 112
Cushee, Leonard, globe maker, 112, 345

Daumas, Maurice, 3
  Les instruments scientifiques, 333, 338
de Bomare, Valmont, *Dictionaire Raisonné universel d'Histoire Naturelle*, 209
de Cassini, Jean-Dominique, 3
  Mémoires ... de l'Observatoire royal de Paris, 333
de Clerq, Peter, 337, 346
de Langlade, Margaret Sophia, marries Dudley Adams, 275
de Magellan, J.H., 188, 191, 250, 251, 252, 259, 352
De Saussure, 231
  Inquiries and Observations on Atmospherical Electricity, 204–5
Deane, William, 14, 26, 52, 53, 54, 56, 340
  The Description of the Copernican System, 337
Dekker, E., *De Leidsche Sphaera*, 337
Demainbray, Dr Stephen, 75, 98, 105, 121, 344
  life, 343–4
Desaguliers, J.T., 17, 26
  A Course on Experimental Philosophy, 101, 337
Desaguliers, Thomas, entry in *Dictionary of National Biography*, 180–81
Devonshire, Duke of, 349
dial
  ring, illustration, 55
  universal inclining, illustration, 297
Dixey, C.W., optician, 145, 348
Dixey, G. & C., opticians, 316
Dixon, Richard, 357
Dodsley, R., publisher, 338
Dollond, John
  optician, 3, 105, 316, 341, 359
  telescope patent, 65
Dollond, P. & J., opticians, 316, 356
Donn, Benjamin
  Description and Use of the Navigation Scale, 94
  Description and Use of the Variation and Tide Instrument, 94
  Essay on Mechanical Geometry, 281
  Map of Devon (1765), 343
  Proposals for Surveying and Making a New and Accurate Map of ... Devon, 92
Donowell, John, engraver, 6
Drakeford, David, 15
draughtman's tool, illustration, 73
Driver, Mary, 269
Dudley, Ann, 42, 339, 356; see also Adams, Ann (wife of George Adams snr)
Dunn, Samuel, teacher of astronomy, 120–21, 346

East India Company, and George Adams snr, 19–20, 336
Edell, S., 345, 347
Edkins, S.S., 131
Edward, Prince (Duke of York), 75, 101
Eldershaw, Joseph, 21
electrical apparatus, illustrations, 203
electrical machines, illustrations, 202
electricity, and medicine, see medical electricity
Encyclopaedia Britannica, 201
eudiometer, 255–6

Fenning, Daniel, *New and Easy Guide to the Use of the Globes*, 108, 345
Ferguson, James, 108, 130, 133, 248, 277, 345, 347, 357
  Introduction to Electricity, 339
  Lectures on Select Subjects in Mechanics, 101, 222, 225, 355
Finchett, Arnold, 171
Finlay, Sir John, 102
Fleet Street
  allocation of numbers, 120, 346, 384
  instrument makers in, 15, 32
  junction with Shoe Lane: engraving (mid 18th century), 6; photograph (1986), 7
  no. 60: description, 310, 312, 313, 314; photograph (1991), 311
Fletcher, bookseller, Oxford, 34
Folkes, Martin, 36

414   Index

Fordham, Sir Herbert George, *John Cary*, 347
Forrest's Coffee House, Charing Cross, 27
Franklin, Benjamin, 68, 69
freeman, becoming a, 4–5
French Revolution, 246
Frison, E., 354

Gage, A.T., *A History of the Linnean Society*, 354
galvanism, 320
geometrical solids, illustrations, 58–9
George I, King, 4
George II, King, 14, 30, 56, 94, 98
George III, King, 55, 68, 85, 103
  collection of scientific instruments, 98, 101, 102, 105, 114, 116, 197, 250, 344
  Dudley Adams, globe maker to, 277
  George Adams snr, mathematical instrument maker to, 94, 146
  view of lightning conductors, 347
George IV, King, 104
Gilbert, John, 60
glass, printing on, 352
globes
  3-inch, 108; cost, 127; description, 127; illustration, 128
  5.4-inch, 127
  6-inch, cost, 127
  9-inch, 129
  12-inch, 109, 120, 127, 285; cost, 130; illustrations, 119, 125, 286, 287, 301, 302
  17-inch, 118
  18-inch, 120, 127, 285; cost, 118, 124; dating of, 126; description, 113–14; illustrations, 87, 116–17, 254–8; manufacture, 260
  21-inch, cost, 130
  28-inch, 112, 118, 313–14
  description, 186
  makers of, 107, 112, 120–21, 130–31
  manufacture, 126
  marketing, 15
  prices, 112–13
  template, illustration, 122
  terrestrial, manufacture, 260
Godfrey, Thomas, 48, 339
Goodison, Nicholas, 355, 359
  English Barometers, 1680–1860, 164, 289, 292, 349
Graham, George, watchmaker, 4, 339, 342
Great Fire of London (1666), 8, 42, 72, 259
Greenwich Observatory, 104, 105
Grocers' Company, 8–9, 11, 31, 44, 135, 184
  archives, 12
  Court of Assistants, 10, 12, 82, 137
  status in City of London, 9
Grocers' Hall
  description, 72, 74
  illustration, 64
Gunther, R.T., 129
  Early Science in Oxford (vol. 2), 347
  Handook of the Museum of the History of Science in the Old Ashmolean Building, 81

Hackmann, W.D., *Electricity from Glass*, 339
Hadley, John
  A Description of a new Instrument, 336
  octant, 18
  quadrants, 25, 45, 48, 60, 63, 82, 230; illustration, 77
  sextants, 229, 230
Haines, George, 177, 179
Ham, Erasmus, 20
Ham, George, 20
Hambly, Maya, *Drawing Instruments*, 194
Hammond, *Practical Surveyor*, 17, 66
Harris, Joseph, *Description and Use of the Globes and Orrery*, 16, 26, 66, 108, 337, 345
Harris, T. & W., globe makers, 316
Harrison, John, 18
Harvard College, 218
Hatton, Thomas
  alliance with George Adams jnr, 172
  An Essay on Gold Coin, 172, 173
  Introduction to Clock and Watch Work, 350
Heald, Tim, *By Appointment*, 360
Hearne, George, 15
Heath & Wing, instrument makers, 12, 66, 336

Heath, J.B., *Some Account of the Worshipful Company of Grocers*, 64, 342, 360
Heath, Thomas, 11, 13, 15, 17, 18, 19, 55, 60, 72, 91, 133, 135, 162, 347, 349
Herschel, Sir John, 218
Hertford, Earl of, 349
Hill, John, electrical experimenter, 205, 206, 213, 216, 354
Hill, John, *The Construction of Timber*, 141, 142, 143, 146, 209
Hill, Nathaniel, globe maker, 15, 112
  5.4-inch globe, 127
  price list, 109
Hindmarsh, Robert, printer, 191, 192, 199, 229
  Rise and Progress of the New Jerusalem Church, 353
  The New Magazine of Knowledge concerning Heaven and Hell, 193, 248, 353
Hinton, Thomas, 68
Hoare's Bank, 13
Hodgson, Alfred, 20
Hooke, Robert, 36
  Micrographia Restaurata, 33
Horwood, Richard, map of London (1799), 276
Huguenots, 3
hygrometers, 231, 292

Ingenhouz, Dr Jan, 252, 256, 261, 354
instrument makers, 12
  in Fleet Street, 15
instruments, drawing
  description, 235
  illustrations, 165, 194–5, 234
  marketing, 15, 17
  mathematical, examples, 11–12
  optical, 12
  surveying, description, 235–6; illustrations, 295

Jack, Richard, 57
  career, 341
  quadrant, patent application, 60–61, 383
Jackson, Joseph, 44, 52, 53, 54
Jenkinson, Robert Banks, Prime Minister, 360
Johnson, Dr Samuel, 113, 118, 345

Joiners' Company, 135, 347
Jones, W. & S., 131, 207, 215, 216, 219, 241, 245, 248–9, 281, 386
  globes: 12-inch, 285; 18-inch, 285; 28-inch, 313–14
  publisher of works of George Adams jnr, 268–9; list, 200
Jones, William, FRS, 216, 247, 357

kaleidoscope, 331
Kanmacher, Frederick, 216
Keith, Thomas, 131
Keller, Helen, 353
Kent, Adolphus, 331
Kepler, Johann, 25
Kettle, Nathaniel, 70, 135
Kew Oservatory, 98, 105
King, H.C.
  Geared to the Stars, 337, 355
  History of the Telescope, 356
King's College, London, 98, 102, 105, 344
Kirby, Joshua, 91
Kirwan, Richard, 188, 189
Knight, Dr Gowin, 70, 134, 204
Koenig, Samuel, 30

La Beaume, Michael, 360
Labaree, Leonard W., *The Papers of Benjamin Franklin*, 341
Lalande, Jerome, 103, 104, 105, 133, 344
  Astronomie, 344
Lane, James, 15
Laurie, R., engraver, 211, 217
Lavoisier, Antoine-Laurent, 247
Legg, Samuel, assignee of Dudley Adams, 308, 313, 359
Leigh & Sotheby, auctioneers, 267
lens, achromatic, 65, 193, 341, 356
Leyden jar, 12, 42, 203, 206, 354
Leyden sphere, 27, 30
lightning conductors, George III's view, 347
Lind, Dr James, 137
Lindsay, George, watchmaker, 337
Linnean Society, 174, 209
Liverpool, Earl of, Prime Minister, 318–19, 326
Lodge, John, engraver, 211, 217
Loft, Matthew, 66
Logan, J., 339

London, map
  (1746), 9
  (1761), illustrations, 96–7
  (1799), 276
longitude, measurement, 18–19
Longitude, Board of, 179, 360
Longitude, Commissioners of, 18
Longitude Prize, 18, 329
Loriners' Company, 5, 72
  appointment of stewards, 9–10
  membership, 8, 334
  minute books, 8
  status in City of London, 9
Loriners' Hall, 9
Loudon, Lord, 137
Louis XVI, King, 246
Lowndes, Francis, 319, 321, 322, 325
  medical electrical apparatus, 320; illustration, 323
  Observations on Medical Electricity, sources, 360
  The Utility of Medical Electricity, sources, 360

Macartney, Lord George, mission to China, 277
McConnell, Dr Anita, 352, 359
Mackenzie, Murdoch, 108
MacLeod, C., *Inventing the Industrial Revolution*, 358
Malie, Thomas, 91
Maling, George, 186
Mann, James, 15, 18, 19
Marsham, Ann, 356
Marsham, Hannah
  marries George Adams jnr, 174, 356; *see also* Adams, Hannah, wife of George Adams jnr
Marsham, Mary, 269
Marsham, Thomas, 174, 209, 350, 354
Martin, Benjamin, 4, 12, 24, 25, 112, 113, 130, 133, 260, 277, 316, 334, 346, 347
  about, 337
  An Essay on the Nature and Superior Use of Globes, 121
  Appendix to the Description and Use of Globes, 121, 123
  coin balances: description, 169; sales, 171
  Essay on Visual Glasses, 355
  George Adams jnr, rivalry with, 171–2, 225, 229
  George Adams snr, rivalry with, 121, 123–4, 138, 199
  globes, 17-inch, 118
  Micrographia Nova, 31, 34, 37
  microscopes, 31–2, 139
  orreries, 121, 218
  Royal Military Academy, Woolwich, supplier to, 176
  Senex globe business, purchase of, 130, 357
  The Monied Man's Vade-Mecum, 171, 172
Maskelyne, Nevil, Astronomer Royal, 174, 176
M'Cormick, Prof John, 248
Mecklenburg-Strelitz, Sophia Charlotte, Princess, 95
medical electricity, 319–20, 322, 326–7
  apparatus, illustration, 323
  efficacy, 326
  theory, 325
Mercers' Company, status in City of London, 9
Merchant Taylors' Company, 44
Merryfield, William, 40
microscopes, 31–2
  compound, 261, 286; illustrations, 23, 28, 86, 190, 215, 298
  Cuff-type: description, 89; illustration, 88
  double, illustration, 22
  lucernal: description, 95, 212–13, 216; illustration, 212
  makers of, 31–2
  pocket: description, 144; illustration, 140
  silver: description, 102, 104; illustration, 103
  simple, illustrations, 29
  solar, 32–3; illustration, 35
  trunnion-mounting: description, 89, 91; illustrations, 84, 86
  variable: description, 144; illustrations, 141, 145
microtome (cutting engine), illustration, 142
Millburn, John R., 347, 356
  Benjamin Martin, 337, 345, 346
  The Library of George Adams, 357

Millburn, John R. (cont'd)
  Retailer of the Sciences, 337, 346, 350
  Wheelwright of the Heavens, 345
Miller, John, 137
Milne, T., artist, 210, 217
money, value in eighteenth century xvii–xix
Monod-Cassidy, Helene, 344
Montagu, Duke of, 54
Moore, Sir John, 74
Morgan, William, 205
Morris, Joseph, 276
Mortimer, Cromwell, 337
Morton, Alan Q., 101
  Public and Private Science, 344, 350
Moxon, James, 107
Moxon, Joseph, 107
Mudge, Lt Col Richard Zachariah, 291
Muller, John, 166
Mynde, J., engraver, 108, 211

Nairne, Edward, 66, 69, 188
Napier's Bones, illustration, 71
Neale, John, watchmaker, 339, 345
Newberry, John, 32
Newton, Isaac, *Principia*, 25
Nicholson, William, 188, 189, 262
  Navigator's Assistant, 230

observatories
  Greenwich, 104, 105
  Kew, 98, 105
octants, 18
  illustrations, 232
Odner, Carl Theophilus, *Annals of the New Church* (vol. 1), 353
Oliver, Richard, schoolteacher, 339
opera glasses, 286
Ordnance, Board of, George Adams snr, supplier, 118
Ordnance, Office of, 14, 52
  Dudley Adams, supplier, 265–6, 292–3; bills, 267, 279
  George Adams jnr, supplier, 167, 176, 177; bills, 267
  George Adams snr, supplier, 52–4, 66, 68, 70, 76, 146, 147, 340
  Jeremiah Sisson, supplier, 162
  Matthew Berge, supplier, 292–3

orreries, 15, 25–7, 30, 121, 218, 246
  compound: description, 218; illustrations, 219, 220, 221, 222, 223
  grand, illustrations, 16, 17
  history, 337
  hybrid: description, 220; illustration, 224

Paine, Thomas, 56
paints, illustrations, 194–5
Palmer, F.W., *Microscopes, to the end of the nineteenth century*, 343
pantograph, 105
Paris, Peace of, 147
Parker, James, 8–9, 11, 335
Parry-Jones, W.L., *The Trade in Lunacy*, 360
Patent Office, 60
Patterson, Prof Robert, 248
Pattrick, Thomas, globe maker, 130, 347
Peel, Robert, 326, 327
Penn, Thomas, 68
perambulator wheel, illustration, 93
  dial, illustration, 244
perpendicular (level), gunner's, illustrations, 150, 175, 288, 289
Philadelphia, Library Company of, 68
Phillips, Hugh, *Mid-Georgian London*, 358
Phillips, Samuel, 351
phlogiston, 247
Pinchbeck, Christopher
  coin balances, 172–3
  toyshop, 171, 172
Pine, John, engraver, 345
plane (plain) tables, illustrations, 80–81, 237
planispheres, *see* globes
Pollock, Dr Alan, 166, 349
Porter, R., *Science and Profit in 18th-century London*, 333, 342
Powder Magazine, Purfleet, supplied by George Adams jnr, 163
The Present State of the Republic of Letters, 19, 336
Prince, Revd John, 213, 216, 248, 354
Pringle, Sir John, royal physician, 104, 105, 344
protractor, brass, illustration, 291
Pyefinch, Henry, 169

418 Index

quadrants, 180, 229, 230
  advertisements for, 61–2
  illustrations, 46, 47, 50, 77, 175, 232, 233
questmen, 5
  meaning, 334

Rackstrow, Benjamin, sculptor, 339
Ramsden, Jesse, 4, 176–7, 179, 180, 231, 245, 255, 355, 384
Randier, Jean, *Marine Navigation Instruments*, 336
Ray, Thomas, 68
Rees, J. Aubrey, *The Worshipful Company of Grocers*, 336
Ribright, Thomas, 56
Rice, James, 75
Richmond, Duke of, 235
Robertson, Captain Archibald, 176
Robinson, Matthew, 330
Rocque, John, map of London (1746), 9
Rogers, Ann, 330, 360, 361
Rogers family, 360–61
Rogers, Hannah, 330, 360, 361
Rogers, Margaret, 12–13
Rogers, Mary Isabel, 360
Rogers, Richard, 360, 361
Rogers, Sarah A., 361
Rössaak, Tor E., 347
Rowley, John, 4, 52, 56
Royal Academy, Portsmouth, 26
royal appointments, misuse of, 359–60
Royal Institute of British Architects, 92
Royal Mathematical School (Christ's Hospital), 56, 68, 75, 76, 124, 147, 153, 161, 266
Royal Microscopical Society, 20, 29, 35
Royal Military Academy, Woolwich, 52, 53, 118, 147, 153, 166–7, 176, 181
Royal Society, 14, 33, 36, 44, 107, 108, 277

Sahiar, A.B., 343
St Bride's
  church, 42; illustration, 43
  parish, 3, 4, 10, 12; officials, 5–6
St Dunstan's
  church, 14, 42
  parish, 78

St James's Chronicle, 95
St Martin in the Fields
  church, 42, 83
  parish, 358
St Thomas's Hospital, 320
Schechner, Sara J., 354
sectors, architectonic
  description, 91
  illustrations, 39, 90, 165
Seely, J., *Stowe*, 358
Senex, John, globe maker, 14, 16, 18, 19, 107, 112, 344–5
  globes, 28-inch, 118
Senex, Mary, 56, 107–8, 109, 345, 357
Seven Years' War, 53, 76, 147, 166, 183
  cost, 198
sextants, 229, 230
'sGravesande, W.J., *Mathematical Elements of Natural Philosophy*, 100, 101
Sharyer, William, 69
Shoe Lane, junction with Fleet Street
  engraving (mid 18th century), 6
  photograph (1986), 7
Short, James, 4, 104, 228
Shuckburgh, Sir George, 231
Simmons, John, 40
Simpson, Dr A.D.C., 348
Sisson, Jeremiah, supplier to Ordnance Office, 161, 167
  bankruptcy, 174, 176
Sisson, Johathan, 4, 15, 104
Smeaton, John
  air-pump, 100, 105, 344
  quadrant, 186, 352
Smith, Caleb
  catadioptical quadrant, 18–19, 25, 229
  The Description and Use of a New Astronomical Instrument, 336
  The Description, Use ... Sea Quadrant, 336
Society for the Encouragement of Arts, Manufactures, and Commerce, 92, 186
Sowerby, Joseph, 17, 56, 57, 340
Spectaclemakers' Company, 314
spectacles
  illustration, 281
  need for, 224–5
spheres, armillary, illustrations, 225

Staples & Barlow, booksellers, York, 33
Staunton, Sir George, *An Authentic Account of an Embassy*, 358
Stedman, Christopher, jnr, apprenticed to George Adams Jnr, 182, 183
Stedman, Christopher, snr, 183
  trade card, illustration, 182
Steggall, Revd W., 205
Sterrop, George, optician, 66
Stewart, Charles, 277
Stirling, James, 15
Stone, Edmund, *The Construction and Uses of Mathematical Instruments*, 355
Sudlow, John, 44
sundials, 72
  illustration, 67
surveyor's level, illustration, 294
Swedenborg, Emanuel
  Apocalypsis Revelata, 192
  Doctrine of the New Jerusalem concerning the Lord, 192
  The True Christian Religion, 191
Swedenborgian movement, 248, 353
  influence in England, 191–2
  influence on George Adams jnr, 192, 353
Symmonds, John, watch-case maker, 338

Tafel, R.L., *Documents concerning the Life and Character of Emanuel Swedenborg*, 353
Tallis, John, publisher of London street views, 310, 312
Tangate, Robert, 135
Tarry, Samuel, 68, 69
taxes, introduction, 197
Taylor, E. Wilfred, *At the Sign of the Orrery*, 339, 340
telescopes, 105, 177
  Dollond's patent, 65
  equatorially-mounted, illustrations, 226–8
  portable, illustrations, 282–3
  reflecting, illustration, 278
  refracting, illustration, 178
Temple Bar, 95, 346
Teyler's Foundation, Harlem, 181, 205, 250, 252
Teyler's Museum, Harlem, 225, 250, 253, 254, 256, 261, 347

theodolites, 177, 236
  4-inch, illustrations, 239, 241, 242, 243
  10-inch, illustration, 239
  illustrations, 187, 238, 240
Theosophical Society, 192
thermometers
  description, 292
  illustration, 274
Thompson, John, 81
Thorne, Richard, creditor of Dudley Adams, 308
Thurlborne, bookseller, Cambridge, 34
timber section, magnified, illustration, 143
time measurement, Atwood's machine, illustration, 253
Toksvig, Signe, *Emanuel Swedenborg, scientist and mystic*, 352
Torrens, Dr H.S., 346, 347
Tower of London, 76
Tracy, John, 15
Tracy, Steven, 27
Trade, Society for Protection of, 343
trade cards, illustrations frontispiece, 134, 182, 275
Trembley, Abraham, 33, 36, 338
Trobridge, G., 352
Troyte, Arthur H. D., 291
trunnion mounting, microscopes, 89
Turner, A.J., *Early Scientific Instruments*, 348
Turner, G. L'E., 343, 344, 348
  Collecting Microscopes, 338
Turton, Thomas, instrument maker, 345
Tyacke, Sarah, *London Map-Sellers, 1660–1720*, 344
Tyro, Magnus, surgeon, 339

Upham, Charles W., 354

van Amstel, Jacob Ploos, 138
van der Kroght, Peter, 115, 346
  Old Globes in the Netherlands, 347, 358
van Leeuwenhoek, Antonie, 34
van Marum, Martinus, 184, 186, 191, 193, 196, 201, 223, 225, 354
  correspondence, 355
  globes supplied by Dudley Adams, 276

van Marum, Martinus (cont'd)
  instruments supplied by George Adams jnr, 250–52, 255–6, 258–62; list, 251
  visit to England, 256, 258
Venus, transit (1769), 98, 121
Vince, Samuel, *Complete System of Astronomy*, 131, 285, 347

Wales, Prince of, George Adams snr appointed mathematical instrument maker to, 75, 83, 85
Walker, Adam, 207, 354
  An easy Introduction to Geography, 347
Wallis, P.J., 346
Wallis, R.V., *Biobibliography*, 346
Wallis, Captain Samuel, 127
Ward, William, 18, 336
Waterloo, Battle of (1815), 305
Watkins, Francis, optician, 36, 62, 65, 66, 68, 104, 341
  L'Exercise du Microscope, 338
  microscopes, 139
Watson, William, 44
Watt's Academy, 25, 26
Watts, William, 15
Webb, William, 21
Wedgwood, Josiah, purchase of globes, 120, 346

Wellcome Institute, London, 115, 346
Wellcome Museum for the History of Science, 20
Wess, Jane A., 101, 344, 350
West, Charles, optician, 135, 348
West, Francis, optician, 314, 316–17, 359
  tract on the human eye, title-page, 315
Wheatland, David P., *The Apparatus of Science at Harvard, 1765–1800*, 228, 355
Whipham & North, silversmiths, 259
Whipple Museum Catalogues, 337
Whipple, R.S., 20, 336
Whitehurst, John, 127, 188
Wickes, bookseller, Norwich, 34
Wigley, J., engraver, 36
William IV, Stadholder, 30
Wilson, J. Simms, 339, 340
Wright, Gabriel, 130, 186
Wright, Thomas, mathematical instrument maker, 14, 25, 26, 44, 52, 53, 56, 94, 108, 347
  death, 349
  orrery, 15; illustration, 16
Wright, Thomas, watch-maker, 192
Wrigley, J., engraver, 211

zograscope, 331, 360
Zuidervaart, Dr H.J., 348